VOLUME THREE HUNDRED AND FIFTY THREE

INTERNATIONAL REVIEW OF CELL AND MOLECULAR BIOLOGY

Cell Death Regulation in Health and Disease - Part C

INTERNATIONAL REVIEW OF CELL AND MOLECULAR BIOLOGY

VOLUME THREE HUNDRED AND FIFTY THREE

INTERNATIONAL REVIEW OF CELL AND MOLECULAR BIOLOGY

Cell Death Regulation in Health and Disease - Part C

Edited by

JOHAN K.E. SPETZ
Harvard T.H. Chan School of Public Health, Boston, MA, USA

LORENZO GALLUZZI
Weill Cornell Medical College, New York, NY, United States

ELSEVIER

ACADEMIC PRESS
An imprint of Elsevier

Academic Press is an imprint of Elsevier
50 Hampshire Street, 5th Floor, Cambridge, MA 02139, United States
525 B Street, Suite 1650, San Diego, CA 92101, United States
The Boulevard, Langford Lane, Kidlington, Oxford OX5 1GB, United Kingdom
125 London Wall, London, EC2Y 5AS, United Kingdom

First edition 2020

ISBN: 978-0-12-820135-0
ISSN: 1937-6448

For information on all Academic Press publications
visit our website at https://www.elsevier.com/books-and-journals

Publisher: Zoe Kruze
Acquisitions Editor: Ashlie M. Jackman
Editorial Project Manager: Leticia M. Lima
Production Project Manager: Denny Mansingh
Cover Designer: Matthew Limbert

Typeset by SPi Global, India

Contents

Contributors

Dieter Adam
Institut für Immunologie, Christian-Albrechts-Universität zu Kiel, Kiel, Germany

Joana D. Amaral
Research Institute for Medicines (iMed.ULisboa), Faculty of Pharmacy, Universidade de Lisboa, Lisbon, Portugal

Ted M. Dawson
Neuroregeneration and Stem Cell Programs, Institute for Cell Engineering; Department of Neurology; Solomon H. Snyder Department of Neuroscience; Department of Pharmacology and Molecular Sciences, Johns Hopkins University School of Medicine, Baltimore, MD; Adrienne Helis Malvin Medical Research Foundation, New Orleans, LA, United States

Valina L. Dawson
Neuroregeneration and Stem Cell Programs, Institute for Cell Engineering; Department of Neurology; Solomon H. Snyder Department of Neuroscience; Department of Physiology, Johns Hopkins University School of Medicine, Baltimore, MD; Adrienne Helis Malvin Medical Research Foundation, New Orleans, LA, United States

Pedro A. Dionísio
Research Institute for Medicines (iMed.ULisboa), Faculty of Pharmacy, Universidade de Lisboa, Lisbon, Portugal

Vivian Gama
Department of Cell & Developmental Biology; Vanderbilt Center for Stem Cell Biology, Vanderbilt University; Neuroscience Program; Vanderbilt Brain Institute, Vanderbilt University Medical Center, Nashville, TN, United States

Michelle Heib
Institut für Immunologie, Christian-Albrechts-Universität zu Kiel, Kiel, Germany

Gergely Imre
Institute of General Pharmacology and Toxicology, University Hospital Frankfurt, Goethe University, Frankfurt am Main, Germany

Tae-In Kam
Neuroregeneration and Stem Cell Programs, Institute for Cell Engineering; Department of Neurology, Johns Hopkins University School of Medicine, Baltimore, MD, United States

Jason Karch
Department of Molecular Physiology and Biophysics; Cardiovascular Research Institute, Baylor College of Medicine, Houston, TX, United States

Hyejin Park
Neuroregeneration and Stem Cell Programs, Institute for Cell Engineering; Department of Neurology, Johns Hopkins University School of Medicine, Baltimore, MD, United States

Pooja Patel
Department of Molecular Physiology and Biophysics, Baylor College of Medicine, Houston, TX, United States

Megan L. Rasmussen
Department of Cell & Developmental Biology, Vanderbilt University, Nashville, TN, United States

Cecília M.P. Rodrigues
Research Institute for Medicines (iMed.ULisboa), Faculty of Pharmacy, Universidade de Lisboa, Lisbon, Portugal

Stefan Rose-John
Institut für Biochemie, Christian-Albrechts-Universität zu Kiel, Kiel, Germany

Preface: Life through death—Key role of cellular suicide for colonial and organismal homeostasis

Johan Spetz[a,b,*], Lorenzo Galluzzi[c,d,e,f,g,*]
[a]John B. Little Center for Radiation Sciences, Harvard School of Public Health, Boston, MA, United States
[b]Laboratory of Systems Pharmacology, Harvard Program in Therapeutic Science, Department of Systems Biology, Harvard Medical School, Boston, MA, United States
[c]Department of Radiation Oncology, Weill Cornell Medical College, New York, NY, United States
[d]Sandra and Edward Meyer Cancer Center, New York, NY, United States
[e]Caryl and Israel Englander Institute for Precision Medicine, New York, NY, United States
[f]Department of Dermatology, Yale School of Medicine, New Haven, CT, United States
[g]Université de Paris, Paris, France
[*]Corresponding authors: e-mail address: spetz@hsph.harvard.edu; deadoc80@gmail.com

The term "regulated cell death" (RCD) is commonly employed to indicate a form of cell death that is controlled by a dedicated molecular machinery, and hence can be modulated (at least to some degree) with specific pharmacological or genetic interventions (Galluzzi et al., 2015, 2018a). RCD can occur in the context of two diametrically opposed scenarios. On the one hand, RCD can represent the ultimate response of cells to microenvironmental perturbations that are excessively intense or prolonged for cells to adapt and recover homeostasis (Spetz et al., 2019). On the other hand, RCD can contribute to purely physiological programs of development and adult tissue homeostasis (Bock and Tait, 2020). The expression "programmed cell death" (PCD) is often used to refer to these latter variants of RCD that occur at a predetermined point in time, irrespective of potentially cytotoxic alterations of the intracellular or extracellular microenvironment (Fuchs and Steller, 2011, 2015).

For multicellular organisms, the advantages provided by genetically encoded programs for cell death are obvious. On the one hand, PCD can be harnessed as a morphogenetic mechanism that shapes first (and maintains then) tissue and organ architecture (Tuzlak et al., 2016). On the other hand, stress-driven RCD constitutes a robust mechanism for the elimination first (and replacement then) of postmitotic cells that are damaged beyond recovery, and hence functionless and/or prone to constitute a threat for the entire organism (e.g., prone to malignant transformation) (Cao and Tait, 2018). Less obvious are the advantages provided by RCD to unicellular organisms, such as bacteria and yeast. However, both prokaryotes and unicellular eukaryotes have rudimentary RCD machineries that allow individual cells to suicide (Carmona-Gutierrez et al., 2018). This happens, as an example, when nutrients for the entire colony are limiting, and the demise of old cells increases the chances of younger, fitter cells to survive dwindling nutrient levels and ultimately reconstitute the colony (Fabrizio et al., 2004; Herker et al., 2004).

In higher eukaryotes, the molecular machinery for RCD has considerably diversified (Galluzzi et al., 2017a; Stockwell et al., 2017; Vanden Berghe et al., 2014; Weinlich et al., 2017) to enable variants of RCD that (1) occur according to diverse kinetics (Buque et al., 2019; Grootjans et al., 2016) and (2) are accompanied by the emission of specific signals that engage or suppress immune reactions (Galluzzi et al., 2017b; Rodriguez-Ruiz et al., 2019; Vanpouille-Box et al., 2018). Thus, dying mammalian cells can deliver signals in the form of damage-associated molecular patterns (DAMPs) or cytokines that are responsible for immune modulation at sites of active RCD (Galluzzi et al., 2018b; Green et al., 2009; Krysko et al., 2012). Many of the enzymes that have long been considered as pure RCD regulators, such as apoptotic caspases, have now been shown to mediate multipronged immunomodulatory functions (Galluzzi et al., 2016; Rodriguez-Ruiz et al., 2019; Rongvaux et al., 2014; White et al., 2014), and to support tissue regeneration, at least in some settings (Cheng et al., 2019; Huang et al., 2011; Ichim and Tait, 2016).

Given the key role of RCD in the maintenance of organismal homeostasis and the control of immune responses against dying cells, it is not surprising that deregulated RCD contributes to the pathology of a variety of human disorders. On the one hand, uncontrolled RCD associated with inflammatory reactions is involved in the pathogenesis of ischemic, cardiac, renal and neurodegenerative conditions, among others (Del Re et al., 2019; Fang et al., 2019; Hetz and Saxena, 2017; Kers et al., 2016; Quinlan et al., 2017). On the other hand, insufficient RCD (as such or in the context of immunosuppressive signaling) is etiologically linked to autoimmunity, proliferative syndromes, and cancer (Cao and Tait, 2018; Czabotar et al., 2014; Khan and Ghazanfar, 2018; Nagata and Tanaka, 2017; Singh et al., 2019; Spetz et al., 2019). In summary, RCD in its various forms stands out as a key mechanism for the preservation of colonial and organismal homeostasis, thus constituting an attractive target for the development of pharmacological interventions against a plethora of human disorders for more than 2 decades (Degterev et al., 2019; Huang and Ichikawa, 2008; Naderer and Fulcher, 2018; Valentin et al., 2018). As of today, however, very few drugs that target the molecular machinery for RCD have successfully entered clinical development, including the BCL2 inhibitor venetoclax (currently approved for the therapy of some hematological tumors) (Jain et al., 2019), and the pan-caspase inhibitor emricasan (which was recently granted fast track status by the US Food and Drug Administration for the therapy of cirrhosis) (Mehta et al., 2018; Shiffman et al., 2019). It is tempting to speculate, yet remains to be formally demonstrated, that translating RCD modulation from the bench to the bedside has been difficult because RCD has been often investigated (especially when the molecular machinery for RCD has begun to emerge) irrespective of its immunological context.

In *Cell Death Regulation in Health and Disease* (a special issue of the *International Review of Cell and Molecular Biology* series) a group of world-renowned experts in the field gathered to discuss mechanistic features and physiopathological implications of RCD in a variety of model organisms. Each of the chapters of *Cell Death Regulation in Health and Disease* addresses a specific aspect of cell death biology, ranging from RCD control in prokaryotes and lower eukaryotes to the contribution of cell death deregulation to various human disorders. The book (which comes in three separate volumes, i.e., parts A, B, and C) is equally addressed to experienced investigators who may want to acquire in-depth insights into a specific aspect of cell death biology, as well as to students or neophytes who are moving their first steps in this broad field of research.

Acknowledgments

L.G. is supported by a Breakthrough Level 2 grant from the US Department of Defense (DoD), Breast Cancer Research Program (BRCP) (#BC180476P1), by the 2019 Laura Ziskin Prize in Translational Research (#ZP-6177, PI: Formenti) from the Stand up to Cancer (SU2C), by a Mantle Cell Lymphoma Research Initiative (MCL-RI, PI: Chen-Kiang) grant from the Leukemia and Lymphoma Society (LLS), by a startup grant from the Department of Radiation Oncology at Weill Cornell Medicine (New York, USA), by a Rapid Response Grant from the Functional Genomics Initiative (New York, USA), by industrial collaborations with Lytix (Oslo, Norway) and Phosplatin (New York, USA), and by donations from Phosplatin (New York, USA), the Luke Heller TECPR2 Foundation (Boston, USA) and Sotio a.s. (Prague, Czech Republic).

Conflict of interest statement

LG received consulting fees from OmniSEQ, Astra Zeneca, Inzen and the Luke Heller TECPR2 Foundation, and he is member of the Scientific Advisory Committee of Boehringer Ingelheim, The Longevity Labs and OmniSEQ.

References

Bock, F.J., Tait, S.W.G., 2020. Mitochondria as multifaceted regulators of cell death. Nat. Rev. Mol. Cell Biol. 21, 85–100.

Buque, A., et al., 2019. Apoptotic caspases cut down the immunogenicity of radiation. Onco. Targets. Ther. 8 (11), e1655364.

Cao, K., Tait, S.W.G., 2018. Apoptosis and cancer: force awakens, phantom menace, or both? Int. Rev. Cell Mol. Biol. 337, 135–152.

Carmona-Gutierrez, D., et al., 2018. Guidelines and recommendations on yeast cell death nomenclature. Microb. Cell 5 (1), 4–31.

Cheng, J., et al., 2019. The caspase-3/PKCdelta/Akt/VEGF-A signaling pathway mediates tumor repopulation during radiotherapy. Clin. Cancer Res. 25 (12), 3732–3743.

Czabotar, P.E., et al., 2014. Control of apoptosis by the BCL-2 protein family: implications for physiology and therapy. Nat. Rev. Mol. Cell Biol. 15 (1), 49–63.

Degterev, A., et al., 2019. Targeting RIPK1 for the treatment of human diseases. Proc. Natl. Acad. Sci. U. S. A. 116 (20), 9714–9722.

Del Re, D.P., et al., 2019. Fundamental mechanisms of regulated cell death and implications for heart disease. Physiol. Rev. 99 (4), 1765–1817.

Fabrizio, P., et al., 2004. Superoxide is a mediator of an altruistic aging program in Saccharomyces cerevisiae. J. Cell Biol. 166 (7), 1055–1067.

Fang, X., et al., 2019. Ferroptosis as a target for protection against cardiomyopathy. Proc. Natl. Acad. Sci. U. S. A. 116 (7), 2672–2680.

Fuchs, Y., Steller, H., 2011. Programmed cell death in animal development and disease. Cell 147 (4), 742–758.

Fuchs, Y., Steller, H., 2015. Live to die another way: modes of programmed cell death and the signals emanating from dying cells. Nat. Rev. Mol. Cell Biol. 16 (6), 329–344.

Galluzzi, L., et al., 2015. Essential versus accessory aspects of cell death: recommendations of the NCCD 2015. Cell Death Differ. 22 (1), 58–73.

Galluzzi, L., et al., 2016. Caspases connect cell-death signaling to organismal homeostasis. Immunity 44 (2), 221–231.

Galluzzi, L., et al., 2017a. Necroptosis: mechanisms and relevance to disease. Annu. Rev. Pathol. 12, 103–130.

Galluzzi, L., et al., 2017b. Immunogenic cell death in cancer and infectious disease. Nat. Rev. Immunol. 17 (2), 97–111.

Galluzzi, L., et al., 2018a. Molecular mechanisms of cell death: recommendations of the nomenclature committee on cell death 2018. Cell Death Differ. 25 (3), 486–541.

Galluzzi, L., et al., 2018b. Linking cellular stress responses to systemic homeostasis. Nat. Rev. Mol. Cell Biol. 19 (11), 731–745.

Green, D.R., et al., 2009. Immunogenic and tolerogenic cell death. Nat. Rev. Immunol. 9 (5), 353–363.

Grootjans, S., et al., 2016. A real-time fluorometric method for the simultaneous detection of cell death type and rate. Nat. Protoc. 11 (8), 1444–1454.

Herker, E., et al., 2004. Chronological aging leads to apoptosis in yeast. J. Cell Biol. 164 (4), 501–507.

Hetz, C., Saxena, S., 2017. ER stress and the unfolded protein response in neurodegeneration. Nat. Rev. Neurol. 13 (8), 477–491.

Huang, D., Ichikawa, K., 2008. Drug discovery targeting apoptosis. Nat. Rev. Drug Discov. 7 (12), 1041.

Huang, Q., et al., 2011. Caspase 3-mediated stimulation of tumor cell repopulation during cancer radiotherapy. Nat. Med. 17 (7), 860–866.

Ichim, G., Tait, S.W., 2016. A fate worse than death: apoptosis as an oncogenic process. Nat. Rev. Cancer 16 (8), 539–548.

Jain, N., et al., 2019. Ibrutinib and venetoclax for first-line treatment of CLL. N. Engl. J. Med. 380 (22), 2095–2103.

Kers, J., et al., 2016. An overview of pathways of regulated necrosis in acute kidney injury. Semin. Nephrol. 36 (3), 139–152.

Khan, U., Ghazanfar, H., 2018. T lymphocytes and autoimmunity. Int. Rev. Cell Mol. Biol. 341, 125–168.

Krysko, D.V., et al., 2012. Immunogenic cell death and DAMPs in cancer therapy. Nat. Rev. Cancer 12 (12), 860–875.

Mehta, G., et al., 2018. A placebo-controlled, multicenter, double-blind, phase 2 randomized trial of the pan-caspase inhibitor emricasan in patients with acutely decompensated cirrhosis. J. Clin. Exp. Hepatol. 8 (3), 224–234.

Naderer, T., Fulcher, M.C., 2018. Targeting apoptosis pathways in infections. J. Leukoc. Biol. 103 (2), 275–285.

Nagata, S., Tanaka, M., 2017. Programmed cell death and the immune system. Nat. Rev. Immunol. 17 (5), 333–340.

Quinlan, S., et al., 2017. MicroRNAs in neurodegenerative diseases. Int. Rev. Cell Mol. Biol. 334, 309–343.

Rodriguez-Ruiz, M.E., et al., 2019. Apoptotic caspases inhibit abscopal responses to radiation and identify a new prognostic biomarker for breast cancer patients. Onco. Targets. Ther. 8 (11), e1655964.

Rongvaux, A., et al., 2014. Apoptotic caspases prevent the induction of type I interferons by mitochondrial DNA. Cell 159 (7), 1563–1577.

Shiffman, M., et al., 2019. Randomised clinical trial: emricasan versus placebo significantly decreases ALT and caspase 3/7 activation in subjects with non-alcoholic fatty liver disease. Aliment. Pharmacol. Ther. 49 (1), 64–73.

Singh, R., et al., 2019. Regulation of apoptosis in health and disease: the balancing act of BCL-2 family proteins. Nat. Rev. Mol. Cell Biol. 20 (3), 175–193.

Spetz, J., et al., 2019. T cells and regulated cell death: kill or be killed. Int. Rev. Cell Mol. Biol. 342, 27–71.

Stockwell, B.R., et al., 2017. Ferroptosis: a regulated cell death nexus linking metabolism, redox biology, and disease. Cell 171 (2), 273–285.

Tuzlak, S., et al., 2016. Interrogating the relevance of mitochondrial apoptosis for vertebrate development and postnatal tissue homeostasis. Genes Dev. 30 (19), 2133–2151.

Valentin, R., et al., 2018. The rise of apoptosis: targeting apoptosis in hematologic malignancies. Blood 132 (12), 1248–1264.

Vanden Berghe, T., et al., 2014. Regulated necrosis: the expanding network of non-apoptotic cell death pathways. Nat. Rev. Mol. Cell Biol. 15 (2), 135–147.

Vanpouille-Box, C., et al., 2018. Cytosolic DNA sensing in organismal tumor control. Cancer Cell 34 (3), 361–378.

Weinlich, R., et al., 2017. Necroptosis in development, inflammation and disease. Nat. Rev. Mol. Cell Biol. 18 (2), 127–136.

White, M.J., et al., 2014. Apoptotic caspases suppress mtDNA-induced STING-mediated type I IFN production. Cell 159 (7), 1549–1562.

CHAPTER ONE

Poly (ADP-ribose) (PAR)-dependent cell death in neurodegenerative diseases

Hyejin Park[a,b,†], Tae-In Kam[a,b,†], Ted M. Dawson[a,b,c,d,e,*], Valina L. Dawson[a,b,c,d,f,*]

[a]Neuroregeneration and Stem Cell Programs, Institute for Cell Engineering, Johns Hopkins University School of Medicine, Baltimore, MD, United States
[b]Department of Neurology, Johns Hopkins University School of Medicine, Baltimore, MD, United States
[c]Adrienne Helis Malvin Medical Research Foundation, New Orleans, LA, United States
[d]Solomon H. Snyder Department of Neuroscience, Johns Hopkins University School of Medicine, Baltimore, MD, United States
[e]Department of Pharmacology and Molecular Sciences, Johns Hopkins University School of Medicine, Baltimore, MD, United States
[f]Department of Physiology, Johns Hopkins University School of Medicine, Baltimore, MD, United States
*Corresponding authors: e-mail address: tdawson@jhmi.edu; vdawson1@jhmi.edu

Contents

† These authors contributed equally.

International Review of Cell and Molecular Biology, Volume 353
ISSN 1937-6448
https://doi.org/10.1016/bs.ircmb.2019.12.009

Abstract

Disruption of cellular functions with aging-induced accumulation of neuronal stressors causes cell death which is a common feature of neurodegenerative diseases. Studies in a variety of neurodegenerative disease models demonstrate that poly (ADP-ribose) (PAR)-dependent cell death, also named parthanatos, is responsible for neuronal loss in neurological diseases, such as Parkinson's disease (PD), Alzheimer's disease (AD), Huntington's disease (HD) and amyotrophic lateral sclerosis (ALS). Parthanatos has distinct features that differ from caspase-dependent apoptosis, necrosis or autophagic cell death. Parthanatos can be triggered by the accumulation of PAR due to overactivation of PAR polymerase-1 (PARP-1). Excess PAR, induces the mitochondrial release apoptosis-inducing factor (AIF), which binds to macrophage migration inhibitory factor (MIF) carrying MIF into the nucleus where it cleaves genomic DNA into large fragments. In this review, we will discuss the molecular mechanisms of parthanatos and their role in neurodegenerative diseases. Furthermore, we will discuss promising therapeutic interventions within the pathological PAR signaling cascade that could be designed to halt the progression of neurodegeneration.

1. Introduction

Neurodegenerative diseases are a broad range of disorders characterized by progressive degeneration and cell death of neurons with symptoms that range from impairment of cognitive function to impairment of movement function. A small fraction of neurodegenerative diseases are caused by mutation of disease-related genes, but most of them are sporadic where environmental factors and aging play a predominant role (Gorman, 2008). Common features of these disorders are abnormal protein dynamics leading to defective degradation and accumulation of aggregated proteins and oxidative stress causing the formation of free radicals. In addition, many of these disorders have mitochondrial dysfunction, axonal transport deficits, and synaptic dysfunction, which contribute to cell death. Disruption of cellular function from aging-induced accumulation of neuronal stressors can contribute to cell death. However, there is relatively little neuronal loss in normal aging, while cell death is a common feature of neurodegenerative diseases. It is likely that several cell death pathways are involved in the neuronal loss in neurodegenerative diseases. Different types of cell death are commonly classified by morphological, enzymological, functional and immunological features (Galluzzi et al., 2012, 2018). This chapter reviews the regulated cell death program known as parthanatos, which is set in motion by excessive poly adenosine diphosphate (ADP)-ribose (PAR) polymerase (PARP) activation. We also discuss the role of parthanatos in neurodegenerative diseases and potential therapeutic approaches to attenuate parthanatos.

2. Overview of parthanatos

Cell death due to overactivation of PARP was known as "PARP-dependent cell death" until around 2009. However, once the byproduct of PARP activity, PAR polymer was recognized as a cell death signal and a mediator of PARP-dependent cell death (Andrabi et al., 2006), this form of cell death was named from the concatenation of "PAR" and "Thanatos." Thanatos being the personification of death in Greek mythology. Thus, "PAR-Thanatos," rendered in writing as "parthanatos," is cell death resulting from the toxic signal of PAR (Fatokun et al., 2014) and is now officially recognized by the Nomenclature Committee on Cell Death (NCCD) (Galluzzi et al., 2018). Parthanatos had previously been misclassified as necrotic due to the depletion of cellular nicotinamide adenine dinucleotide (NAD^+) and adenosine triphosphate (ATP), similar to necrosis. The loss of NAD+ and ATP was thought be due to PARP overactivation and consumption of NAD+ and the subsequent loss of ATP (Ha and Snyder, 1999). However, it is now known that the loss of NAD+ and ATP is due to PAR binding to hexokinase and inhibition of glycolysis (Andrabi et al., 2014; Fouquerel et al., 2014). Parthanatic cell death exhibits several biochemical features that make it distinct from apoptosis and necrosis or any other cell death. Unlike caspase-dependent apoptosis, parthanatos does not depend on a caspase cascade and therefore cannot be prevented by a pan-caspase inhibitor such as carbobenzoxy-valyl-alanyl-aspartyl-[O-methyl]-fluoromethylketone (*Z*-VAD-dmk). Knockout of the essential caspase mediators, Apoptotic protease activating factor 1 (APAF-1) or (BCL2 Associated X, Apoptosis Regulator) BAX also does not impair parthanatos (Cregan et al., 2002). It also does not involve apoptotic bodies, and is characterized by large-scale, rather than small-scale, DNA fragmentation (Fatokun et al., 2014; Wang et al., 2009). Although both parthanatos and necrosis involve the loss of cell-membrane integrity, parthanatos does not share the cell swelling that is characteristic of necrosis (Fan et al., 2017; Fatokun et al., 2014). Table 1 highlights the morphological

Table 1 Differential features of parthanatos, apoptosis, necroptosis and autophagic cell death.

Forms of death		Parthanatos	Apoptosis	Necroptosis	Autophagic cell death
Trigger		Genotoxic stress, Excitotoxicity	Death receptor (extrinsic), Intrinsic	Activation of death receptors, Inhibition of caspase-8	Starvation, Rapamycin
Cell morphology	Membrane	Loss of membrane integrity	Cell membrane blebbing Apoptotic body	Cellular swelling, Membrane rupture	Cell membrane blebbing
	Nucleus	Chromatin condensation, Large DNA fragmentation	Chromatin condensation, Small DNA fragmentation	–	–
	Cytosol	Mitochondrial depolarization, Cell shrinkage	Mitochondrial depolarization, Cell shrinkage	Mitochondrial swelling	Autophagic vacuoles, Increased lysosomal activity
Key regulators		PARP-1, AIF, MIF	Caspases, Bax, Bak, APAF1, cytochrome *c*	RIP1/3, MLKL	Atg proteins, Beclin-1
Inhibitors		PARP inhibitors, MIF nuclease inhibitors (putative), PARG, Iduna	Caspases inhibitors, Bcl-2	Necrostatin-1	3-Methyladenine, Wortmannin
Energy (ATP) requirement		No	Yes	No	Yes
Cellular markers		PARP1 activation, PAR accumulation, AIF and MIF translocation to nucleus	Caspase activation	RIPK3 activation	Autophagosome formation, Lysosomal degradation

and biochemical characteristics and neurodegenerative disease relevance that differentiates parthanatos from other well-known regulated cell death types including apoptosis, necroptosis and autophagic cell death.

Parthanatos can be triggered by oxidative stress, nitric oxide production or extensive DNA damage. When DNA damage is induced by ionizing radiation, exposure to *N*-methyl-D-aspartate (NMDA) or DNA-alkylating agent *N*-methyl-*N′*-nitro-*N*-nitrosoguanidine (MNNG), oxygen glucose deprivation (OGD) or prolonged treatment with hydrogen peroxide and other reactive oxygen species (ROS), PARP-1 overactivates and produces long-chained and branched PAR polymers (Dawson and Dawson, 2017; Wang et al., 2004; Yu et al., 2002, 2006). PAR generated from overactivation of PARP-1 is released from the nucleus into cytosol where it binds apoptosis-inducing factor (AIF) causing it release from the mitochondria (Wang et al., 2011). There is small pool of AIF that resides on the outer surface of the mitochondria and this is the releasable pool of AIF (Yu et al., 2009). Once AIF is released from the outer surface of the mitochondria it binds to macrophage migration inhibitory factor (MIF) and together they translocate to the nucleus (Wang et al., 2016). MIF is a DNA nuclease that cleaves genomic DNA into large (approximately 50 kb fragments) leading to chromatin and nuclear condensation fragmentation, which is likely the final step in the parthanatic death cascade (Fig. 1). The translocation of mitochondrial AIF into the nucleus can be prevented by the PAR degrading enzyme, PAR glycohydrolase (PARG) or ADP-ribosyl-acceptor hydrolase 3 (ARH3), but not by caspase inhibition (Mashimo et al., 2013; Yu et al., 2006).

When cells die, the specific cell death mechanism depends on the nature, duration, and degree of the stress. Cell death that is due to parthanatos should be completely prevented by inhibitors or deletion of PARP-1 (Fan et al., 2017). Parthanatos plays a role in most neurologic and neurodegenerative diseases including stroke (Eliasson et al., 1997), Parkinson's disease (PD) (Kam et al., 2018; Mandir et al., 1999), Alzheimer's disease (AD) (Love et al., 1999), Huntington's disease (HD) (Vis et al., 2005) and amyotrophic lateral sclerosis (ALS) (Hivert et al., 1998) among others (Yu et al., 2003).

3. Regulation of parthanatos

The biochemical events of parthanatos are distinct and occur with precision. Events begin with the rapid overactivation of PARP, the resulting synthesis and accumulation of PAR, then the translocation of mitochondrial

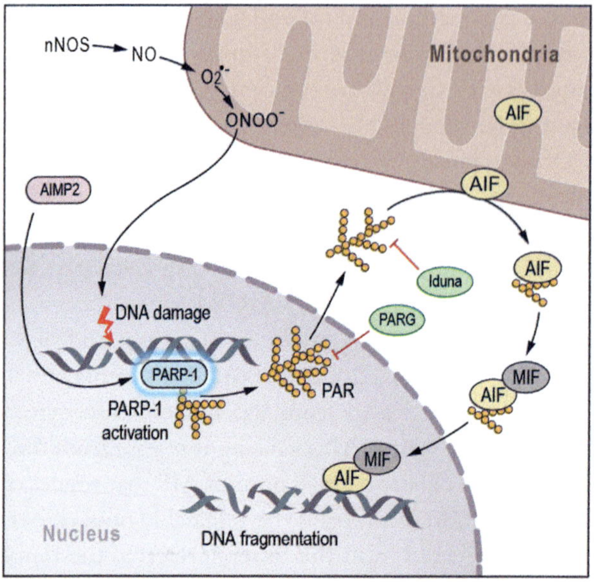

Fig. 1 Mechanisms of parthanatos. The first step of parthanatos is overactivation of PARP-1 due to direct or indirect DNA damage. Stressors such as oxidative stress, ischemia, NMDA, or α-synuclein PFF activate neuronal nitric oxide synthase (nNOS) that produces nitric oxide (NO). NO reacts with superoxide ($O_2^{\cdot -}$) in the mitochondria to generate peroxynitrite ($ONOO^-$), resulting in DNA damage. PARP-1 is also activated by direct DNA damage, including reactive oxygen species (ROS), alkylating agents [e.g., *N*-methyl-*N*′-nitro-*N*-nitrosoguanidine (MNNG)], and UV and ionizing radiation. Nuclear translocation and direct interaction of PARP-1 with AIMP2 activates PARP-1 in a DNA damage-independent manner (Lee et al., 2013). Overactivated PARP-1 leads to poly (ADP-ribose) (PAR) polymer synthesis and accumulation. PAR polymers then translocate from the nucleus to the cytosol and mitochondria where it binds to apoptosis inducing factor (AIF), inducing mitochondrial release and translocation of AIF to the nucleus. Interaction of AIF with macrophage migration inhibitory factor (MIF) is required for recruitment of MIF to the nucleus, where MIF cleaves genomic DNA into large-scale fragments via its nuclease activity, which causes cell death. PARG or Iduna degrade PAR or PAR-binding proteins, respectively, to prevent parthanatos.

AIF into the cytosol where it binds to MIF and the MIF/AIF complex translocates to the nucleus. These events conclude by inducing large-scale DNA fragmentation (Dawson and Dawson, 2017; Fatokun et al., 2014). The major events and their participants will be explored below.

3.1 Poly (ADP-ribose) polymerase-1 (PARP-1)

PARP-1 (EC 2.4.2.30) is the most abundant and well characterized among the 17-mammalian PARP family proteins classified on the basis of enzymatic function and protein domain homology (D'Amours et al., 1999). It is a nuclear enzyme that regulates cellular homeostasis and genomic stability (Krietsch et al., 2013). PARP-1 is a 116-kDa protein with three major functional domains. There is an N-terminal DNA-binding domain (42 kDa) that has three zinc finger motifs to sense DNA strand breaks and a nuclear localization sequence (NLS) in a helix-turn-helix motif used for DNA-binding. In the center is an automodification domain (16 kDa) that contains sites for self-ADP-ribosylation and a BRCA1 C-terminal (BRCT) motif for protein-protein interaction. At the C-terminus there is a catalytic domain (55 kDa) harboring an NAD^+-binding site and a conserved PARP signature motif (PSM) for PAR synthesis (D'Amours et al., 1999; Luo and Kraus, 2012) (Fig. 2). PARP-1 is highly conserved in eukaryotes; there is close to 92% homology between human and mouse PARP-1 (Virag and Szabo, 2002).

Under physiological conditions, PARP-1 is known to regulate mitochondrial function, gene expression, cell differentiation, cell division, DNA replication and repair, and cell death (D'Amours et al., 1999; Ji and Tulin, 2010; Kraus, 2008; Kraus and Lis, 2003; Krietsch et al., 2013; Schreiber et al., 2006). PARP-1, after sensing DNA breaks, activates a DNA base-excision repair system through the synthesis of PAR polymer. When PARP-1 recognizes mild DNA damage, its activity increases 500-fold (D'Amours et al., 1999). It synthesizes negatively charged PAR polymers by using oxidized NAD^+ produced from ATP and links the polymers using glycosidic bonds (D'Amours et al., 1999) (Fig. 3). About 20–500 PAR polymer residues are then covalently attached to PARP itself

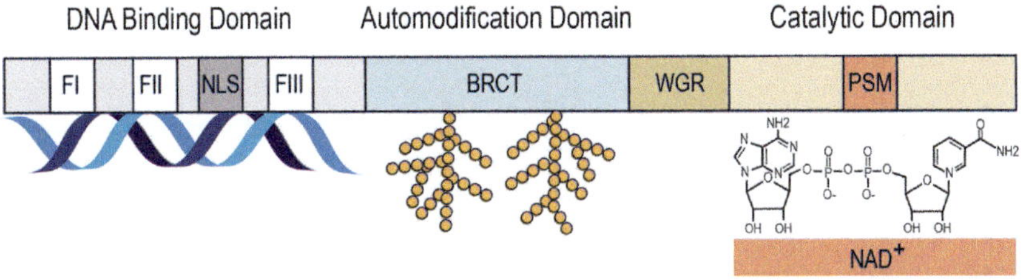

Fig. 2 PARP-1 protein domains. PARP-1 has an amino-terminal DNA binding domain containing three zinc finger motifs that are required for DNA binding and DNA damage detection and a nuclear localization signal (NLS). The central automodification domain contains a BRCT motif for protein-protein interaction and glutamate and lysine residues as the target of covalent auto-poly (ADP-ribosyl)ation to regulate PARP-1 activity. PARP-1 also has a carboxyl-terminal catalytic domain with enzymatic activity in the PARP signature motif (PSM) that catalyzes ribosylation reactions using nicotinamide adenine dinucleotide (NAD+) as a donor of ADP-ribose groups.

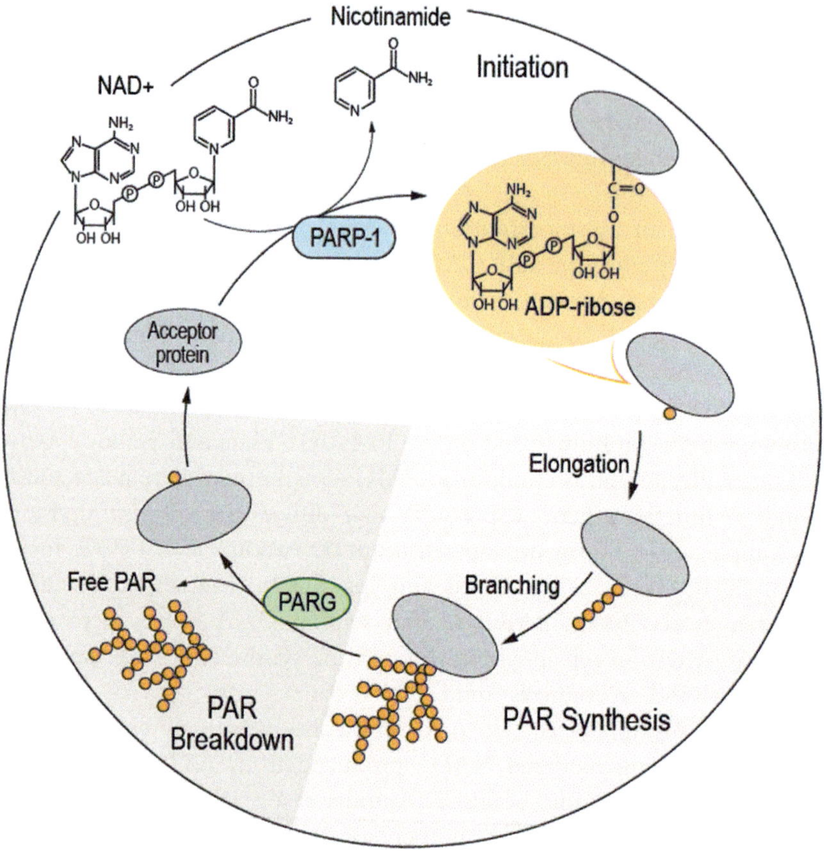

Fig. 3 Metabolism of poly (ADP-ribose). PARP-1 catalyzes the three steps of poly (ADP-ribose) anabolism; (i) Hydrolyzing NAD+ to ADP-ribose unit on the acceptor proteins, with the concomitant release of nicotinamide (Initiation), (ii) Elongation, and (iii) Branching of the polymer (PAR synthesis). PARG cleaves glycosidic bonds between ADP-ribose subunits located at the ends and within the poly (ADP-ribose) chains, which causes the degradation and release of poly (ADP-ribose) from the acceptor proteins (PAR breakdown). The final ADP-ribose on the acceptor proteins is removed by mono-ADP-ribosyl-protein lyase.

(autoribosylation) or to other nuclear proteins (heteroribosylation) such as histones, transcription factors, DNA ligase-2, topoisomerase I and II, DNA polymerase, high-mobility group proteins, as well as many other proteins (Gibson and Kraus, 2017). Through this poly ADP-ribosylation and interaction with BRCT, PARP-1 can recruit DNA repair proteins to the damaged site to facilitate DNA repair (Li et al., 2013).

When DNA damage moves past mild into severe, PARP-1 becomes excessively activated leading to the production of a large amount of PAR. This is the first step of the toxic parthanatic cascade followed by migration of AIF from the mitochondria into the cytosol where it binds MIF and then together they enter the nucleus followed by large scale DNA fragmentation by MIF, chromatin condensation and cell death. Thus, it is very important to distinguish between the degrees of DNA damage that activates PARP-1. Many toxic stimuli are shown to induce excessive activation of PARP-1; these include oxidative stress caused by ROS, ischemia, ischemia-reperfusion, hypoxia, reactive nitrogen species including peroxynitrite ($ONOO^-$), formed from reaction of nitric oxide with the superoxide anion, and neurotoxic aggregated proteins such as α-synuclein and amyloid-β (Aβ) (Angelova and Abramov, 2014; Kam et al., 2018; Wang et al., 2004; Yu et al., 2002, 2006). Ischemia and reperfusion lead to large-scale release of the excitatory neurotransmitter glutamate. This excessively stimulates glutamate receptors, the NMDA receptor in particular, and causes a large calcium influx. This calcium influx causes the activation of neuronal nitric oxide synthase (nNOS) and cell death via parthanatos (Dawson and Dawson, 2017; Dawson et al., 1991; Zhang et al., 1994).

3.2 Poly (ADP-ribose) (PAR) polymer

PAR is a polymer of two or more ADP-ribose units that can be attached as a post-translational modification. Multiple ADP-ribose can be covalently attached to proteins at different amino acids including lysine, arginine, glutamate, aspartate, cysteine, diphthamide, serine, threonine, phosphoserine and asparagine (Teloni and Altmeyer, 2016). To make the PAR polymer, subunits are linked to each other via ribose-ribose bonds. The ribose group of one ADP-ribose unit is usually attached to the adenosine of the adjacent ADP-ribose unit. Branching of the polymer can occur when a non-adenosine ribose from a neighboring ADP-ribose unit is linked (Fig. 4) (Leung, 2014; Miwa et al., 1979). Both protein-bound and free PAR polymers can non-covalently bind to other proteins. Binding occurs through their conserved PAR-binding motifs including PAR-binding motifs (PBMs), macrodomains, WWE domains, and PAR-binding zinc finger (PBZF) domains as well as other binding domains (Krietsch et al., 2013; Luo and Kraus, 2012). Bound proteins experience changes in their structure, function and localization as the binding motifs often overlap with other functional domains.

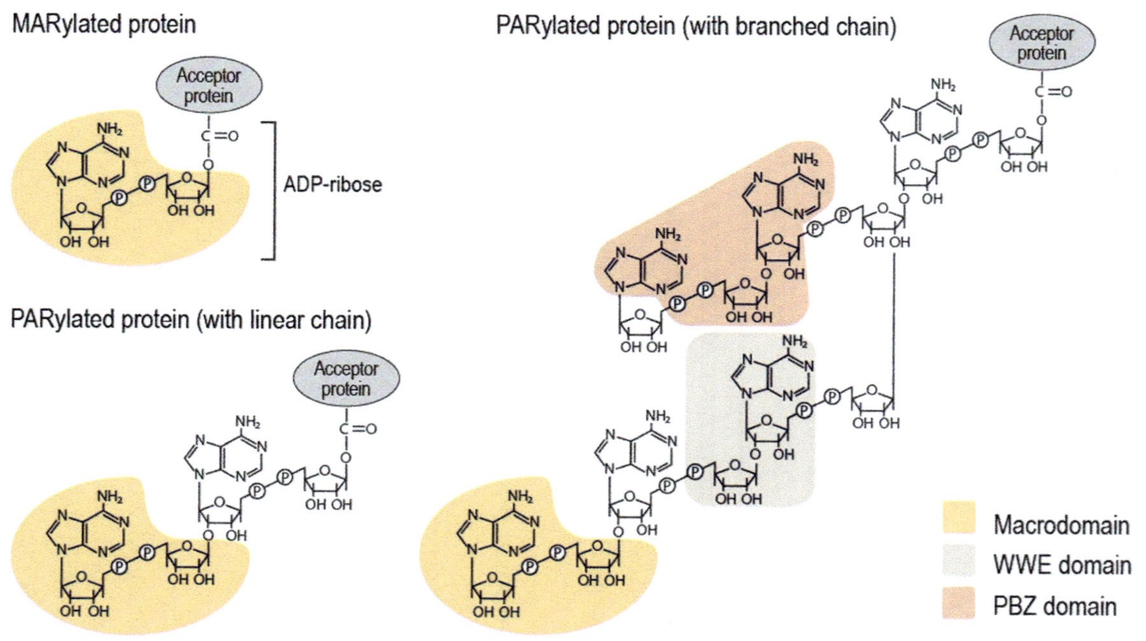

Fig. 4 Protein ADP-ribosylation and ADP-ribose recognition by protein domains. Schematic illustrations of proteins with mono (ADP-ribosyl) ation (MARylated protein) and poly (ADP-ribosyl)ation (PARylated protein). A linear chain is formed when adenine-linked ribose is joined to a non-adenine ribose, while a branched chain is formed when two non-adenine riboses are linked together. Shown are examples of PAR recognition domains with PAR moieties. The Macro domain interacts with the terminal ADP ribose unit. The WWE domain interacts with the isoADP-ribose unit. The PBZ domain interacts with the adenine in two tandem ADP-ribose units.

As already discussed, PAR is involved in the initial stage of parthanatos through PARP overactivation. Experimental evidence that indicates PAR mediates parthanatos comes from the protection from cell death induced by NMDA treatment in mouse primary neurons when cytosolic PAR is neutralized with PAR-specific antibodies (Andrabi et al., 2006). It was further demonstrated that PAR itself was toxic to neurons when exogenous PAR administered via Bioporter to facilitate PAR entry into cells was shown to directly induce cell death (Andrabi et al., 2006). PAR toxicity was further shown to be dependent on the extent of the dose and/or the complexity of the polymer (Andrabi et al., 2006). This study suggests that PAR polymer has the ability to exit the nucleus and translocate into the cytoplasm and induce cell death. It is not clear how the large, highly negatively charged PAR polymer escapes from nucleus and arrives at the mitochondria, but it is hypothesized that a binding partner binds to PAR to facilitate the translocation.

3.3 Poly (ADP-ribose) glycohydrolase (PARG) and ADP-ribosyl hydrolase 3 (ARH3)

PARG is an endogenous enzyme that regulates PAR levels by degrading the polymers after they are synthesized by PARPs (Davidovic et al., 2001). PARG-catalyzed hydrolysis of PAR polymers yields free ADP-ribose units through its endoglycosidic and exoglycosidic activity (O'Sullivan et al., 2019). PARG exists in alternatively spliced isoforms with full-length PARG (110 kDa) localized in the nucleus and the splice variants in diverse localization patterns (Bonicalzi et al., 2003; Haince et al., 2006; Meyer et al., 2007; Meyer-Ficca et al., 2004). Many studies have demonstrated the importance of PARG in parthanatos. Overexpression of PARG has been found to protect against PAR-mediated cell death (Andrabi et al., 2006), hydrogen peroxide-induced cell death (Blenn et al., 2006), excitotoxicity and stroke (Andrabi et al., 2006; Cozzi et al., 2006), while reducing its levels leads to increased toxicity through the accumulation of PAR in cells (Andrabi et al., 2006). Trophoblasts from the early stage of PARG-knockout embryos cannot survive without the PARP inhibitor, benzamide, and these cells were more sensitive to MNNG-induced toxicity (Koh et al., 2004). Furthermore, whole-body knockout of PARG is embryonically lethal around day E3.5 due to the excessive accumulation of the synthesized PAR polymer (Koh et al., 2004). Interestingly, PARG catalytic domain-deficient drosophila exhibit similar lethality in the larval stages, but mutants that live at higher temperature show progressive neurodegeneration with reduced locomotor activity and a decreased lifespan along with extensive accumulation of PAR in the central nervous system (Hanai et al., 2004). These data suggest that PARG plays a critical role in the normal cell function in neurons and that parthanatos is responsible for neurodegeneration (Zhou et al., 2011).

A different enzyme showing some PARG-like activity, named ADP-ribosyl hydrolase 3 (ARH3 or ADPRHL2) was also identified (Mueller-Dieckmann et al., 2006). ARH3 hydrolyzes the O-glycosidic bond of PAR and O-acetyl-ADP-ribose (OAADPr), in both instances generating mono (ADP-ribose) (MAR) (Kasamatsu et al., 2011; Ono et al., 2006). Mouse fibroblasts derived from ARH3-knockout embryos show increased PAR content in the nucleus and cytoplasm, release of AIF from the mitochondria into the nucleus, chromatin condensation, and nuclear shrinkage, resulting in susceptibility to hydrogen peroxide-induced cell death through parthanatos (Mashimo et al., 2013). Recently, a family was identified with an ARH3 gene mutation that results in a truncated and inactive protein. The patient exhibited a progressive neurodegeneration phenotype (Mashimo et al., 2019). Furthermore, the patient's fibroblasts were more sensitive to hydrogen peroxide-induced PAR accumulation and cell death, suggesting that ARH3 regulates parthanatos and neurodegeneration.

3.4 Apoptosis-inducing factor (AIF)

AIF is a flavoprotein that is primarily localized to the intermembrane space of the mitochondria, although 20–30% of it is loosely associated with the cytosolic side of the outer mitochondrial membrane (Yu et al., 2009). AIF translocates from the mitochondria to the nucleus in parthanatos (Yu et al., 2002, 2006). AIF contains 16 exons and is located on chromosome X (Susin et al., 1999). AIF is first synthesized in the cytoplasm as a 67kDa-precursor protein and is then imported into mitochondria using its mitochondrial localization sequence (MLS) in its N-terminus where it is processed into the mature 62kDa form (Otera et al., 2005). AIF is ubiquitously expressed and has five isoforms in humans (Delettre et al., 2006a,b). The crystal structures of mouse and human AIFs have been solved revealing the positive electrostatic potential at their surface, which may function to bind DNA (Ye et al., 2002). The crystal structure of AIF reveals three putative functional domains as well: (1) a N-terminal FAD-binding domain, (2) a central reduced NAD (NADH)-binding domain and (3) a C-terminal domain. The first two domains are responsible for the oxidoreductase function of AIF, which confers electron transfer activity to the protein.

The C-terminal domain mainly contributes to the cell death (parthanatos) properties of AIF (Wang et al., 2011). The oxidoreductase activity of AIF is not required for it to function in its cell death capacity (Ye et al., 2002). However, mutation of AIF's PAR-binding site (separate from its DNA-binding site), which has NADH oxidase activity and binds FAD or DNA, renders AIF unable to translocate into the nucleus and mediate cell death when PARP-1 is activated (Wang et al., 2011).

AIF is a bifunctional protein supporting cell survival with vital mitochondria functionality that also displays a lethal function as a cell death mediator in parthanatos. AIF may be involved in cortical development by supporting normal mitochondrial respiration or the maintenance of mitochondrial structure (Cheung et al., 2005). Loss of AIF in muscle leads to mitochondrial dysfunction, muscle atrophy and dilated cardiomyopathy (Joza et al., 2005) while it is associated with a protective function in diabetes and obesity (Pospisilik et al., 2007). In addition, forebrain-specific AIF null mice show defection in cortical development and die at E17 (Cheung et al., 2005). Harlequin (Hq) mice, which have about an 80% down-regulation of AIF expression, serve as a very useful model to study AIF-mediated cell death. These mice develop oxidative stress-mediated neurodegeneration (Klein et al., 2002). All of these data suggest that AIF supports cell survival. On the other hand, AIF is also a key regulator that mediates PARP-1 dependent cell death (Cregan et al., 2002; Wang et al., 2004; Yu et al., 2002, 2006). Exposure to toxins that induce DNA damage, such as MNNG, NMDA or oxidative stress, leads to AIF translocation from the mitochondria to nucleus resulting in large-scale DNA fragmentation and cell death (Yu et al., 2002). Cortical neurons from Harlequin mice are resistant to NMDA toxicity and treatment with PAR polymer (Yu et al., 2006) and these mice have reduced lesions compared to wild-type mice in excitotoxicity or stroke conditions (Culmsee et al., 2005; Yu et al., 2006). However, forced AIF expression raised NMDA-induced cell death to the same level as wild type neurons in the Hq neurons (Yu et al., 2006). In mouse embryonic fibroblasts this process is not blocked by the broad-spectrum caspase inhibitors boc-aspartyl-fmk or *Z*-VAD-fmk (Yu et al., 2002). Therefore, AIF is the commitment point (point of no return) in parthanatic cell death, which is independent of caspase activation.

Upon entry of AIF into the nucleus, AIF binds to DNA and it is thought to be required for the precipitation of cell death, however, other mechanisms may also exist because unlike human AIF, mouse AIF does not have any structural motifs recognizable as DNA binding motifs (Mate et al., 2002).

3.5 Macrophage migration inhibitory factor (MIF)

MIF is a pro-inflammatory cytokine that was first identified in 1966 as an inhibitor of macrophage random migration when released from T lymphocytes during a delayed-type hypersensitivity response (Bloom and Bennett, 1966; David, 1966; Schindler et al., 2018). The human MIF gene is located on chromosome 22q11.2 and consists of 3 exons and 2 introns encoding a protein with 114 amino acid residues (12.5kDa) (Paralkar and Wistow, 1994). All mammalian MIFs (mouse, rat, cattle and human) have ~90% homology. Such a high-level of conservation of MIF implies that it might have critical biological functions. Crystallographic studies of MIF indicate that it is a homotrimer showing a unique molecular structure that is different from other cytokines (Lolis and Bucala, 1996; Sun et al., 1996). Initial three-dimensional structure comparison of MIF indicated that it resembles three microbial enzymes having enzymatic activities of tautomerase, thiol-protein oxidoreductase and ketoenol isomerase. The amino-terminal proline residue is critical for this catalytic activity; however, it is unclear to what extent this functional enzyme activity of MIF contributes to its biological function (Bendrat et al., 1997; Hermanowski-Vosatka et al., 1999; Kleemann et al., 2000; Lubetsky et al., 2002). Originally, the main cellular source of MIF in the immune system was thought to be T cells, but current research shows MIF is constitutively expressed by monocytes, macrophages and lymphocytes in contrast to most cytokines. Besides the immune system, MIF is also broadly expressed in other cell types throughout the body, including neurons in the brain (Baugh and Bucala, 2002; Calandra and Roger, 2003; Lue et al., 2002).

Currently, MIF is recognized as a pleiotropic inflammatory cytokine. MIF is found in the cytoplasm of resting immune cells, but is released upon stimulation via a non-classical secretory pathway (Merk et al., 2009). Secreted MIF activates signaling by binding to transmembrane receptors including CD74. Following receptor activation, MIF facilitates cell proliferation, cell survival and inhibition of apoptosis via ERK1/2 MAP kinase, PI3K/AKT pathway and down-regulation of p53 (Fingerle-Rowson et al., 2003; Morand et al., 2006). MIF has been recognized as an important regulator of the innate immune response, where it stimulates the production of pro-inflammatory mediators such as TNF-α, interferon-γ and interleukins (IL-1β, IL-2, IL-6, and IL-8) thereby triggering cytokine production (Calandra and Roger, 2003). Given its broad pro-inflammatory activities, MIF is thought to be involved in the pathogenesis inflammatory diseases as rheumatoid asthma, arthritis, diabetes, sepsis, cancer, atherosclerosis and other cardiovascular diseases (Arjona et al., 2007; Calandra and Roger, 2003; Lue et al., 2002; Mizue et al., 2005; Morand et al., 2006; Sanchez et al., 2006).

Apart from the known roles of MIF, it was also identified in a screen for AIF interacting proteins as a new member of a large nuclease superfamily (Wang et al., 2016). MIF interacts with AIF, translocates to the nucleus from the cytoplasm MIF's nuclease activity is responsible for the large-scale DNA fragmentation during parthanatos (Wang et al., 2016). MIF does not show any sequence homology to other human nucleases. However, alignments of the PD-D/E(x)K motif in MIF with bacterial restriction endonuclease, including *Eco*RI, *Eco*RV, ExoIII and *Pvu*II, reveal similar orientations of the four beta strands in the beta sheet against the alpha helices, as observed in a typical PD-D/E(x)K motif active site, suggesting a structural similarity (Wang et al., 2016). MIF has both 3′exonuclease and endonuclease activity in the presence of Mg^{2+} or Ca^{2+}. Furthermore, depletion of MIF reduced chromatinolysis and NMDA induced cell death in neurons or focal stroke in mice (Wang et al., 2016). Inhibition of MIF's nuclease activity by mutation of key amino acid residues in its nuclease domain or prevention of its translocation to the nucleus by mutation of the AIF/MIF interaction site markedly protects against parthanatos and attenuates histological and behavioral defects in the focal ischemia model of stroke (Wang et al., 2016). This indicates that MIF is a PARP-1-dependent AIF-associated nuclease (PAAN) that is required for parthanatos (Jonas, 2016; Wang et al., 2016).

4. Parthanatos in neurodegenerative diseases

Neuronal cell death is one of the main hallmarks of neurodegenerative diseases. Regions of the brain that suffer neuronal loss differ among diseases and the disease progresses other regions or neuronal subtypes can become involved. Another common hallmark of neurodegenerative diseases is aberrant protein aggregation. While the localization and types of protein aggregation are different, it is generally accepted that these aggregates or the precursors of the aggregates are responsible for the pathology and neuronal cell death. Thus, an increased understating of the molecular mechanisms of cell death, especially parthanatos, and their involvement in different types of neurodegenerative diseases is critical to the development of effective treatments for these diseases.

4.1 Alzheimer's disease (AD)

AD the most common type of dementia, is characterized by extracellular deposition of Aβ in senile plaques and paired helical filaments (PHFs) containing hyperphosphorylated microtubule-associated tau proteins in neurofibrillary tangles (NFT) (O'Brien and Wong, 2011). The upstream regulator of parthanatos, PARP-1, was reported to be overactive in AD human brains as evidenced by higher levels of PAR (Love et al., 1999). PARP-1 activation was also indicated at the early stage in the hippocampus and cortex of TgCRND8 mice containing the double Swedish (K670N/M671L) and Indiana (V717F) amyloid precursor protein (APP) mutations under the hamster prion (PrP) promoter (Martire et al., 2013). Treatment with Aβ caused PAR accumulation, which could be prevented by co-incubation with the PARP-1 inhibitor PJ34 in cultured cells (Abeti et al., 2011; Angelova and Abramov, 2014; Kauppinen et al., 2011). Moreover, hAPP-J20 mice having the Swedish (K670N/M671L) and Indiana (V717F) APP mutations under platelet-derived growth factor (PDGF) promoter crossed with PARP-1 knockout mice (hAPP-J20/PARP-1$^{-/-}$) showed reduced brain pathology including microglial activation, hippocampal synaptic integrity and cognitive function (Kauppinen et al., 2011). Injection of Aβ into the brain of PARP-1 knockout mice or treatment with PJ34 attenuated microglial activation (Turunc Bayrakdar et al., 2014). These data suggest a protective role for PARP-1 in AD models through genetic or pharmacological inhibition. In addition to PARP-1, translocation of AIF was also described in TgCRND8 mice (Yu et al., 2014). Additionally, there is a marked increase in MIF levels in the cerebrospinal fluid (CSF) of AD patients. Taken together, these results suggest the involvement of parthanatos in AD (Bacher et al., 2010; Popp et al., 2009).

4.2 Parkinson's disease (PD)

PD is the most common movement disorder disease. It is characterized by progressive, age-dependent dopaminergic neuronal loss in the substrata nigra pars compacta. This neurodegeneration is caused by fibrillary α-synuclein (Goedert, 2001). PD shares a common pathway with AD, as oxidative stress is an activator of PARP-1, which sets in motion parthanatos. Injection of 1-methyl-4-phenyl-1,2,3,6-tetrahydropyridine (MPTP), a toxin to induce PD-like symptoms in mice, induces PARP-1 activation (Cosi et al., 1996; Wang et al., 2003; Wu et al., 2014). PARP-1 knockout mice and MPTP-injected mice treated with the PARP inhibitor benzamide were resistant to toxicity and showed a reduced level of dopamine (DA) (Mandir et al., 1999; Yokoyama et al., 2010). Similarly, PARP-1 knockout mice prevented dopaminergic neurodegeneration and PD-related symptoms by suppressing AMP protein kinase (AMPK) and the nuclear translocation of AIF following injection of 6-hydroxydopamine (6-OHDA), a toxin that also induces PD-like symptoms in mice (Kim et al., 2013). Recently, the importance of PARP-1 was emphasized in the PD mouse model using α-synuclein preformed fibrils (PFFs). Injection of α-synuclein PFF into the striatum led to PAR-induced aggregation of α-synuclein and neurotoxicity that was mitigated by genetic deletion of PARP-1 or treatment with the PARP inhibitor, ABT-888 (Kam et al., 2018). These studies suggest a feedforward mechanism whereby the parthanatos induced by fibrillary α-synuclein hyperactivates PARP-1, which leads to accumulation of PAR that can interact with α-synuclein promoting more aggregation and toxicity (Fig. 5). Interestingly, elevated levels of PAR were found in the CSF of PD patients suggesting that PAR could be a potential biomarker for PD (Kam et al., 2018). In addition to PARP-1 activation, nuclear translocation of AIF has also been found in tyrosine-hydroxylase (TH) positive neurons in primary neurons of PD patients (Burguillos et al., 2011).

4.3 Huntington's disease (HD)

HD is an autosomal dominant neurodegenerative disorder. Symptoms include the progressive loss of motor and cognitive function, loss of spatial and self-awareness, and can include depression and dementia. It is caused by a CAG trinucleotide repeat expansion in exon 1 of the huntingtin gene. In HD, the number of CAG repeats is expanded from the normal range of 10–35 repeats to 40–80 repeats (Labbadia and Morimoto, 2013; Munoz-Sanjuan and Bates, 2011). This results in an elongated polyglutamine (polyQ) tract at the N-terminus of the huntingtin (htt) protein. This polyQ expansion is responsible for htt aggregation and is associated with neuronal intranuclear inclusions and toxicity (Williams and Paulson, 2008). The htt protein localizes to the sites of DNA damage and scaffolds DNA repair proteins, suggesting a relationship with parthanatos (Maiuri et al., 2017). Indeed, in the caudate nucleus of severely affected HD brains, accumulation of damaged DNA and increased PARP-1 expression were reported, while only weak caspase-3 activation was detected, suggesting that caspase-mediated cell death has little contribution to pathology of HD

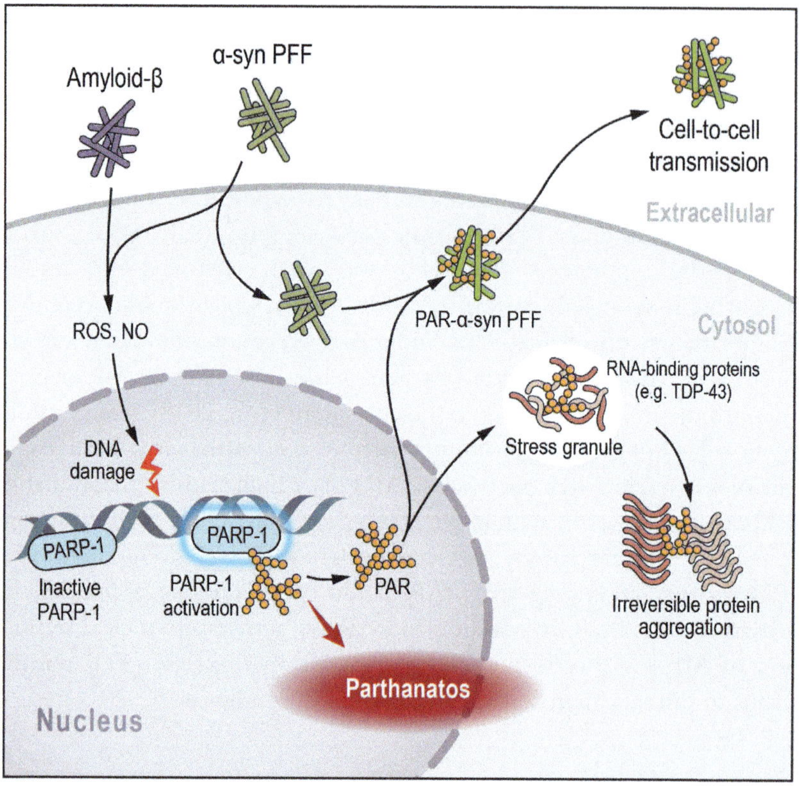

Fig. 5 Role of poly (ADP-ribose) and PARP-1 in neurodegenerative diseases. In Alzheimer's disease (AD) or Parkinson's disease (PD), aggregated proteins (amyloid-β or α-synuclein) cause the hyperactivation of PARP-1 through accumulation of oxidative stress (i.e., reactive oxygen species or nitric oxide) and increased levels of cellular PAR, which together lead to neuronal cell death by parthanatos. In a feedforward loop, PAR interacts with α-synuclein and causes the formation of a more toxic α-synuclein strain, resulting in accelerated pathologic α-synuclein cell-to-cell transmission and toxicity in PD models (Kam et al., 2018). PAR also binds to RNA-binding proteins (e.g., TDP-43) and promotes its accumulation in stress granules that causes irreversible aggregation in amyotrophic lateral sclerosis (ALS) models (McGurk et al., 2018).

(Askeland et al., 2018; Vis et al., 2005). Moreover, when HD model R6/2 mutant mice expressing a portion of the human HD gene are treated with the PARP-1 inhibitor INO-1001, they survive longer and display less signs of neuropathological dysfunction than vehicle-treated mice. This suggests a neuroprotective effect of PARP-1 in HD (Cardinale et al., 2015; Paldino et al., 2017).

4.4 Amyotrophic lateral sclerosis (ALS)

Amyotrophic Lateral Sclerosis (ALS) is a fatal neurodegenerative disease where the upper and lower motor neurons degenerate causing muscle atrophy, paralysis and death within ~2–5 years of disease onset (Hobson and McDermott, 2016; McGurk et al., 2018). The primary cytological features of ALS are cytoplasmic inclusions that contain primarily TAR DNA-binding protein (TDP-43). Sometimes other proteins are found within the inclusion, but there is a strong correlation between TDP-43 presence and neuronal loss (Hobson and McDermott, 2016). In the ALS brain, increased expression of PARP-1 is also detected in the motor cortex, parietal cortex and cerebellum, and is localized to a subset of TDP-43 inclusions (Kim et al., 2004; McGurk et al., 2014). Recently, elevated PAR levels were also reported in the nuclei of motor neurons in the spinal cord of ALS patients and the TDP-43 induced toxicity in those neurons could be mitigated by treatment with the PARP-1 inhibitor veliparib (McGurk et al., 2018) (Fig. 5). AIF release from the mitochondria and subsequent translocation into the nucleus were also observed in the spinal cord of ALS model mice expressing symptomatic superoxide dismutase 1 (SOD1) G93A mutation, suggesting a role for AIF in causing motor neuronal death (Oh et al., 2006).

5. Parthanatos and therapeutic targets in neurodegenerative diseases

Activation of PARP-1, accumulation of PAR polymers, the release of AIF from the mitochondria and subsequent nuclear translocation, and MIF-mediated DNA fragmentation and chromatin condensation are the key steps in the parthanatos pathway that provide therapeutic opportunities for the treatment of neurodegenerative diseases. Some PARP inhibitors have already been developed for the treatment of cancers, and it is time to repurpose them for treatment of other diseases (Berger et al., 2018).

5.1 PARP inhibitors

The development of PARP inhibitors for therapy of human disorders has a long history (Berger et al., 2018; Sachdev et al., 2019). There are four Food and Drug Administration (FDA) and European Medicine Agency (EMA) approved PARP inhibitors for the treatment of ovarian cancer and include olaparib, niraparib, rucaparib and talazoparib (McCann, 2019). Most PARP inhibitor development has focused on promoting cell death in cancer cells rather than preventing cell death in neurodegenerative diseases. In cancer cells, inhibition of PARP and/or trapping of PARP promotes cell death by blocking the DNA repair system required for their survival. Development of PARP inhibitors to treat neurodegenerative diseases may be more challenging due to potential side effects associated with prolonged treatment in neurons. Nevertheless, the beneficial effects of PARP inhibitors have been well described in several neurodegenerative disease models. For example, the treatment of PARP-1 inhibitors, ABT-888 (veliparib), AG-014699 (rucaparib) and BMN-673 (talazoparib) showed reduced pathologic α-synuclein aggregation and increased cell viability in models of PD (Kam et al., 2018), veliparib reduced TDP-43 mediated cell death in spinal cord cultures (McGurk et al., 2018) and olaparib attenuated TDP-43 induced cytotoxicity in motor neuron cells in models of ALS (Duan et al., 2019). There are many well-characterized PARP inhibitors that have yet to be tested for use in neurodegenerative disease and they should be considered for the treatment of neurologic diseases (Berger et al., 2018).

5.2 Inhibitors of parthanatos mediators

Apart from the PARP inhibitors, most potential interventions of parthanatos are still in the experimental stage. The next possible therapeutic target following PARP-1 activation is to clear the excessive accumulation of PAR polymer. Iduna is a PAR-dependent E3 ubiquitin ligase that ubiquitinates poly ADP-ribosylated substrates and targets them for degradation (Kang et al., 2011). It protects against glutamate NMDA receptor-mediated excitotoxicity both in vitro and in vivo, as well as MCAO-induced stroke in mice because of its ability to bind with PAR (Andrabi et al., 2011). Substrates of Iduna that are involved in parthanatos have not been reported. Strategies aimed at enhancing Iduna activity or enhancing the degradation of PAR could be therapeutic targets. Identification of Iduna substrates is priority.

Another attractive therapeutic target following PARP activation is to prevent the release of AIF from the mitochondria and its translocation to the nucleus. To date, no AIF inhibitor has been identified to prevent mitochondrial translocation of AIF to the nucleus. A high-throughput screen of small molecules using fluorescently tagged AIF for visualization could potentially reveal some candidates. Preventing PAR from binding AIF is a promising target. Another potential target is preventing the AIF-MIF interaction or the nuclease activity of MIF. As noted above, MIF's nuclease activity is required for large-scale DNA fragmentation is likely the final step of parthanatos.

MIF nuclease activity is an attractive therapeutic target. MIF does not share homology with other human DNases and its nuclease activity is separate from its tautomerase activity making it possible to develop compounds that should have a high safety profile. Thus, identification of selective MIF blockers that inhibit only MIF nuclease activity but not its normal enzymatic activity such as its tautomerase and oxidoreductase activity have a high likelihood of providing disease-modifying therapeutic agents with minimal side-effects in parthanatos driven neurodegenerative diseases.

6. Conclusion

PARP-1 dependent cell death, parthanatos, is a unique cell death pathway that is different from other types of cell death such as apoptosis, necroptosis or autophagic cell death. The molecular mechanism of parthanatos involves excessive activation of PARP-1 leading to an elevation of PAR polymer that acts as a cell death signal. PAR binds to AIF leading to its translocation from the mitochondria into the cytosol where it binds MIF and the AIF/MIF complex enters the nucleus where MIF causes large-scale DNA fragmentation and chromatin condensation. Parthanatic cell death plays an important role in neurodegenerative disorders including AD, PD, HD and ALS. Although cell death is the last stage in neurodegeneration, genetic depletion or pharmacological inhibition of parthanatos' regulators show beneficial effects in cellular and animal models of neurodegenerative diseases. Thus, further investigation into the parthanatos pathway, as well as identifying and developing inhibitors of parthanatos' components, are promising therapeutic approaches in the treatment of neurodegenerative diseases.

Acknowledgments

We would like to thank Dr. Jennifer Roshek and Ms. Elizabeth Wilmarth for editorial assistance. This work was supported by US National Institutes of Health grants NS38377, NS67525 and by the JPB Foundation. The authors acknowledge the joint participation by the Adrienne Helis Malvin Medical Research Foundation through its direct engagement in the continuous active conduct of medical research in conjunction with the Johns Hopkins Hospital and the Johns Hopkins University School of Medicine and the Foundation's Parkinson's Disease Programs M-2016. The authors thank I-Hsun Wu for assistance with the illustrations. T.M.D. is the Leonard and Madlyn Abramson Professor in Neurodegenerative Diseases.

References

Abeti, R., Abramov, A.Y., Duchen, M.R., 2011. Beta-amyloid activates PARP causing astrocytic metabolic failure and neuronal death. Brain 134, 1658–1672.

Andrabi, S.A., Kim, N.S., Yu, S.W., Wang, H., Koh, D.W., Sasaki, M., Klaus, J.A., Otsuka, T., Zhang, Z., Koehler, R.C., Hurn, P.D., Poirier, G.G., Dawson, V.L., Dawson, T.M., 2006. Poly(ADP-ribose) (PAR) polymer is a death signal. Proc. Natl. Acad. Sci. U. S. A. 103, 18308–18313.

Andrabi, S.A., Kang, H.C., Haince, J.F., Lee, Y.I., Zhang, J., Chi, Z., West, A.B., Koehler, R.C., Poirier, G.G., Dawson, T.M., Dawson, V.L., 2011. Iduna protects the brain from glutamate excitotoxicity and stroke by interfering with poly(ADP-ribose) polymer-induced cell death. Nat. Med. 17, 692–699.

Andrabi, S.A., Umanah, G.K., Chang, C., Stevens, D.A., Karuppagounder, S.S., Gagne, J.P., Poirier, G.G., Dawson, V.L., Dawson, T.M., 2014. Poly(ADP-ribose) polymerase-dependent energy depletion occurs through inhibition of glycolysis. Proc. Natl. Acad. Sci. U. S. A. 111, 10209–10214.

Angelova, P.R., Abramov, A.Y., 2014. Interaction of neurons and astrocytes underlies the mechanism of Abeta-induced neurotoxicity. Biochem. Soc. Trans. 42, 1286–1290.

Arjona, A., Foellmer, H.G., Town, T., Leng, L., Mcdonald, C., Wang, T., Wong, S.J., Montgomery, R.R., Fikrig, E., Bucala, R., 2007. Abrogation of macrophage migration inhibitory factor decreases West Nile virus lethality by limiting viral neuroinvasion. J. Clin. Invest. 117, 3059–3066.

Askeland, G., Dosoudilova, Z., Rodinova, M., Klempir, J., Liskova, I., Kusnierczyk, A., Bjoras, M., Nesse, G., Klungland, A., Hansikova, H., Eide, L., 2018. Increased nuclear DNA damage precedes mitochondrial dysfunction in peripheral blood mononuclear cells from Huntington's disease patients. Sci. Rep. 8, 9817.

Bacher, M., Deuster, O., Aljabari, B., Egensperger, R., Neff, F., Jessen, F., Popp, J., Noelker, C., Reese, J.P., Al-Abed, Y., Dodel, R., 2010. The role of macrophage migration inhibitory factor in Alzheimer's disease. Mol. Med. 16, 116–121.

Baugh, J.A., Bucala, R., 2002. Macrophage migration inhibitory factor. Crit. Care Med. 30, S27–S35.

Bendrat, K., Al-Abed, Y., Callaway, D.J., Peng, T., Calandra, T., Metz, C.N., Bucala, R., 1997. Biochemical and mutational investigations of the enzymatic activity of macrophage migration inhibitory factor. Biochemistry 36, 15356–15362.

Berger, N.A., Besson, V.C., Boulares, A.H., Burkle, A., Chiarugi, A., Clark, R.S., Curtin, N.J., Cuzzocrea, S., Dawson, T.M., Dawson, V.L., Hasko, G., Liaudet, L., Moroni, F., Pacher, P., Radermacher, P., Salzman, A.L., Snyder, S.H., Soriano, F.G., Strosznajder, R.P., Sumegi, B., Swanson, R.A., Szabo, C., 2018. Opportunities for the repurposing of PARP inhibitors for the therapy of non-oncological diseases. Br. J. Pharmacol. 175, 192–222.

Blenn, C., Althaus, F.R., Malanga, M., 2006. Poly(ADP-ribose) glycohydrolase silencing protects against H2O2-induced cell death. Biochem. J. 396, 419–429.

Bloom, B.R., Bennett, B., 1966. Mechanism of a reaction in vitro associated with delayed-type hypersensitivity. Science 153, 80–82.

Bonicalzi, M.E., Vodenicharov, M., Coulombe, M., Gagne, J.P., Poirier, G.G., 2003. Alteration of poly(ADP-ribose) glycohydrolase nucleocytoplasmic shuttling characteristics upon cleavage by apoptotic proteases. Biol. Cell 95, 635–644.

Burguillos, M.A., Hajji, N., Englund, E., Persson, A., Cenci, A.M., Machado, A., Cano, J., Joseph, B., Venero, J.L., 2011. Apoptosis-inducing factor mediates dopaminergic cell death in response to LPS-induced inflammatory stimulus: evidence in Parkinson's disease patients. Neurobiol. Dis. 41, 177–188.

Calandra, T., Roger, T., 2003. Macrophage migration inhibitory factor: a regulator of innate immunity. Nat. Rev. Immunol. 3, 791–800.

Cardinale, A., Paldino, E., Giampa, C., Bernardi, G., Fusco, F.R., 2015. PARP-1 inhibition is neuroprotective in the R6/2 mouse model of Huntington's disease. PLoS One 10, e0134482.

Cheung, E.C., Melanson-Drapeau, L., Cregan, S.P., Vanderluit, J.L., Ferguson, K.L., Mcintosh, W.C., Park, D.S., Bennett, S.A., Slack, R.S., 2005. Apoptosis-inducing factor is a key factor in neuronal cell death propagated by BAX-dependent and BAX-independent mechanisms. J. Neurosci. 25, 1324–1334.

Cosi, C., Chopin, P., Marien, M., 1996. Benzamide, an inhibitor of poly(ADP-ribose) polymerase, attenuates methamphetamine-induced dopamine neurotoxicity in the C57B1/6N mouse. Brain Res. 735, 343–348.

Cozzi, A., Cipriani, G., Fossati, S., Faraco, G., Formentini, L., Min, W., Cortes, U., Wang, Z.Q., Moroni, F., Chiarugi, A., 2006. Poly(ADP-ribose) accumulation and enhancement of postischemic brain damage in 110-kDa poly(ADP-ribose) glycohydrolase null mice. J. Cereb. Blood Flow Metab. 26, 684–695.

Cregan, S.P., Fortin, A., Maclaurin, J.G., Callaghan, S.M., Cecconi, F., Yu, S.W., Dawson, T.M., Dawson, V.L., Park, D.S., Kroemer, G., Slack, R.S., 2002. Apoptosis-inducing factor is involved in the regulation of caspase-independent neuronal cell death. J. Cell Biol. 158, 507–517.

Culmsee, C., Zhu, C., Landshamer, S., Becattini, B., Wagner, E., Pellecchia, M., Blomgren, K., Plesnila, N., 2005. Apoptosis-inducing factor triggered by poly(ADP-ribose) polymerase and Bid mediates neuronal cell death after oxygen-glucose deprivation and focal cerebral ischemia. J. Neurosci. 25, 10262–10272.

D'Amours, D., Desnoyers, S., D'silva, I., Poirier, G.G., 1999. Poly(ADP-ribosyl)ation reactions in the regulation of nuclear functions. Biochem. J. 342 (Pt. 2), 249–268.

David, J.R., 1966. Delayed hypersensitivity in vitro: its mediation by cell-free substances formed by lymphoid cell-antigen interaction. Proc. Natl. Acad. Sci. U. S. A. 56, 72–77.

Davidovic, L., Vodenicharov, M., Affar, E.B., Poirier, G.G., 2001. Importance of poly(ADP-ribose) glycohydrolase in the control of poly(ADP-ribose) metabolism. Exp. Cell Res. 268, 7–13.

Dawson, T.M., Dawson, V.L., 2017. Mitochondrial mechanisms of neuronal cell death: potential therapeutics. Annu. Rev. Pharmacol. Toxicol. 57, 437–454.

Dawson, V.L., Dawson, T.M., London, E.D., Bredt, D.S., Snyder, S.H., 1991. Nitric oxide mediates glutamate neurotoxicity in primary cortical cultures. Proc. Natl. Acad. Sci. U. S. A. 88, 6368–6371.

Delettre, C., Yuste, V.J., Moubarak, R.S., Bras, M., Lesbordes-Brion, J.C., Petres, S., Bellalou, J., Susin, S.A., 2006a. AIFsh, a novel apoptosis-inducing factor (AIF) pro-apoptotic isoform with potential pathological relevance in human cancer. J. Biol. Chem. 281, 6413–6427.

Delettre, C., Yuste, V.J., Moubarak, R.S., Bras, M., Robert, N., Susin, S.A., 2006b. Identification and characterization of AIFsh2, a mitochondrial apoptosis-inducing factor (AIF) isoform with NADH oxidase activity. J. Biol. Chem. 281, 18507–18518.

Duan, Y., Du, A., Gu, J., Duan, G., Wang, C., Gui, X., Ma, Z., Qian, B., Deng, X., Zhang, K., Sun, L., Tian, K., Zhang, Y., Jiang, H., Liu, C., Fang, Y., 2019. PARylation regulates stress granule dynamics, phase separation, and neurotoxicity of disease-related RNA-binding proteins. Cell Res. 29, 233–247.

Eliasson, M.J., Sampei, K., Mandir, A.S., Hurn, P.D., Traystman, R.J., Bao, J., Pieper, A., Wang, Z.Q., Dawson, T.M., Snyder, S.H., Dawson, V.L., 1997. Poly(ADP-ribose) polymerase gene disruption renders mice resistant to cerebral ischemia. Nat. Med. 3, 1089–1095.

Fan, J., Dawson, T.M., Dawson, V.L., 2017. Cell death mechanisms of neurodegeneration. Adv. Neurobiol. 15, 403–425.

Fatokun, A.A., Dawson, V.L., Dawson, T.M., 2014. Parthanatos: mitochondrial-linked mechanisms and therapeutic opportunities. Br. J. Pharmacol. 171, 2000–2016.

Fingerle-Rowson, G., Petrenko, O., Metz, C.N., Forsthuber, T.G., Mitchell, R., Huss, R., Moll, U., Muller, W., Bucala, R., 2003. The p53-dependent effects of macrophage migration inhibitory factor revealed by gene targeting. Proc. Natl. Acad. Sci. U. S. A. 100, 9354–9359.

Fouquerel, E., Goellner, E.M., Yu, Z., Gagne, J.P., Barbi De Moura, M., Feinstein, T., Wheeler, D., Redpath, P., Li, J., Romero, G., Migaud, M., Van Houten, B., Poirier, G.G., Sobol, R.W., 2014. ARTD1/PARP1 negatively regulates glycolysis by inhibiting hexokinase 1 independent of NAD+ depletion. Cell Rep. 8, 1819–1831.

Galluzzi, L., Vitale, I., Abrams, J.M., Alnemri, E.S., Baehrecke, E.H., Blagosklonny, M.V., Dawson, T.M., Dawson, V.L., El-Deiry, W.S., Fulda, S., Gottlieb, E., Green, D.R., Hengartner, M.O., Kepp, O., Knight, R.A., Kumar, S., Lipton, S.A., Lu, X., Madeo, F., Malorni, W., Mehlen, P., Nunez, G., Peter, M.E., Piacentini, M., Rubinsztein, D.C., Shi, Y., Simon, H.U., Vandenabeele, P., White, E., Yuan, J., Zhivotovsky, B., Melino, G., Kroemer, G., 2012. Molecular definitions of cell death subroutines: recommendations of the Nomenclature Committee on Cell Death 2012. Cell Death Differ. 19, 107–120.

Galluzzi, L., Vitale, I., Aaronson, S.A., Abrams, J.M., Adam, D., Agostinis, P., Alnemri, E.S., Altucci, L., Amelio, I., Andrews, D.W., Annicchiarico-Petruzzelli, M., Antonov, A.V., Arama, E., Baehrecke, E.H., Barlev, N.A., Bazan, N.G., Bernassola, F., Bertrand, M.J.M., Bianchi, K., Blagosklonny, M.V., Blomgren, K., Borner, C., Boya, P., Brenner, C., Campanella, M., Candi, E., Carmona-Gutierrez, D., Cecconi, F., Chan, F.K., Chandel, N.S., Cheng, E.H., Chipuk, J.E., Cidlowski, J.A., Ciechanover, A., Cohen, G.M., Conrad, M., Cubillos-Ruiz, J.R., Czabotar, P.E., D'angiolella, V., Dawson, T.M., Dawson, V.L., De Laurenzi, V., De Maria, R., Debatin, K.M., Deberardinis, R.J., Deshmukh, M., Di Daniele, N., Di Virgilio, F., Dixit, V.M., Dixon, S.J., Duckett, C.S., Dynlacht, B.D., El-Deiry, W.S., Elrod, J.W., Fimia, G.M., Fulda, S., Garcia-Saez, A.J., Garg, A.D., Garrido, C., Gavathiotis, E., Golstein, P., Gottlieb, E., Green, D.R., Greene, L.A., Gronemeyer, H., Gross, A., Hajnoczky, G., Hardwick, J.M., Harris, I.S., Hengartner, M.O., Hetz, C., Ichijo, H., Jaattela, M., Joseph, B., Jost, P.J., Juin, P.P., Kaiser, W.J., Karin, M., Kaufmann, T., Kepp, O., Kimchi, A., Kitsis, R.N., Klionsky, D.J., Knight, R.A., Kumar, S., Lee, S.W., Lemasters, J.J., Levine, B., Linkermann, A., Lipton, S.A., Lockshin, R.A., Lopez-Otin, C., Lowe, S.W., Luedde, T., Lugli, E., Macfarlane, M., Madeo, F., Malewicz, M., Malorni, W., Manic, G., et al., 2018. Molecular mechanisms of cell death: recommendations of the Nomenclature Committee on Cell Death 2018. Cell Death Differ. 25, 486–541.

Gibson, B.A., Kraus, W.L., 2017. Identification of protein substrates of specific PARP enzymes using analog-sensitive PARP mutants and a "clickable" NAD(+) analog. Methods Mol. Biol. 1608, 111–135.

Goedert, M., 2001. Alpha-synuclein and neurodegenerative diseases. Nat. Rev. Neurosci. 2, 492–501.

Gorman, A.M., 2008. Neuronal cell death in neurodegenerative diseases: recurring themes around protein handling. J. Cell. Mol. Med. 12, 2263–2280.

Ha, H.C., Snyder, S.H., 1999. Poly(ADP-ribose) polymerase is a mediator of necrotic cell death by ATP depletion. Proc. Natl. Acad. Sci. U. S. A. 96, 13978–13982.

Haince, J.F., Ouellet, M.E., Mcdonald, D., Hendzel, M.J., Poirier, G.G., 2006. Dynamic relocation of poly(ADP-ribose) glycohydrolase isoforms during radiation-induced DNA damage. Biochim. Biophys. Acta 1763, 226–237.

Hanai, S., Kanai, M., Ohashi, S., Okamoto, K., Yamada, M., Takahashi, H., Miwa, M., 2004. Loss of poly(ADP-ribose) glycohydrolase causes progressive neurodegeneration in Drosophila melanogaster. Proc. Natl. Acad. Sci. U. S. A. 101, 82–86.

Hermanowski-Vosatka, A., Mundt, S.S., Ayala, J.M., Goyal, S., Hanlon, W.A., Czerwinski, R.M., Wright, S.D., Whitman, C.P., 1999. Enzymatically inactive macrophage migration inhibitory factor inhibits monocyte chemotaxis and random migration. Biochemistry 38, 12841–12849.

Hivert, B., Cerruti, C., Camu, W., 1998. Hydrogen peroxide-induced motoneuron apoptosis is prevented by poly ADP ribosyl synthetase inhibitors. Neuroreport 9, 1835–1838.

Hobson, E.V., McDermott, C.J., 2016. Supportive and symptomatic management of amyotrophic lateral sclerosis. Nat. Rev. Neurol. 12, 526–538.

Ji, Y., Tulin, A.V., 2010. The roles of PARP1 in gene control and cell differentiation. Curr. Opin. Genet. Dev. 20, 512–518.

Jonas, E., 2016. The MIFstep in parthanatos. Science 354, 36–37.

Joza, N., Oudit, G.Y., Brown, D., Benit, P., Kassiri, Z., Vahsen, N., Benoit, L., Patel, M.M., Nowikovsky, K., Vassault, A., Backx, P.H., Wada, T., Kroemer, G., Rustin, P., Penninger, J.M., 2005. Muscle-specific loss of apoptosis-inducing factor leads to mitochondrial dysfunction, skeletal muscle atrophy, and dilated cardiomyopathy. Mol. Cell. Biol. 25, 10261–10272.

Kam, T.I., Mao, X., Park, H., Chou, S.C., Karuppagounder, S.S., Umanah, G.E., Yun, S.P., Brahmachari, S., Panicker, N., Chen, R., Andrabi, S.A., Qi, C., Poirier, G.G., Pletnikova, O., Troncoso, J.C., Bekris, L.M., Leverenz, J.B., Pantelyat, A., Ko, H.S., Rosenthal, L.S., Dawson, T.M., Dawson, V.L., 2018. Poly(ADP-ribose) drives pathologic alpha-synuclein neurodegeneration in Parkinson's disease. Science 362, eaat8407.

Kang, H.C., Lee, Y.I., Shin, J.H., Andrabi, S.A., Chi, Z., Gagne, J.P., Lee, Y., Ko, H.S., Lee, B.D., Poirier, G.G., Dawson, V.L., Dawson, T.M., 2011. Iduna is a poly(ADP-ribose) (PAR)-dependent E3 ubiquitin ligase that regulates DNA damage. Proc. Natl. Acad. Sci. U. S. A. 108, 14103–14108.

Kasamatsu, A., Nakao, M., Smith, B.C., Comstock, L.R., Ono, T., Kato, J., Denu, J.M., Moss, J., 2011. Hydrolysis of O-acetyl-ADP-ribose isomers by ADP-ribosylhydrolase 3. J. Biol. Chem. 286, 21110–21117.

Kauppinen, T.M., Suh, S.W., Higashi, Y., Berman, A.E., Escartin, C., Won, S.J., Wang, C., Cho, S.H., Gan, L., Swanson, R.A., 2011. Poly(ADP-ribose)polymerase-1 modulates microglial responses to amyloid beta. J. Neuroinflammation 8, 152.

Kim, S.H., Engelhardt, J.I., Henkel, J.S., Siklos, L., Soos, J., Goodman, C., Appel, S.H., 2004. Widespread increased expression of the DNA repair enzyme PARP in brain in ALS. Neurology 62, 319–322.

Kim, T.W., Cho, H.M., Choi, S.Y., Suguira, Y., Hayasaka, T., Setou, M., Koh, H.C., Hwang, E.M., Park, J.Y., Kang, S.J., Kim, H.S., Kim, H., Sun, W., 2013. (ADP-ribose) polymerase 1 and AMP-activated protein kinase mediate progressive dopaminergic neuronal degeneration in a mouse model of Parkinson's disease. Cell Death Dis. 4, e919.

Kleemann, R., Rorsman, H., Rosengren, E., Mischke, R., Mai, N.T., Bernhagen, J., 2000. Dissection of the enzymatic and immunologic functions of macrophage migration inhibitory factor. Full immunologic activity of N-terminally truncated mutants. Eur. J. Biochem. 267, 7183–7193.

Klein, J.A., Longo-Guess, C.M., Rossmann, M.P., Seburn, K.L., Hurd, R.E., Frankel, W.N., Bronson, R.T., Ackerman, S.L., 2002. The harlequin mouse mutation downregulates apoptosis-inducing factor. Nature 419, 367–374.

Koh, D.W., Lawler, A.M., Poitras, M.F., Sasaki, M., Wattler, S., Nehls, M.C., Stoger, T., Poirier, G.G., Dawson, V.L., Dawson, T.M., 2004. Failure to degrade poly(ADP-ribose) causes increased sensitivity to cytotoxicity and early embryonic lethality. Proc. Natl. Acad. Sci. U. S. A. 101, 17699–17704.

Kraus, W.L., 2008. Transcriptional control by PARP-1: chromatin modulation, enhancer-binding, coregulation, and insulation. Curr. Opin. Cell Biol. 20, 294–302.

Kraus, W.L., Lis, J.T., 2003. PARP goes transcription. Cell 113, 677–683.

Krietsch, J., Rouleau, M., Pic, E., Ethier, C., Dawson, T.M., Dawson, V.L., Masson, J.Y., Poirier, G.G., Gagne, J.P., 2013. Reprogramming cellular events by poly(ADP-ribose)-binding proteins. Mol. Aspects Med. 34, 1066–1087.

Labbadia, J., Morimoto, R.I., 2013. Huntington's disease: underlying molecular mechanisms and emerging concepts. Trends Biochem. Sci. 38, 378–385.

Lee, Y., Karuppagounder, S.S., Shin, J.H., Lee, Y.I., Ko, H.S., Swing, D., Jiang, H., Kang, S.U., Lee, B.D., Kang, H.C., Kim, D., Tessarollo, L., Dawson, V.L., Dawson, T.M., 2013. Parthanatos mediates AIMP2-activated age-dependent dopaminergic neuronal loss. Nat. Neurosci. 16, 1392–1400.

Leung, A.K., 2014. Poly(ADP-ribose): an organizer of cellular architecture. J. Cell Biol. 205, 613–619.

Li, M., Lu, L.Y., Yang, C.Y., Wang, S., Yu, X., 2013. The FHA and BRCT domains recognize ADP-ribosylation during DNA damage response. Genes Dev. 27, 1752–1768.

Lolis, E., Bucala, R., 1996. Crystal structure of macrophage migration inhibitory factor (MIF), a glucocorticoid-induced regulator of cytokine production, reveals a unique architecture. Proc. Assoc. Am. Physicians 108, 415–419.

Love, S., Barber, R., Wilcock, G.K., 1999. Increased poly(ADP-ribosyl)ation of nuclear proteins in Alzheimer's disease. Brain 122 (Pt. 2), 247–253.

Lubetsky, J.B., Dios, A., Han, J., Aljabari, B., Ruzsicska, B., Mitchell, R., Lolis, E., Al-Abed, Y., 2002. The tautomerase active site of macrophage migration inhibitory factor is a potential target for discovery of novel anti-inflammatory agents. J. Biol. Chem. 277, 24976–24982.

Lue, H., Kleemann, R., Calandra, T., Roger, T., Bernhagen, J., 2002. Macrophage migration inhibitory factor (MIF): mechanisms of action and role in disease. Microbes Infect. 4, 449–460.

Luo, X., Kraus, W.L., 2012. On PAR with PARP: cellular stress signaling through poly(ADP-ribose) and PARP-1. Genes Dev. 26, 417–432.

Maiuri, T., Mocle, A.J., Hung, C.L., Xia, J., Van Roon-Mom, W.M., Truant, R., 2017. Huntingtin is a scaffolding protein in the ATM oxidative DNA damage response complex. Hum. Mol. Genet. 26, 395–406.

Mandir, A.S., Przedborski, S., Jackson-Lewis, V., Wang, Z.Q., Simbulan-Rosenthal, C.M., Smulson, M.E., Hoffman, B.E., Guastella, D.B., Dawson, V.L., Dawson, T.M., 1999. Poly(ADP-ribose) polymerase activation mediates 1-methyl-4-phenyl-1, 2,3,6-tetrahydropyridine (MPTP)-induced parkinsonism. Proc. Natl. Acad. Sci. U. S. A. 96, 5774–5779.

Martire, S., Fuso, A., Rotili, D., Tempera, I., Giordano, C., De Zottis, I., Muzi, A., Vernole, P., Graziani, G., Lococo, E., Faraldi, M., Maras, B., Scarpa, S., Mosca, L., D'erme, M., 2013. PARP-1 modulates amyloid beta peptide-induced neuronal damage. PLoS One 8, e72169.

Mashimo, M., Kato, J., Moss, J., 2013. ADP-ribosyl-acceptor hydrolase 3 regulates poly (ADP-ribose) degradation and cell death during oxidative stress. Proc. Natl. Acad. Sci. U. S. A. 110, 18964–18969.

Mashimo, M., Bu, X., Aoyama, K., Kato, J., Ishiwata-Endo, H., Stevens, L.A., Kasamatsu, A., Wolfe, L.A., Toro, C., Adams, D., Markello, T., Gahl, W.A., Moss, J., 2019. PARP1 inhibition alleviates injury in ARH3-deficient mice and human cells. JCI Insight 4, 124519.

Mate, M.J., Ortiz-Lombardia, M., Boitel, B., Haouz, A., Tello, D., Susin, S.A., Penninger, J., Kroemer, G., Alzari, P.M., 2002. The crystal structure of the mouse apoptosis-inducing factor AIF. Nat. Struct. Biol. 9, 442–446.

McCann, K.E., 2019. Advances in the use of PARP inhibitors for BRCA1/2-associated breast cancer: talazoparib. Future Oncol. 15, 1707–1715.
McGurk, L., Lee, V.M., Trojanowksi, J.Q., Van Deerlin, V.M., Lee, E.B., Bonini, N.M., 2014. Poly-A binding protein-1 localization to a subset of TDP-43 inclusions in amyotrophic lateral sclerosis occurs more frequently in patients harboring an expansion in C9orf72. J. Neuropathol. Exp. Neurol. 73, 837–845.
McGurk, L., Mojsilovic-Petrovic, J., Van Deerlin, V.M., Shorter, J., Kalb, R.G., Lee, V.M., Trojanowski, J.Q., Lee, E.B., Bonini, N.M., 2018. Nuclear poly(ADP-ribose) activity is a therapeutic target in amyotrophic lateral sclerosis. Acta Neuropathol. Commun. 6, 84.
Merk, M., Baugh, J., Zierow, S., Leng, L., Pal, U., Lee, S.J., Ebert, A.D., Mizue, Y., Trent, J.O., Mitchell, R., Nickel, W., Kavathas, P.B., Bernhagen, J., Bucala, R., 2009. The Golgi-associated protein p115 mediates the secretion of macrophage migration inhibitory factor. J. Immunol. 182, 6896–6906.
Meyer, R.G., Meyer-Ficca, M.L., Whatcott, C.J., Jacobson, E.L., Jacobson, M.K., 2007. Two small enzyme isoforms mediate mammalian mitochondrial poly(ADP-ribose) glycohydrolase (PARG) activity. Exp. Cell Res. 313, 2920–2936.
Meyer-Ficca, M.L., Meyer, R.G., Coyle, D.L., Jacobson, E.L., Jacobson, M.K., 2004. Human poly(ADP-ribose) glycohydrolase is expressed in alternative splice variants yielding isoforms that localize to different cell compartments. Exp. Cell Res. 297, 521–532.
Miwa, M., Saikawa, N., Yamaizumi, Z., Nishimura, S., Sugimura, T., 1979. Structure of poly(adenosine diphosphate ribose): identification of 2′-[1″-ribosyl-2″-(or 3″-)(1‴-ribosyl)]adenosine-5′,5″,5‴-tris(phosphate) as a branch linkage. Proc. Natl. Acad. Sci. U. S. A. 76, 595–599.
Mizue, Y., Ghani, S., Leng, L., Mcdonald, C., Kong, P., Baugh, J., Lane, S.J., Craft, J., Nishihira, J., Donnelly, S.C., Zhu, Z., Bucala, R., 2005. Role for macrophage migration inhibitory factor in asthma. Proc. Natl. Acad. Sci. U. S. A. 102, 14410–14415.
Morand, E.F., Leech, M., Bernhagen, J., 2006. MIF: a new cytokine link between rheumatoid arthritis and atherosclerosis. Nat. Rev. Drug Discov. 5, 399–410.
Mueller-Dieckmann, C., Kernstock, S., Lisurek, M., Von Kries, J.P., Haag, F., Weiss, M.S., Koch-Nolte, F., 2006. The structure of human ADP-ribosylhydrolase 3 (ARH3) provides insights into the reversibility of protein ADP-ribosylation. Proc. Natl. Acad. Sci. U. S. A. 103, 15026–15031.
Munoz-Sanjuan, I., Bates, G.P., 2011. The importance of integrating basic and clinical research toward the development of new therapies for Huntington disease. J. Clin. Invest. 121, 476–483.
O'Brien, R.J., Wong, P.C., 2011. Amyloid precursor protein processing and Alzheimer's disease. Annu. Rev. Neurosci. 34, 185–204.
Oh, Y.K., Shin, K.S., Kang, S.J., 2006. AIF translocates to the nucleus in the spinal motor neurons in a mouse model of ALS. Neurosci. Lett. 406, 205–210.
Ono, T., Kasamatsu, A., Oka, S., Moss, J., 2006. The 39-kDa poly(ADP-ribose) glycohydrolase ARH3 hydrolyzes O-acetyl-ADP-ribose, a product of the Sir2 family of acetyl-histone deacetylases. Proc. Natl. Acad. Sci. U. S. A. 103, 16687–16691.
O'Sullivan, J., Tedim Ferreira, M., Gagne, J.P., Sharma, A.K., Hendzel, M.J., Masson, J.Y., Poirier, G.G., 2019. Emerging roles of eraser enzymes in the dynamic control of protein ADP-ribosylation. Nat. Commun. 10, 1182.
Otera, H., Ohsakaya, S., Nagaura, Z., Ishihara, N., Mihara, K., 2005. Export of mitochondrial AIF in response to proapoptotic stimuli depends on processing at the intermembrane space. EMBO J. 24, 1375–1386.
Paldino, E., Cardinale, A., D'angelo, V., Sauve, I., Giampa, C., Fusco, F.R., 2017. Selective sparing of striatal interneurons after poly (ADP-Ribose) polymerase 1 inhibition in the R6/2 mouse model of Huntington's disease. Front. Neuroanat. 11, 61.

Paralkar, V., Wistow, G., 1994. Cloning the human gene for macrophage migration inhibitory factor (MIF). Genomics 19, 48–51.

Popp, J., Bacher, M., Kolsch, H., Noelker, C., Deuster, O., Dodel, R., Jessen, F., 2009. Macrophage migration inhibitory factor in mild cognitive impairment and Alzheimer's disease. J. Psychiatr. Res. 43, 749–753.

Pospisilik, J.A., Knauf, C., Joza, N., Benit, P., Orthofer, M., Cani, P.D., Ebersberger, I., Nakashima, T., Sarao, R., Neely, G., Esterbauer, H., Kozlov, A., Kahn, C.R., Kroemer, G., Rustin, P., Burcelin, R., Penninger, J.M., 2007. Targeted deletion of AIF decreases mitochondrial oxidative phosphorylation and protects from obesity and diabetes. Cell 131, 476–491.

Sachdev, E., Tabatabai, R., Roy, V., Rimel, B.J., Mita, M.M., 2019. PARP inhibition in cancer: an update on clinical development. Target. Oncol. 14, 657–679.

Sanchez, E., Gomez, L.M., Lopez-Nevot, M.A., Gonzalez-Gay, M.A., Sabio, J.M., Ortego-Centeno, N., De Ramon, E., Anaya, J.M., Gonzalez-Escribano, M.F., Koeleman, B.P., Martin, J., 2006. Evidence of association of macrophage migration inhibitory factor gene polymorphisms with systemic lupus erythematosus. Genes Immun. 7, 433–436.

Schindler, L., Dickerhof, N., Hampton, M.B., Bernhagen, J., 2018. Post-translational regulation of macrophage migration inhibitory factor: basis for functional fine-tuning. Redox Biol. 15, 135–142.

Schreiber, V., Dantzer, F., Ame, J.C., De Murcia, G., 2006. Poly(ADP-ribose): novel functions for an old molecule. Nat. Rev. Mol. Cell Biol. 7, 517–528.

Sun, H.W., Bernhagen, J., Bucala, R., Lolis, E., 1996. Crystal structure at 2.6-A resolution of human macrophage migration inhibitory factor. Proc. Natl. Acad. Sci. U. S. A. 93, 5191–5196.

Susin, S.A., Lorenzo, H.K., Zamzami, N., Marzo, I., Snow, B.E., Brothers, G.M., Mangion, J., Jacotot, E., Costantini, P., Loeffler, M., Larochette, N., Goodlett, D.R., Aebersold, R., Siderovski, D.P., Penninger, J.M., Kroemer, G., 1999. Molecular characterization of mitochondrial apoptosis-inducing factor. Nature 397, 441–446.

Teloni, F., Altmeyer, M., 2016. Readers of poly(ADP-ribose): designed to be fit for purpose. Nucleic Acids Res. 44, 993–1006.

Turunc Bayrakdar, E., Uyanikgil, Y., Kanit, L., Koylu, E., Yalcin, A., 2014. Nicotinamide treatment reduces the levels of oxidative stress, apoptosis, and PARP-1 activity in Abeta(1-42)-induced rat model of Alzheimer's disease. Free Radic. Res. 48, 146–158.

Virag, L., Szabo, C., 2002. The therapeutic potential of poly(ADP-ribose) polymerase inhibitors. Pharmacol. Rev. 54, 375–429.

Vis, J.C., Schipper, E., De Boer-Van Huizen, R.T., Verbeek, M.M., De Waal, R.M., Wesseling, P., Ten Donkelaar, H.J., Kremer, B., 2005. Expression pattern of apoptosis-related markers in Huntington's disease. Acta Neuropathol. 109, 321–328.

Wang, H., Shimoji, M., Yu, S.W., Dawson, T.M., Dawson, V.L., 2003. Apoptosis inducing factor and PARP-mediated injury in the MPTP mouse model of Parkinson's disease. Ann. N. Y. Acad. Sci. 991, 132–139.

Wang, H., Yu, S.W., Koh, D.W., Lew, J., Coombs, C., Bowers, W., Federoff, H.J., Poirier, G.G., Dawson, T.M., Dawson, V.L., 2004. Apoptosis-inducing factor substitutes for caspase executioners in NMDA-triggered excitotoxic neuronal death. J. Neurosci. 24, 10963–10973.

Wang, Y., Dawson, V.L., Dawson, T.M., 2009. Poly(ADP-ribose) signals to mitochondrial AIF: a key event in parthanatos. Exp. Neurol. 218, 193–202.

Wang, Y., Kim, N.S., Haince, J.F., Kang, H.C., David, K.K., Andrabi, S.A., Poirier, G.G., Dawson, V.L., Dawson, T.M., 2011. Poly(ADP-ribose) (PAR) binding to apoptosis-inducing factor is critical for PAR polymerase-1-dependent cell death (parthanatos). Sci. Signal. 4, ra20.

Wang, Y., An, R., Umanah, G.K., Park, H., Nambiar, K., Eacker, S.M., Kim, B., Bao, L., Harraz, M.M., Chang, C., Chen, R., Wang, J.E., Kam, T.I., Jeong, J.S., Xie, Z., Neifert, S., Qian, J., Andrabi, S.A., Blackshaw, S., Zhu, H., Song, H., Ming, G.L., Dawson, V.L., Dawson, T.M., 2016. A nuclease that mediates cell death induced by DNA damage and poly(ADP-ribose) polymerase-1. Science 354, aad6872.

Williams, A.J., Paulson, H.L., 2008. Polyglutamine neurodegeneration: protein misfolding revisited. Trends Neurosci. 31, 521–528.

Wu, X.L., Wang, P., Liu, Y.H., Xue, Y.X., 2014. Effects of poly (ADP-ribose) polymerase inhibitor 3-aminobenzamide on blood-brain barrier and dopaminergic neurons of rats with lipopolysaccharide-induced Parkinson's disease. J. Mol. Neurosci. 53, 1–9.

Ye, H., Cande, C., Stephanou, N.C., Jiang, S., Gurbuxani, S., Larochette, N., Daugas, E., Garrido, C., Kroemer, G., Wu, H., 2002. DNA binding is required for the apoptogenic action of apoptosis inducing factor. Nat. Struct. Biol. 9, 680–684.

Yokoyama, H., Kuroiwa, H., Tsukada, T., Uchida, H., Kato, H., Araki, T., 2010. Poly(ADP-ribose)polymerase inhibitor can attenuate the neuronal death after 1-methyl-4-phenyl-1,2,3,6-tetrahydropyridine-induced neurotoxicity in mice. J. Neurosci. Res. 88, 1522–1536.

Yu, S.W., Wang, H., Poitras, M.F., Coombs, C., Bowers, W.J., Federoff, H.J., Poirier, G.G., Dawson, T.M., Dawson, V.L., 2002. Mediation of poly(ADP-ribose) polymerase-1-dependent cell death by apoptosis-inducing factor. Science 297, 259–263.

Yu, S.W., Wang, H., Dawson, T.M., Dawson, V.L., 2003. Poly(ADP-ribose) polymerase-1 and apoptosis inducing factor in neurotoxicity. Neurobiol. Dis. 14, 303–317.

Yu, S.W., Andrabi, S.A., Wang, H., Kim, N.S., Poirier, G.G., Dawson, T.M., Dawson, V.L., 2006. Apoptosis-inducing factor mediates poly(ADP-ribose) (PAR) polymer-induced cell death. Proc. Natl. Acad. Sci. U. S. A. 103, 18314–18319.

Yu, S.W., Wang, Y., Frydenlund, D.S., Ottersen, O.P., Dawson, V.L., Dawson, T.M., 2009. Outer mitochondrial membrane localization of apoptosis-inducing factor: mechanistic implications for release. ASN Neuro 1, e00021.

Yu, W., Bonnet, M., Farso, M., Ma, K., Chabot, J.G., Martin, E., Torriglia, A., Guan, Z., Mclaurin, J., Quirion, R., Krantic, S., 2014. The expression of apoptosis inducing factor (AIF) is associated with aging-related cell death in the cortex but not in the hippocampus in the TgCRND8 mouse model of Alzheimer's disease. BMC Neurosci. 15, 73.

Zhang, J., Dawson, V.L., Dawson, T.M., Snyder, S.H., 1994. Nitric oxide activation of poly(ADP-ribose) synthetase in neurotoxicity. Science 263, 687–689.

Zhou, Y., Feng, X., Koh, D.W., 2011. Activation of cell death mediated by apoptosis-inducing factor due to the absence of poly(ADP-ribose) glycohydrolase. Biochemistry 50, 2850–2859.

Further reading

D'Arcy, M.S., 2019. Cell death: a review of the major forms of apoptosis, necrosis and autophagy. Cell Biol. Int. 43, 582–592.

CHAPTER TWO

Molecular mechanisms of necroptosis and relevance for neurodegenerative diseases

Pedro A. Dionísio, Joana D. Amaral, Cecília M.P. Rodrigues*
Research Institute for Medicines (iMed.ULisboa), Faculty of Pharmacy, Universidade de Lisboa, Lisbon, Portugal
*Corresponding author: e-mail address: cmprodrigues@ff.ulisboa.pt

Contents

Abstract

Necroptosis is a regulated cell death pathway morphologically similar to necrosis that depends on the kinase activity of receptor interacting protein 3 (RIP3) and the subsequent activation of the pseudokinase mixed lineage kinase domain-like protein (MLKL), being also generally dependent on RIP1 kinase activity. Necroptosis can be recruited during pathological conditions, usually following the activation of death receptors under specific cellular contexts. In this regard, necroptosis has been implicated in the pathogenesis of multiple disorders, including acute and chronic neurodegenerative diseases, such as Parkinson's and Alzheimer's diseases, and multiple sclerosis. Here, we summarize the molecular mechanisms regulating the induction of necroptosis and downstream effectors of this form of cell death, besides exploring non-necroptotic roles for necroptosis-related proteins that may impact on alternative cell death pathways and inflammatory mechanisms in disease. Finally, we outline the recent evidence implicating necroptosis in neurodegenerative conditions and the emerging therapeutic perspectives targeting necroptosis in these diseases.

International Review of Cell and Molecular Biology, Volume 353
ISSN 1937-6448
https://doi.org/10.1016/bs.ircmb.2019.12.006

1. Introduction

Traditionally, apoptosis was considered a highly regulated process, as opposed to passive necrosis, which could result from extreme microenvironmental insults resulting in accidental cell death (Galluzzi et al., 2018). Passive necrosis results in rapid cytoplasmic and organelle swelling (oncosis), along with plasma membrane permeabilization and subsequent leakage of intracellular molecules containing damage associated molecular patterns (DAMPs), which results in inflammation (Galluzzi et al., 2018; Yuan et al., 2019a). However, mounting evidence from the last decades has unraveled multiple forms of regulated necrotic modalities that are morphologically similar to passive necrosis, such as mitochondrial pore transition (MPT)-dependent necrosis, necroptosis, ferroptosis, parthanatos or pyroptosis. Importantly, these necrotic pathways depend on specific molecular mechanisms that are recruited during distinct pathological conditions and can be pharmacologically or genetically targeted (Galluzzi et al., 2018). Necroptosis is probably the most well-studied regulated necrotic subroutine, since it shares multiple molecular components with the extrinsic apoptotic pathway and can act as a fail-safe mechanism to ensure cell death progression when apoptosis is abnormally inhibited, being also relevant during several pathological conditions (Choi et al., 2019; Yuan et al., 2019a).

The term necroptosis was originally coined in 2005 to indicate a non-apoptotic cell death pathway triggered by death receptor (DR) activation upon caspase inhibition that was potently inhibited by necrostatin-1 (Nec-1), a widely used inhibitor of the kinase activity of receptor interacting protein (RIP)1 (also known as RIPK1) (Degterev et al., 2005). Recently, necroptosis was defined, according to the Nomenclature Committee on Cell Death, as an RCD that critically relies on RIP3 (also known as RIPK3) and the pseudokinase mixed lineage kinase domain-like protein (MLKL), being also dependent on RIP1 kinase activity in several scenarios (Galluzzi et al., 2018).

2. Upstream regulators of necroptosis

Necroptosis can be elicited under specific intracellular conditions following the formation of signaling complexes containing proteins that possess a RIP homotypic interacting motif (RHIM) domain, such as RIP1, TIR-domain-containing adapter-inducing interferon-β (TRIF) and Z-DNA-binding protein 1 (ZBP1), also known as DNA-dependent activator of interferon regulatory factors (IRFs) (DAI), which allow for the recruitment of RIP3 via RHIM homotypic interactions. Usually, formation of these necroptotic complexes occurs in specific cellular conditions downstream of DRs, such as tumor necrosis factor-α (TNF-α) receptor 1 (TNFR1), TNF-α-related apoptosis-inducing ligand receptors (TRAILRs) and Fas, or pattern recognition receptors (PRRs) involved in the immune response, including TLR3 and TLR4, which recruit TRIF when activated, and ZBP1 itself (Choi et al., 2019; Galluzzi et al., 2018).

Necroptosis initiation downstream of DRs is mostly dependent on RIP1 recruitment and activation of its kinase activity (Yuan et al., 2019a). Although Fas and TRAILRs can also induce necroptosis via non-canonical recruitment of RIP1, necroptotic signaling downstream of TNFR1 activation is the most thoroughly studied case (Seo et al., 2019). In fact, TNF-α signaling can exert multiple cellular responses, ranging from cell survival and expression of pro-inflammatory genes to several different forms of cell death, depending on cellular context (Fig. 1) (Seo et al., 2019). In this regard, downstream signaling is mostly determined by RIP1, which regulates the crosstalk between survival and pro-inflammatory response or cell death, depending on post-translational modifications and binding partners (Seo et al., 2019; Yuan et al., 2019a). RIP1 is composed of an N-terminal Ser/Thr kinase domain which is essential for its pro-death functions, an intermediate domain involved in NF-κB signaling that also contains a RHIM domain, and a C-terminal DD (Fig. 2) (Yuan et al., 2019a). Initially, binding of TNF-α to TNFR1 stabilizes receptor trimerization, which recruits TRADD and RIP1 via death domain (DD) homotypic interactions for subsequent pro-survival complex I assembly at the plasma membrane (Galluzzi et al., 2018; Sedger and McDermott, 2014).

TRADD becomes able to recruit TNFR1-associated factor 2 (TRAF2), which binds cellular inhibitor of apoptosis (cIAP)1/2 (Annibaldi et al., 2018). In turn, cIAPs ubiquitinate themselves, RIP1 and TRAF2 with K63, K11 and K48 poly-Ubiquitin (Ub) chains (Annibaldi et al., 2018; Bertrand et al., 2008; Seo et al., 2019; Varfolomeev et al., 2008). K63 poly-Ub chains then recruit the linear ubiquitin assembly chain complex (LUBAC), constituted by haem-oxidized iron-regulatory protein 2 ubiquitin ligase-1 (HOIL-1), HOIL-1 interacting protein (HOIP) and SHANK-associated RH domain-interacting protein (SHARPIN). LUBAC further ubiquitinates multiple complex I components with M1-linear poly-Ub chains, including RIP1. These K63 and M1 poly-Ub chains then serve as scaffolding platforms

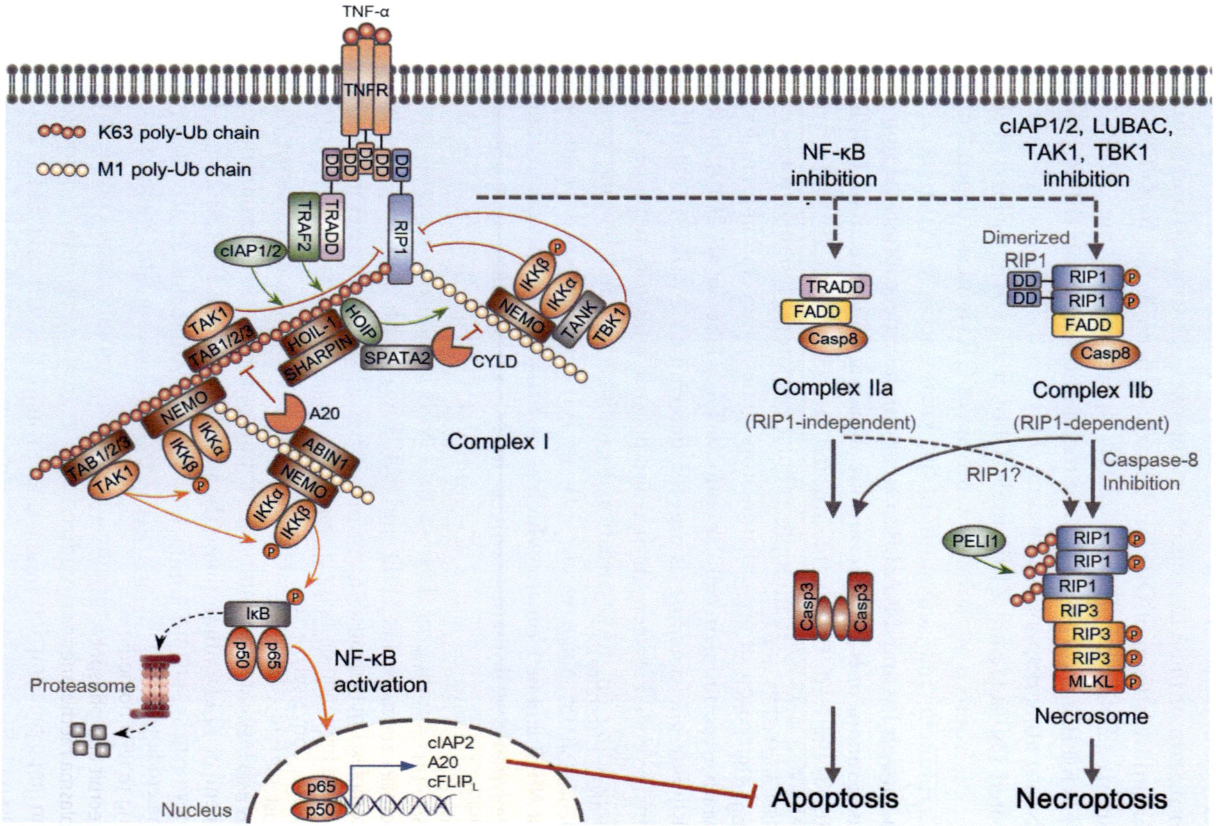

Fig. 1 See legend on opposite page.

for recruitment of the complexes TAK1 binding protein (TAB) 1/2/3/ TAK1 and NF-κB essential modulator (NEMO)/inhibitor of NF-κB (IκB) kinase (IKK)α/IKKβ (Gerlach et al., 2011; Haas et al., 2009; Seo et al., 2019). There, TAK1 kinase activity contributes to activation of downstream MAPKs and IKKβ, which phosphorylates IκB (Blonska et al., 2005; Shim et al., 2005). IκB phosphorylation leads to its proteasomal degradation, thus releasing NF-κB for pro-survival and pro-inflammatory gene expression (Seo et al., 2019; Zandi et al., 1998). However, upon internalization of activated TNFR1, complex I may dissociate, mostly due to the action of multiple

Fig. 1 TNF-α signaling pathways. Upon TNFR1 activation, TRADD and RIP1 are recruited through DD interactions. TRADD then binds TRAF2, which further recruits cIAPs, leading to K63 and K11 poly-ubiquitination of RIP1. These K63 poly-Ub chains serve as scaffolding platforms for the binding of signaling complexes, such as TAB1/2/3/TAK1, NEMO/IKKα/IKKβ and LUBAC, composed by HOIP, HOIL, and SHARPIN. LUBAC ubiquitinates RIP1 and NEMO with M1 poly-Ub chains, which further recruits NEMO/IKKα/IKKβ, leading to the formation of a large signaling complex termed complex I. TAK1 activates IKKβ, which in turn phosphorylates IκB, thus contributing to its proteasomal degradation and activation of NF-κB, resulting in the expression of pro-survival genes that inhibit subsequent TNF-α-mediated apoptosis. TAK1, IKKβ and TBK1, which can be recruited to complex I downstream of NEMO-bound TRAF family member-associated NF-κB activator (TANK), phosphorylate RIP1, thus inhibiting pro-death RIP1 dimerization and kinase activation. NF-κB-dependent expression of A20 acts as a negative feedback regulator of NF-κB and RIP1 activation via recruitment of A20 to complex I by ABIN1, where A20 deubiquitinates RIP1-linked K63 poly-Ub chains and in turn mediates K48 poly-ubiquitination, leading to RIP1 proteasomal degradation. Conversely, CYLD is recruited to complex I through LUBAC-bound adaptor SPATA2, where it removes K63 and M1 poly-Ub chains from several complex I components, including RIP1, which is crucial for RIP1 activation and transition into cytosolic death complexes. Upon NF-κB inhibition, complex I may dissociate, allowing TRADD, FADD and caspase-8 to interact, leading to late formation of the cytosolic complex IIa and induction of RIP1-independent apoptosis. Conversely, inhibition or depletion of cIAPs, LUBAC, TAK1 and TBK1 allows for fast RIP1 homodimerization via DD interactions and transphosphorylation, leading to activation of its kinase activity and formation of complex IIb following the recruitment of FADD-caspase-8, which may result in RIP1 kinase-dependent apoptosis. However, in conditions where caspase-8 is inhibited or depleted, active RIP1 can heterodimerize with RIP3, forming an amyloid-like structure called the necrosome, leading to RIP3 homodimerization and activating *cis*-autophosphorylation. Active RIP3 can recruit and phosphorylate MLKL, leading to MLKL oligomerization and translocation to plasma membranes to induce necroptosis. RIP1 K63 poly-ubiquitination by PELI1 is also an important contributor to necroptotic induction. Complex IIa can also lead to necroptosis following caspase-8 inhibition, although with a much lower kinetics than complex IIb, which may depend on cytosolic RIP1 recruitment. *Based on Seo, J., Kim, M.W., Bae, K.H., Lee, S.C., Song, J., Lee, E.W., 2019. The roles of ubiquitination in extrinsic cell death pathways and its implications for therapeutics. Biochem. Pharmacol. 162, 21–40.*

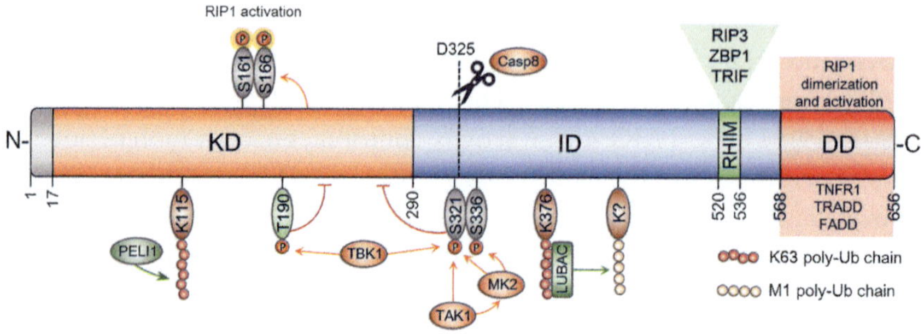

Fig. 2 Activation of RIP1 is tightly regulated by multiple post-translational modifications. The domain architecture of murine RIP1 consists of an N-terminal kinase domain (KD), an intermediate domain (ID), where the RHIM is contained, and a C-terminal DD. K63 poly-ubiquitination at K376 can be mediated by cIAPs and is essential for the recruitment of multiple signaling complexes resulting in NF-κB activation and suppression of RIP1 activity, including LUBAC, which can mediate RIP1 M1 poly-ubiquitination. Recruitment of TAK1 and TBK1 into complex I can suppress RIP1 activation by mediating inhibitory phosphorylations on S231 and T190/S231, respectively, while TAK1-downstream activation of MK2 leads to phosphorylation on S231/S336 in cytosolic RIP1. Active caspase-8 can also cleave RIP1 at D235, which suppresses RIP1 activity within death complexes. The RHIM motif is crucial for RIP1/RIP3 interaction and subsequent induction of necroptosis, but also contributes to RIP1 interaction with TRIF and ZBP1 to regulate pro-inflammatory responses and necroptosis downstream of these proteins. PELI1-mediated K63 poly-Ub chain formation on K155 of active RIP1 contributes to necrosome assembly and necroptosis. The DD is crucial not only for interaction with multiple adaptor proteins, such as TNFR1, TRADD and FADD, but also to mediate RIP1 homodimerization and subsequent transphosphorylation in activating residues, especially S161/S166.

DUBs that remove poly-Ub chains involved in complex I stabilization, which can lead to the formation of the cytosolic complex IIa, where TRADD/RIP1 can recruit Fas associated via DD (FADD) for subsequent binding of procaspase-8 and the pseudocaspase-8 cellular FLICE/caspase-8-like inhibitory protein (cFLIP), or the complex IIb, where RIP1 is the main recruiter of FADD-procaspase-8-cFLIP for induction of RIP1 kinase-dependent apoptosis or, eventually, necroptosis (Seo et al., 2019; Wang et al., 2008; Yuan et al., 2019a). RIP1 activation and subsequent FADD recruitment depend on RIP1 homodimerization via DD and transphosphorylation on activating residues, most notably S166/S161 (Meng et al., 2018; Ofengeim et al., 2015; Xu et al., 2018; Zhang et al., 2017). Formation and activation of either complex II variants are finely regulated by early checkpoints following TNFR1 activation, which modulate RIP1 activation through post-translational modifications such as poly-ubiquitination and phosphorylation, and late checkpoints that rely on NF-κB-dependent expression of pro-survival genes (Fig. 2) (Seo et al., 2019).

RIP1 pro-survival poly-ubiquitination by TRAF2, cIAPs and LUBAC at the complex I is the first early checkpoint against TNF-induced cell death, and their deletion or inhibition induce fast complex IIb formation and cell death (Haas et al., 2009; Seo et al., 2019; Wang et al., 2008). M1 and K63 poly-Ub modifications inhibit RIP1 activation and cell death by stabilizing complex I due to their scaffolding properties, by recruiting complexes TAB1/2/3/TAK1 and NEMO/IKKα/β and, in the case of cIAP1-mediated K48 poly-Ub chain formation, by directly inhibiting RIP1 activation and targeting it for proteasomal degradation (Annibaldi et al., 2018; Bertrand et al., 2011; Haas et al., 2009; Seo et al., 2019). In this regard, K63 poly-Ub chain formation at K376 of RIP1 by cIAPs and TRAF2 is essential for subsequent LUBAC recruitment, canonical NF-κB activation, and cell survival, although its pro-survival functions at early time-points are NF-κB-independent (Seo et al., 2019; Tang et al., 2019; Wang et al., 2018). Recruitment of the kinases TAK1, IKKα/β and TANK-binding kinase 1 (TBK1)/IKKε downstream of M1 and K63 poly-Ub chain formation in complex I is also a crucial step to keep RIP1 kinase activity in check, since these kinases have been shown to directly phosphorylate RIP1 and thus inhibit its activity (Dondelinger et al., 2015, 2017; Geng et al., 2017). TAK1 inhibition is sufficient to induce massive RIP1-dependent cell death following TNFR1 activation, even in the presence of properly poly-ubiquitinated RIP1 (Dondelinger et al., 2013). TAK1 has been shown to directly phosphorylate murine RIP1 at S321, besides activating IKKα/β, which can also mediate inhibitory phosphorylations of complex I-bound RIP1, and downstream p38-MK2 that can phosphorylate cytosolic RIP1 at S321 and S336 (Dondelinger et al., 2015, 2017; Geng et al., 2017; Jaco et al., 2017; Menon et al., 2017). Moreover, TBK1/IKKε have also been shown recently to be recruited to complex I in a LUBAC-dependent manner, where they can phosphorylate multiple murine RIP1 residues such as S231 and T190, thus preventing RIP1-dependent cell death (Lafont et al., 2018; Xu et al., 2018). Interestingly, it was recently shown that TAK1 pharmacological or genetic inhibition, or depletion of cIAPs induced by Smac mimetics, leads to the fast formation of an intermediate complex composed by insoluble and highly M1- and K63-poly-ubiquitinated RIP1 (iuRIP1) bound to complex I, which can further drive cytosolic complex IIb formation and subsequent caspase-8 activation. The E3 Ub ligase cellular-Casitas B-lineage lymphoma (c-Cbl) is responsible for marking RIP1 with pro-death K63 poly-Ub chains within complex I, although the specific residues targeted by this enzyme and subsequent mode of action remain unknown. LRRK2 can also interact with RIP1 within complex I early during iuRIP1 complex formation, and its knockdown or genetic deletion inhibits iuRIP1 formation. Interestingly, this effect was independent on LRRK2 kinase activity, rather suggesting a role for its scaffolding functions in the process, while several autosomal dominant gain-of-function LRRK2 mutations lead to hyper-kinase activity in Parkinson's disease (PD) (Amin et al., 2018; Martin et al., 2014).

Complex IIa formation, on the other hand, occur later following TNFR1 stimulation upon inhibition of NF-κB-dependent expression of pro-survival genes, such as cIAP2, anti-apoptotic signaling protein 20 (A20) and, especially, the long isoform of cFLIP ($cFLIP_L$), which is most commonly achieved experimentally by exposure to the translational inhibitor cycloheximide (Seo et al., 2019; Wang et al., 2008). In normal conditions, NF-κB-dependent overexpression of $cFLIP_L$ inhibits caspase-8 activation following complex IIa formation (Seo et al., 2019). In turn, A20 is recruited to complex I by the M1 poly-Ub chain binding protein A20-binding inhibitor of NF-κB activation 1 (ABIN1), where it acts as an inhibitor of NF-κB signaling and RIP1 activation, mostly by replacing pro-survival K63 poly-Ub chains in RIP1 with K48 poly-Ub chains, which targets RIP1 to the proteasome (Dziedzic et al., 2018; Wertz et al., 2004). Similarly, OPTN, an Ub-binding scaffold protein recruited downstream of M1 poly-Ub chain assembly in complex I, may also contribute to K48 poly-ubiquitination of RIP1 and subsequent proteasomal degradation (Ito et al., 2016; Nakazawa et al., 2016).

Complex I dissociation is also regulated be several DUBs, most notably cylindromatosis (CYLD), which is recruited to the LUBAC component HOIP in complex I by the bridging adaptor spermatogenesis-associated 2 (SPATA2) (Kupka et al., 2016; Seo et al., 2019). Upon recruitment, CYLD removes pro-survival M1 and K63 poly-Ub chains in several complex I components, including RIP1, and this function is crucial to permit RIP1 activation and transition into cytosolic death complexes (Draber et al., 2015; Seo et al., 2019; Wang et al., 2008).

In conditions that allow for the formation of complex IIa or IIb downstream of TNFR1 activation, necroptosis is suppressed by caspase-8 activity, which cleaves RIP1, RIP3 and CYLD (Feng et al., 2007; Newton et al., 2019; Oberst et al., 2011; O'Donnell et al., 2011; Pop et al., 2011).

In agreement with a physiological role for caspase-8 in inhibiting necroptosis downstream of TNFR1 signaling, genetic ablation of caspase-8 or FADD leads to early embryonic lethality in mice, which is rescued by RIP3 co-deletion (Dillon et al., 2012; Oberst et al., 2011; Yeh et al., 1998). Importantly, pro-caspase-8 and cFLIP$_L$ heterodimers possess limited proteolytic activity that is sufficient to mediate RIP1 cleavage and inhibit necroptosis without inducing apoptosis (Oberst et al., 2011; Pop et al., 2011). In fact, cFLIP$_L$ is an essential inhibitor of apoptosis and necroptosis, as demonstrated by the fact that cFLIP$_L$ ablation results in embryonic lethality that can only be rescued by co-deletion of FADD and RIP3 (Dillon et al., 2012). It was also recently demonstrated that, while knock-in mice for a catalytically inactive form of caspase-8 also die as embryos due to excessive necroptosis, knock-in mice for an uncleavable form of caspase-8 are viable, thus supporting the view that procaspase-8/cFLIP$_L$ heterodimers are sufficient to limit necroptosis (Newton et al., 2019).

Therefore, induction of RIP1-dependent necroptosis depends on caspase-8 inhibition, achieved either genetically or pharmacologically (for example with the widely used pan-caspase inhibitor zVAD.fmk [zVAD]), or by the expression of viral proteins that inhibit caspases. Interestingly, cIAPs depletion in cancer cell lines may lead to the spontaneous formation of an intracellular complex containing RIP1, RIP3, FADD, procaspase-8 and cFLIP isoforms independently of transmembrane receptor stimulation due to decreased proteasomal degradation of RIP1, and is thus termed Ripoptosome. In this complex, high cFLIP$_S$ levels can inhibit apoptosis but induce necroptosis (Feoktistova et al., 2011; Mandal et al., 2014; Tenev et al., 2011). Conversely, excessive cFLIP$_L$ expression and defective caspase-8 activation are detected in cortical lesions of multiple sclerosis (MS) patients and animal models, which may contribute to disease progression by limiting caspase-8 activity and cleavage of necroptotic proteins (Ofengeim et al., 2015). Therefore, unbalanced recruitment of procaspase-8 and cFLIP isoforms into death complexes and subsequent downregulation of caspase-8 activity may sensitize cells to the induction of necroptosis in pathological settings.

Lipopolysaccharide (LPS)-driven activation of TLR4 or dsRNA/polyinosinic-polycytidilic acid (polyI:C)-mediated activation of TLR3 can also induce necroptosis upon caspase inhibition via downstream TRIF signaling, which recruits RIP1 and/or RIP3 through RHIM homotypic interactions (He et al., 2011; Kaiser et al., 2013; Kim and Li, 2013). Under normal conditions, RIP1 recruitment and poly-ubiquitination is essential to induce TRIF-dependent NF-κB activation (Cusson-Hermance et al., 2005;

Meylan et al., 2004; Moriwaki and Chan, 2017). RIP1 can further recruit FADD, procaspase-8 and cFLIP isoforms, which may act as a terminating signaling mechanism by cleaving RIP1 and RIP3, but can eventually result in apoptosis, or necroptosis when caspase-8 is inhibited (Feoktistova et al., 2011; Kaiser and Offermann, 2005). Interestingly, RIP1 kinase activity contributes to TLR3/4-mediated necroptosis in macrophages and microglia but is unnecessary in several cell lines (Fricker et al., 2013; Kaiser et al., 2013; Kearney et al., 2015; Kim and Li, 2013). In fact, RIP1 silencing or genetic ablation can induce massive TLR3/4-driven necroptosis even in the absence of caspase inhibition, indicating that RIP1 is involved in limiting TLR-TRIF-mediated necroptosis, although this function appears to be cell- and context-dependent (Dillon et al., 2014; Kaiser et al., 2013). Similarly, activation of ZBP1 by cytosolic dsDNA, which is usually associated to viral infection, can lead to RHIM-dependent recruitment of RIP3 and induction of necroptosis, while RIP1 binding to ZBP1 counteracts this process (Ingram et al., 2019; Upton et al., 2012). In fact, RIP1 acts as an essential basal inhibitor of ZBP1-mediated necroptosis during both development and adult tissue homeostasis (Lin et al., 2016; Newton et al., 2016b). Therefore, while TNFR1-mediated necroptosis is mostly dependent on RIP1 kinase activity, RIP1 scaffolding functions are vitally essential to ensure pro-survival signaling via NF-κB, to properly regulate extrinsic apoptosis, and to generally inhibit necroptosis downstream of TRIF and ZBP1 (Dillon et al., 2014; Kaiser et al., 2013; Lin et al., 2016; Tang et al., 2019). Moreover, RIP1 is also essential to suppress spontaneous RIP3 oligomerization and activation in the cytosol (Orozco et al., 2014). In agreement, RIP1 kinase dead (K45A or D138N) knock-in mice are fully viable and remain protected from TNF-α-induced shock, but RIP1ko mice die early postnatally due to excessive caspase-8 activation and RIP3-dependent necroptosis. Moreover, this early postnatal lethality is fully rescued by concomitant co-deletion of RIP3 and either caspase-8 or FADD (Dillon et al., 2014; Kaiser et al., 2014; Polykratis et al., 2014). Similarly, RIP1 with a mutant RHIM knock-in mice also die perinatally, which can be prevented by RIP3, MLKL or ZBP1 co-deletion (Lin et al., 2016).

3. Effectors of necroptosis

Although the discovery of RIP1 involvement in TNF-α-dependent necroptosis dates to 2005, it was only in 2009 that the essential role of RIP3 for the induction of TNF-α-mediated necroptosis following pharmacological or viral caspase inhibition was described (Cho et al., 2009; Degterev et al., 2005; He et al., 2009; Zhang et al., 2009). RIP3 is a 528 amino acid protein with a conserved RIP Ser/Thr kinase domain in the N-terminus and a RHIM domain close to the C-terminus, and both are crucial for its necroptotic activity (Fig. 3) (Cho et al., 2009). During permissive conditions for RIP1-dependent necroptosis, active RIP1 can further recruit RIP1 and RIP3 through RHIM interactions, leading to the formation and oligomerization of heterodimers of these proteins, which originates a large hydrophobic amyloid signaling complex termed necrosome (Li et al., 2012; Mompean et al., 2018; Wu et al., 2014). Necrosome formation also appears to be assisted by heat shock protein 90 (Hsp90) and the cochaperone cell cycle division cycle 37 (Li et al., 2015; Zhao et al., 2016). Importantly, although RIP1-RIP3 heterodimerization is important for necroptotic signaling downstream of RIP1 activation, these heterodimers cannot in itself induce necroptosis, being dependent on the subsequent formation of RIP3-RIP3 homodimers that allows for *cis*-autophosphorylation of T231/S232 in mouse and S227 in human RIP3 and necroptotic induction (Raju et al., 2018; Sun et al., 2012; Wu et al., 2014; Xie et al., 2013b). Therefore, RIP3-RIP3 homodimerization can occur spontaneously or downstream

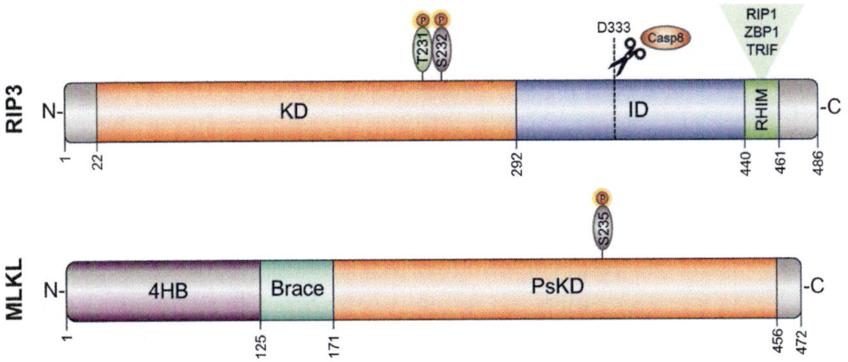

Fig. 3 Domain architecture and post-translational modifications of murine RIP3 and MLKL. RIP3 is constituted by a N-terminal kinase domain (KD) and an intermediate domain, where the RHIM motif is contained, which is essential for RIP3 recruitment and induction of necroptosis by RIP1, TRIF and ZBP1. RIP3 KD homodimerization downstream of RHIM interactions leads to *cis*-autophosphorylation of T231/S232, which is essential to recruit and phosphorylate MLKL. MLKL consists of an N-terminal 4HB domain, involved in membrane interactions and subsequent permeabilization, a central brace region composed by two helices and a C-terminal pseudokinase domain (PsKD), which usually inhibits 4HB executioner functions. However, phosphorylation on S235 of the PsKD of MLKL by RIP3 allows for the release of 4HB and formation of active oligomers through the central brace region.

of other RHIM-containing proteins, independently of RIP1 (Mompean et al., 2018; Orozco et al., 2014; Wu et al., 2014). It was also recently shown that the RIP3 kinase domain can homodimerize, probably as a secondary mechanism downstream of RHIM-dependent dimerization, which is essential to mediate conformational changes that allow for RIP3 autophosphorylation (Raju et al., 2018). The steady-state levels of RIP3 and RIP1 are maintained by the E3 Ub ligase carboxyl terminus of Hsp70-interacting protein (CHIP), which ubiquitinates these proteins and targets them for lysosomal degradation (Seo et al., 2016). Similarly to RIP1, RIP3 can be poly-ubiquitinated by cIAPs, which may contribute to RIP3 proteasomal degradation (Bertrand et al., 2011).

When active, the kinase domain of active RIP3 can bind the C-terminal pseudokinase domain of MLKL and phosphorylate T357/S358 in human MLKL or S345 in mouse MLKL (Rodriguez et al., 2016; Sun et al., 2012; Xie et al., 2013b). This event, initially identified in 2012, is critical for necroptotic induction, and MLKL still remains the sole identified *bona fide* effector of this RCD (Petrie et al., 2019a; Sun et al., 2012; Zhao et al., 2012). In accordance, MLKL-deficient cells and mice are fully protected from necroptotic engagement (Murphy et al., 2013; Wu et al., 2013). Structurally, human and mouse MLKL have similar domain compositions, possessing an N-terminal 4-helix bundle (4HB) essential for its executioner function that remains basally inhibited by the pseudokinase domain, and an intermediate brace region composed by two helices (Fig. 3) (Murphy et al., 2013; Petrie et al., 2018, 2019a). RIP3-mediated phosphorylation on the activation loop of the MLKL pseudokinase domain appears to be essential to induce the release of the 4HB domain and formation of homo-oligomers through the central brace region (Davies et al., 2018; Hildebrand et al., 2014; Petrie et al., 2019a; Rodriguez et al., 2016; Wang et al., 2014a). MLKL oligomers can then translocate to the inner leaflet of the plasma membrane through low-affinity phosphatidylinositol phosphate (PIP)-binding sites contained in the 4HB, which is followed by a "rolling over" of the 4HB in order to expose further higher-affinity PIP-binding sites that ensure a robust interaction between MLKL oligomers and the plasma membrane (Dondelinger et al., 2014; Quarato et al., 2016). Importantly, human and mouse MLKL orthologs display divergent activating mechanisms, and both full-length MLKL and killer constructs are negligible inducers of necroptosis when expressed in MLKL-deficient cells from the other species (Tanzer et al., 2016). In the case of mouse MLKL, RIP3-mediated phosphorylation at S345 is sufficient to fully activate mouse

MLKL, since phosphomimetic mutants can induce necroptosis in RIP3-deficient cells (Jacobsen et al., 2016; Petrie et al., 2019a; Tanzer et al., 2015). Conversely, activation of human MLKL is more tightly regulated (Petrie et al., 2019a). In fact, recruitment to RIP3 is an essential prerequisite to the induction of necroptosis, since phosphomimetic MLKL mutants for the RIP3 substrate sites T357/S358, which negate RIP3 engagement, completely impair necroptosis in MLKL-deficient human cells, which suggest that RIP3, or perhaps other necrosome-contained protein, may have additional scaffolding functions essential for proper MLKL oligomerization and function (Petrie et al., 2018, 2019a). Interestingly, although recombinant human 4HB domain can rapidly induce extensive membrane permeabilization in *in vitro* liposome assays, a significant delay occurs between MLKL translocation to the plasma membrane and subsequent membrane permeabilization and cell death (Dondelinger et al., 2014; Petrie et al., 2019a; Tanzer et al., 2015; Wang et al., 2014a). These observations, allied to the specificity of the species of the host cells to MLKL orthologs, strongly suggest the existence of regulatory mechanisms downstream of MLKL activation that may promote or inhibit necroptosis, which are now starting to be unraveled (Petrie et al., 2019a; Tanzer et al., 2016). In this regard, HSP90 is essential to proper human and mouse MLKL oligomerization and translocation to plasma membranes besides its function in the necrosome (Bigenzahn et al., 2016; Jacobsen et al., 2016; Zhao et al., 2016). ATP binding to the pseudokinase domain of human MLKL has also been shown to destabilize oligomerization, suggesting that ATP depletion may favor MLKL activation (Petrie et al., 2018). TRAF2 was also reported to inhibit necroptosis by constitutively associating with MLKL, which is reversed by CYLD-dependent TRAF2 deubiquitination during TNF-α-driven necroptosis (Petersen et al., 2015). Recently, it was also shown that human MLKL can bind highly phosphorylated soluble inositol phosphates (IPs) in three distinct sites within the 4HB, which removes auto-inhibitory mechanisms for full-blown MLKL activation. Importantly, deletion of the lipid kinases involved in the synthesis of these IPs strongly protected human cells from necroptosis (Dovey et al., 2018; McNamara et al., 2019). Necroptosis can also be attenuated or prevented downstream of MLKL activation and translocation to the plasma membrane with concomitant PS exposure, but prior to full membrane permeabilization, due to the formation of extracellular vesicles containing phosphorylated MLKL (p-MLKL), which depends on endosomal sorting complex required for transport (ESCRT)-III machinery. This process allows necroptotic cells to live longer and produce inflammatory mediators prior to cell demise, or survive altogether in the presence of a mild necroptotic stimulus (Gong et al., 2017; Yoon et al., 2017; Zargarian et al., 2017).

Nevertheless, the precise mechanisms involved in membrane permeabilization following MLKL oligomerization are still a matter of intense debate. Biophysical analyses performed *in vitro* suggest that mouse MLKL forms preferably trimers, while human MLKL forms tetramers, although assembly into higher order oligomeric species can probably occur in living cells (Davies et al., 2018; Huang et al., 2017; Petrie et al., 2018, 2019a). MLKL oligomers have been suggested to disrupt the plasma membrane by partial insertion, or whole insertion, leading to the formation of pores, or cation channels, which would allow for Ca^{2+} and Na^{+} influx and subsequent loss of ionic and osmotic homeostasis, leading to cell lysis and release of intracellular contents, including DAMPs (Dondelinger et al., 2014; He and Wang, 2018; Hildebrand et al., 2014; Huang et al., 2017; Wang et al., 2014a; Xia et al., 2016). MLKL may also rely on auxiliary recruitment of ion channels, such as the transient receptor potential melastatin related 7 (TRPM7), for Ca^{2+} influx (Cai et al., 2014). Conversely, a recent study suggests that MLKL forms small plasma membrane nanopores that are sufficient to induce plasma membrane permeabilization independently of Ca^{2+} influx (Ros et al., 2017).

Overproduction of reactive oxygen species (ROS) during necroptotic signaling also appears to be a universal mechanism, but their relevance to necroptotic cell death is widely controversial, since they are dispensable for necroptosis in several cell types (He et al., 2009; Tait et al., 2013; Yang et al., 2018 ; Zhang et al., 2017). Specifically, necrosomal RIP3 appears to be essential for ROS production during necroptosis. For example, while RIP3 was necessary for cell death-dependent ROS production during necroptosis in FADD-deficient Jurkat cells, MLKL silencing only partially protected from ROS generation but halted cell death, suggesting other targets for RIP3-dependent ROS production (Schenk and Fulda, 2015; Zhang et al., 2009). ROS production probably acts as a positive feedback mechanism to ensure proper necrosome formation, since it was shown that ROS mediate the oxidation of specific cysteines in RIP1, which facilitates RIP1-RIP1 interactions via formation of disulfide bond-linked high molecular weight complexes and subsequently increases RIP1 transphosphorylation at S161. In these studies, ROS quenching with several antioxidants fully protected L929 cells and FADD-deficient Jurkat cells from necroptotic cell death, and strongly decreased necrosome formation (Schenk and Fulda,

2015; Zhang et al., 2017). Overall, these results suggest that ROS can not only contribute to necroptosis as an amplification device but may also reduce the threshold for necroptotic induction in other settings due to the ROS sensing properties of RIP1.

Mitochondria are generally regarded as the probable culprit involved in ROS production during necroptosis, although other mechanisms of ROS generation have also been proposed, such as activation of the NADPH oxidase NOX1 downstream of RIP1 during TNF-α-dependent necroptosis in L929 cells (Kim et al., 2007; Yang et al., 2018; Zhang et al., 2017). An earlier study implicated the mitochondrial protein phosphoglycerate mutase family member 5 (PGAM5) as a required executioner of necroptosis downstream of RIP3 activation by inducing mitochondrial fission via dynamin-related protein 1 (DRP1), but ample evidence from subsequent studies employing genetic deletion or silencing of both proteins disproved this hypothesis (Lu et al., 2016; Moriwaki et al., 2016; Remijsen et al., 2014; Wang et al., 2012b). Interestingly, mitophagy-dependent mitochondrial depletion reduced ROS production during TNF-α/zVAD-induced necroptosis in NIH3T3 fibroblast cells, supporting a role for mitochondria in necroptotic-mediated ROS generation, but necroptotic cell death still proceeded normally in this and other cell lines (Tait et al., 2013). Conversely, mitochondria-depleted L929 cells are partially protected from TNF-α/zVAD-induced necroptosis (Yang et al., 2018; Zhang et al., 2017). Moreover, TNF-α-dependent necroptosis in L929 cells following cIAPs or TAK1 inhibition was suggested to depend on mitochondrial ROS production downstream of RIP1/3 activation, but independently of NOX1, and ROS scavengers significantly protected from necroptotic cell death (Vanlangenakker et al., 2011). Ample evidence indicate that aerobic respiration is increased during necroptosis in multiple cell lines, especially L929 cells, and this effect may be the driving force behind necroptotic ROS generation (Goossens et al., 1999; Schulze-Osthoff et al., 1992; Yang et al., 2018). A recent study reported that the active necrosome can be translocated to the mitochondria in a MLKL-dependent fashion, where RIP3 directly phosphorylates the pyruvate dehydrogenase complex (PDC) E3 subunit at T135, leading to PDC activation and a subsequent increase in aerobic respiration and mitochondrial ROS production (Yang et al., 2018). An earlier work from the same group has also showed that RIP3 can stimulate the activity of several metabolic enzymes involved in glycogenolysis, glutaminolysis and glycolysis, which increases the availability of substrates for the Krebs cycle and further amplifies aerobic respiration and ROS generation (Zhang et al., 2009). In agreement, mitochondrial complex I inhibition with amytal or rotenone has been shown to reduce TNF-α-dependent necroptosis in L929 cells (Goossens et al., 1999; Schulze-Osthoff et al., 1992; Yang et al., 2018; Zhang et al., 2017). Conversely, mitochondrial complex I inhibition was reported to induce necroptotic features in melanoma cell lines by increasing ROS production via PINK1- and DRP1-dependent mitophagy, which was attenuated by knockdown of these or necrosome-related proteins, ROS quenching and Nec-1 (Basit et al., 2017). A recent study has also reported that RIP3 induces mitochondrial ROS production independently of MLKL by directly interacting with NOX4 and promoting its mitochondrial translocation, and by inhibiting mitochondrial complexes I and III, in *in vivo* and *in vitro* models of sepsis-induced acute kidney injury, which are dependent on TLR signaling (Sureshbabu et al., 2018). These controversial results indicate that the relevance of mitochondrial electron chain inhibition for the induction of necroptosis depends on cell type, metabolic conditions, and type of stimulus used. PINK1-dependent mitophagy also appears to contribute to necroptotic induction in *in vivo* and *in vitro* models of chronic obstructive pulmonary disease, while deletion of PINK1 or DRP1 is irrelevant for TNF-α-dependent necroptosis in L929 cells (Mizumura et al., 2014; Remijsen et al., 2014).

Importantly, Nec-1 stable (Nec-1s) exposure potentiates the neural differentiation of iPSCs from PD patients carrying dominant PD-causing mutations in the gene encoding optic atrophy 1 (OPA1), a GTPase involved in mitochondrial dynamics. These PD-derived neural cells present abnormal mitochondrial morphologies, aerobic respiration impairment, ATP deficits and higher ROS production, along with higher protein levels of RIP1, RIP3 and MLKL (Iannielli et al., 2018). Overall, it is possible that an "intrinsic" form of necroptosis can be induced by mitochondrial dysfunction/mitophagy and subsequent ROS production independently of receptor activation in some settings, possibly via the ROS sensing mechanism of RIP1, although autocrine death receptor activation remains to be evaluated in these models (Basit et al., 2017; Iannielli et al., 2018; Mizumura et al., 2014).

Some studies have also suggested the involvement of BAX/BAK oligomerization and subsequent mitochondrial outer membrane permeabilization (MOMP) as a downstream event of necroptotic cell death (Irrinki et al., 2011; Karch et al., 2015; Rohde et al., 2017; Tischner et al., 2012). Double BAX/BAK ko fully protects mouse fibroblasts from

TNF-α/zVAD-induced necroptosis, which was reported to depend on MLKL mitochondrial translocation and BAX/BAK homo- or heterodimerization, along with reduced expression of the anti-apoptotic protein Mcl-1 (Karch et al., 2015). Other study showed that RIP3/MLKL-dependent ROS production was essential for BAK-mediated MOMP during Smac mimetic/glucocorticoid-induced necroptosis, while BAK silencing delayed cell death and decreased MOMP induction (Rohde et al., 2017). Interestingly, cyclophilin D (CYPD) deletion or pharmacological inhibition, an essential component of MPT-dependent necrosis, which result in the permeabilization of outer and inner mitochondrial membranes, was also shown to partially protect from necroptosis, suggesting that this RCD may contribute to the necroptotic phenotype (Karch et al., 2015; Tischner et al., 2012). However, concomitant pharmacological or genetic inhibition of necroptosis and CYPD further prolonged the survival of mice following kidney ischemia/reperfusion than inhibition of either necrotic pathway alone, suggesting that, in the eventuality of participating in necroptosis, CYPD may do so in restricted settings (Linkermann et al., 2013). Overall, although MLKL remains the only confirmed executioner of necroptosis, ROS production and crosstalk with death mechanisms of other RCDs, such as apoptosis or MPT-driven necrosis, probably drives necroptotic cell death in multiple experimental and pathological conditions. More research is needed to unravel these mechanisms in disease, which may eventually lead to the development of novel therapies that can singlehandedly block or attenuate multiple forms of cell demise.

4. Post-translational and epigenetic regulation of necroptosis

As aforementioned, necroptotic signaling downstream of DR or PRR activation is tightly regulated by several post-translational modifications, such as ubiquitination, phosphorylation and proteolytic cleavage, which mainly converge on RIP1 before necrosome assembly. However, post-translational modifications can also promote or inhibit necroptotic cell death during or following necrosome assembly. In this regard, RIP1 was shown to be heavily modified with M1- and K63-linked poly-Ub chains during TNF-α/zVAD/Smac mimetic-mediated necroptosis in HT29 cells in several lysine residues, and these modifications where dependent on RIP1 kinase activity. Moreover, mutation of RIP1 at K115 reduced necroptotic cell death and limited necrosome assembly (de Almagro et al., 2017). Importantly, although it cannot be excluded that the TNF-α-dependent pro-death ubiquitination of RIP1 at complex I upon cIAPs or TAK1 inhibition is also represented in this study, K115 poly-ubiquitination was only detected upon exposure to TNF-α/zVAD/Smac mimetic but not TNF-α/Smac mimetic, suggesting that this important post-translational modification happens downstream of necrosome formation (Amin et al., 2018; de Almagro et al., 2017). An independent study then showed that the E3 Ub ligase Pellino-1 (PELI1) was the one responsible for K63-linked poly-ubiquitination on K115 in RIP1 during necroptosis following RIP1 kinase activation, and that this event was essential for the downstream recruitment of RIP3/MLKL (Wang et al., 2017). Conversely, PELI1 has also been reported to mediate K48-linked poly-ubiquitination of kinase active RIP3, which targets active RIP3 for proteasomal degradation and attenuates necroptotic induction (Choi et al., 2018). Further research is thus warranted to dissect the different roles of PELI1 during necroptosis. K63-linked poly-ubiquitination on K5 in RIP3 by an unknown E3 Ub ligase also promotes necrosome assembly and cell death, which is reverted by the deubiquitinating activity of A20 (Onizawa et al., 2015). CYLD can also contribute to necroptosis by removing pro-survival poly-ubiquitin chains from RIP1 within the necrosome (Moquin et al., 2013). The E3 Ub ligase Triad3a was also shown to mediate K48-linked poly-ubiquitination of RIP1 upon RIP3 recruitment downstream of LPS/ZVAD-mediated necroptosis in macrophages, which induced RIP1 proteasomal degradation and early necrosome disassembly, thus limiting necroptosis (Alturki et al., 2018). Interestingly, parkin was recently implicated as a negative regulator of necroptosis when activated by phosphorylation on S9 by AMP-activated protein kinase (AMPK), which leads to K33-linked poly-ubiquitination of RIP3 and inhibition of necrosome formation (Lee et al., 2019). Protein phosphatase 1B (PPM1B) can also suppress spontaneous and TNF-α-dependent necroptosis by dephosphorylating kinase active RIP3 (Chen et al., 2015). Further research is bound to identify several other ubiquitin-modifying enzymes and kinases/phosphatases that are involved in the intricate regulation of necroptotic signaling.

RIP3 is weakly expressed throughout the human and mouse bodies, with the unsurprising exception of immunologic organs, such as the spleen and thymus, and in lymphoid and myeloid cells (Belizario et al., 2015; He et al., 2009; Sun et al., 1999). However, RIP3 expression massively increases in multiple pathological conditions, which may predispose the affected regions to necroptosis (He et al., 2009; Lin et al., 2019). Inhibition of RIP3 expression and subsequently of necroptotic signaling is also typical in cancer cell lines, which has been associated with gene promotor methylation (Koo et al., 2015). Importantly, stimulation of the receptors for type I interferon (IFN)α/β and type II IFNγ can also induce necroptosis through transcriptional mechanisms, such as MLKL transcriptional upregulation, in several cell types, but especially macrophages (McComb et al., 2014; Sarhan et al., 2019; Thapa et al., 2013). Intriguingly, IFNγ-dependent expression of MLKL was linked to hepatocellular necroptosis in an inflammation-driven hepatitis murine model, and MLKL activation was shown to proceed independently of RIPK3, albeit being dependent on RIP1 kinase activity (Gunther et al., 2016). These observations suggest that murine MLKL activation can happen independently of RIP3 in some contexts, but such mechanisms are probably absent in humans due to the requirement of RIP3-MLKL interaction for full activation of human MLKL (Petrie et al., 2019a). Recently, it was also shown that the bromodomain protein 4 can also interact with IRF1 independently of IFN signaling to mediate MLKL transcription (Xiong et al., 2019). It is probable that many more transcriptional mechanisms of necroptotic regulation exist, which may constitute interesting therapeutic targets to treat pathologies related to excessive or deficient necroptosis.

MicroRNAs (miRNAs), which are small non-coding RNAs that are involved in post-transcriptional gene silencing, may also contribute to the regulation of necroptosis. For example, miR-874 expression increases in cellular and animal models of myocardial infarction, and its inhibition attenuates myocardial necroptosis by preventing miR-874-mediated downregulation of caspase-8 expression (Wang et al., 2013). In turn, miR-155 prevent necroptosis in cardiomyocyte progenitor cells by targeting RIP1 (Liu et al., 2011).

Recently, miR-425 was shown to be downregulated in the SN of mice exposed to the sub-chronic MPTP regimen, which correlated with higher levels of RIP1, RIP3 and p-MLKL in dopaminergic neurons and neuronal loss. This miRNA appears to target RIP1 transcripts, and its genetic knockdown worsened MPTP-driven dopaminergic loss with a concomitant increase of several necroptotic markers, including RIP1. Accordingly, intracerebral administration of a miR-425 mimic reduced RIP1 expression and ameliorated MPTP-driven pathology. Importantly, miR-425 deficiency and an increase in necroptotic markers were detected in the SN of PD patients, supporting a role for this miRNA and necroptosis in PD progression (Hu et al., 2019).

5. Non-necroptotic roles for necroptosis-related proteins

Multiple studies showed that RIP1 activation, with or without concomitant RIP3 recruitment but with no MLKL requirement, can mediate the expression of pro-inflammatory genes independently of cell death, especially in macrophages and microglia (Christofferson et al., 2012; Degterev et al., 2019; Ito et al., 2016; Najjar et al., 2016; Ofengeim et al., 2017; Saleh et al., 2017; Zhu et al., 2018). Importantly, genetic or pharmacological RIP1 inhibition attenuated inflammation-related phenotypes in mouse models of Alzheimer's disease (AD) and amyotrophic lateral sclerosis (ALS) independently of neurodegeneration, which was attributed to decreased microglial expression of M1- or disease-linked pro-inflammatory genes, supporting a key role for RIP1 in microglial-driven neuroinflammation in neurodegenerative diseases (Ito et al., 2016; Ofengeim et al., 2017). In itself, RIP3 may also modulate the pro-inflammatory response independently of necroptosis in specific settings. For example, dendritic cells from RIP3ko mice developed a blunted pro-inflammatory response when exposed to LPS, which was partially dependent on defective NF-κB signaling downstream of IκB degradation. In a model of colitis, RIP3ko mice also exhibited an impaired induction of interleukin (IL)-1β, IL-23, and IL-22 cytokine cascade in dendritic cells, which promote tissue repair in this model (Moriwaki et al., 2014, 2017). In fact, RIP3 deletion affords more protection than MLKL deficiency in several mouse models of disease, such as renal ischemia-reperfusion injury, metastatic tumor, TNF-α challenge, IL-1β-dependent arthritis, *Staphylococcus aureus* infection and sepsis-induced acute kidney injury, supporting other non-necroptotic roles for RIP3, especially in inflammation (Hanggi et al., 2017; Kitur et al., 2016; Lawlor et al., 2015; Newton et al., 2016a; Sureshbabu et al., 2018).

Moreover, extensive research has implicated the recruitment of RIP1, RIP3, MLKL and caspase-8 downstream of TLR3/4-TRIF-signaling as an important contributor to inflammasome activation in immune cells, such as dendritic cells and macrophages, in a cell death-independent manner (Conos et al., 2017; Kang et al., 2015; Lawlor et al., 2015). Inflammasomes are multiprotein complexes formed upon exposure to pro-inflammatory stimuli in which the effector caspase-1 is activated to cleave pro-IL-1β into the mature cytokine, although caspase-8 can also contribute to this process by cleaving procaspase-1 or pro-IL-1β directly (Moriwaki and Chan, 2017). The precise mechanisms linking the necroptotic machinery to inflammasome activation are still a matter of debate, but usually depend on the inhibition or absence of cIAPs, A20 and/or caspase-8 (Duong et al., 2015; Lawlor et al., 2015; Moriwaki and Chan, 2017; Vince et al., 2012). In the absence of cIAPs in macrophages, TLR4 stimulation can mediate NLRP3 inflammasome activation via the recruitment of RIP3 and subsequent kinase activity-independent activation of caspase-8, which result in NLRP3-caspase-1 activation. However, if caspase-8 is also inhibited, RIP3 kinase activity and MLKL become essential to mediate NLRP3 activation (Lawlor et al., 2015). MLKL-dependent inflammasome activation appears to rely on MLKL translocation to the plasma membrane and subsequent K^+ efflux but is independent on cell permeabilization. In fact, NLRP3 inflammasome inhibition prevents pro-IL-1β maturation but does not affect MLKL-driven necroptosis, suggesting that inflammasome activation during necroptosis is a secondary mechanism to potentiate release of pro-inflammatory mediators without affecting the fate of cells (Conos et al., 2017).

RIP3 may also participate in the regulation of other RCDs, especially apoptosis. In fact, earlier studies preceding the discovery of necroptosis attributed an apoptotic role for RIP3 when overexpressed, which was suggested to depend on its RHIM-containing C-terminal domain (Sun et al., 1999; Yu et al., 1999). Interestingly, while RIP3ko mice display no obvious phenotype, expression of the catalytic inactive $RIP3^{D161N}$ induce perinatal lethality in mice due to excessive activation of RIP1- and caspase-8-dependent apoptosis, but these animals can survive until birth if RIP1 is deleted (Newton et al., 2014; Raju et al., 2018). However, mice expressing other form of kinase dead $RIP3^{K51A}$ are viable and fertile (Mandal et al., 2014). Cellular expression of mutant $RIP3^{D161N}$ or exposure to some ATP-competitive RIP3 inhibitors at high concentrations were shown to induce caspase-8-dependent apoptosis by mediating the spontaneous formation of a Ripoptosome-like complex, containing RIP3, RIP1, FADD and $cFLIP_L$, but not MLKL. Intriguingly, activation of caspase-8 in this complex is independent of the kinase activities of RIP1 and RIP3, but dependent on $cFLIP_L$ recruitment (Mandal et al., 2014). Recently, it was reported that the $RIP3^{D161N}$ mutation locks RIP3 into an active conformation that, despite lacking catalytic activity, can still act as an allosteric activator of Wt RIP3 or RIP1 by homo- or heterodimerizing with their kinase domains (Raju et al., 2018). Since caspase-8 activation in this context is independent of the kinase activity of RIP1, RIP1/RIP3^{D161N} kinase dimerization probably enhances the scaffolding functions of RIP1 for the subsequent recruitment of Ripoptosome components. However, forcible dimerization of Wt RIP3 can still result in caspase-8 and RIP1-dependent apoptosis when MLKL is absent (Cook et al., 2014). Curiously, it was also shown that, during conditions of cIAPs or TAK1 inhibition, RIP3-deficient mouse fibroblasts are protected from TNF-α-mediated RIP1 kinase- and caspase-8-dependent apoptosis, but apoptosis could be reinstated by expression of a RIP3^{K51A} kinase dead or a RHIM domain mutant RIP3, suggesting that RIP3 may have additional roles during RIP1 kinase-dependent apoptosis. Moreover, RIP3 deletion strongly limited ROS production during this form of apoptosis but, while ROS scavengers protected from cell death and prevented complex IIb formation, RIP3 deletion had no effect on complex IIb formation, suggesting specific downstream RIP3-dependent mechanisms of ROS generation (Dondelinger et al., 2013). Given the possible recruitment of MOMP, a classical apoptotic component, during necroptosis in a ROS-dependent fashion, it would be interesting to determine whether this RIP3-dependent ROS production during apoptosis eventually targets mitochondria (Rohde et al., 2017). Overall, these observations support the existence of an intricate network between apoptotic and necroptotic components, implicating that the induction of either apoptosis or necroptosis in a specific setting is dependent on complex interactions between the cellular levels, as well as the kinase and scaffolding activities, of intervening components such as RIP1, RIP3, MLKL, caspase-8, FADD and $cFLIP_L$.

6. Evolutionary and physiological roles of necroptosis

As discussed earlier, RIP3- and MLKL-deficient mice develop normally and mature into fertile adults devoid of aberrant phenotypes, indicating that necroptosis is not required during development or adult homeostasis (Newton et al., 2004; Wu et al., 2013). The general consensus is that necroptosis evolved as a mechanism to fight viral infections capable of halting apoptosis in the host cell to fulfill their life cycles. Since necroptosis is a pro-inflammatory cell death pathway, induction of this RCD in infected cells can not only terminate viral replication but also contribute to the development of an immune response against the invading pathogen by releasing pro-inflammatory molecules such as cytokines, DAMPs and pathogen associated molecular patterns (PAMPs) (Dondelinger et al., 2016).

In fact, a recent study showed that cytokine production can proceed in necroptotic cells following membrane permeabilization, which boosts the immune response toward these cells (Orozco et al., 2019). The fact that all RHIM-containing proteins capable of activating RIP3, namely, RIP1, TRIF and ZBP1, are intimately involved in the recognition of PAMPs further strengthens this hypothesis. In accordance, cellular infection with poxvirus vaccinia virus induces TNF-α-dependent necroptosis due to the expression of the viral pan-caspase inhibitor B13R/Spi2 and subsequent caspase-8 inhibition. Moreover, RIP3ko mice are highly susceptible to vaccinia virus infection (Cho et al., 2009). The NS1 protein of human influenza A virus has also been recently demonstrated to induce necroptosis by directly interacting with MLKL and increasing its oligomerization downstream of RIP3-dependent necroptotic signaling (Gaba et al., 2019). However, multiple viruses have also evolved anti-necroptotic mechanisms. For example, the murine cytomegalovirus encodes for the anti-apoptotic protein viral inhibitor of caspase-8 activation along with the anti-necroptotic viral inhibitor of RIP activation (vIRA), which contains an N-terminal RHIM domain capable of interacting with the RHIM of RIP3 (Upton et al., 2010, 2012). The human herpes simplex virus type 1 and type 2 are also capable of inhibiting both apoptotic and necroptotic pathways by the single expression of ICP6 or ICP10, respectively, which can bind caspase-8 through the N-terminal domain and RIP1/3 through the C-terminal RHIM (Guo et al., 2015). Very recently, it was also discovery that some poxviruses express homologs of the MLKL pseudokinase domain, which were shown to sequester RIP3 and therefore prevent necroptosis (Petrie et al., 2019b).

Multiple lines of evidence also indicate that necroptosis can play a role in the activation of the innate and adaptive immune system against cancer cells. It has been shown that necroptotic cells are highly immunogenic not only due to the release of DAMPs, such as ATP and high mobility group box 1 (HMGB1), but also due to NF-κB-dependent transcription before cell death. This promotes efficient cross-priming of $CD8^+$ T cells and production of IFNγ and IL-12, which elicits an adaptive immune response against tumor cells *in vivo* (Aaes et al., 2016; Yang et al., 2016; Yatim et al., 2015). Accordingly, RIP3 expression is highly downregulated in breast cancer tissue, suggesting that absence of RIP3 is positively selected during tumorigenesis, possibly in order to evade necroptotic induction upon gain of resistance to apoptosis (Koo et al., 2015). In contrast, other studies have suggested that necroptosis and necroptosis-driven inflammation can contribute to tumorigenesis and potentiate tumor extravasation (Lee et al., 2019; Strilic et al., 2016).

Necroptosis may also act as a backup when apoptosis is inhibited during physiological conditions. For example, BIM-deficient mice develop lymphoproliferative disorder and autoimmunity due to limited induction of apoptosis in T cells, which can be reverted by conditional deletion of caspase-8 in these cells and subsequent necroptotic engagement (Bohgaki et al., 2011). Likewise, caspase-8 and RIP3 or FADD and MLKL double ko mice are viable and fertile but develop a progressive lymphoproliferative disease with accumulation of T cells in several tissues (Oberst et al., 2011; Zhang et al., 2016). Intriguingly, although necroptosis is a pro-inflammatory cell death pathway, *in vivo* and *in vitro* studies suggest that clearance of TNFR1- or TLR-stimulated immune cells by necroptosis can attenuate rather than promote inflammation (Kearney et al., 2015; Li et al., 2019b). Of note, systemic zVAD administration can alleviate the pathogenesis of endotoxin shock after LPS challenge in mice by inducing necroptosis of activated macrophages (Li et al., 2019b). Importantly, induction of necroptosis in TLR3/4-stimulated microglia was shown to protect neurons against neurotoxic inflammation (Fricker et al., 2013). Conversely, microglial necroptosis has been detected in several neurodegenerative diseases, such as AD, ALS and intracerebral hemorrhage (Caccamo et al., 2017; Ito et al., 2016; Lu et al., 2019).

7. Dysregulation of necroptosis in CNS diseases

Abnormal activation of necroptosis has also been extensively reported to be involved in the etiology of multiple pathologies with an inflammatory component, including acute (Table 1) and chronic (Table 2) neurodegenerative diseases (Choi et al., 2019; Yuan et al., 2019a).

7.1 Acute neurodegenerative diseases

The first Nec-1 therapeutic application was reported in a mouse model of ischemic brain injury (Table 1), where Nec-1 decreased infarct volume following middle cerebral artery occlusion (MCAO), while zVAD co-administration resulted in a synergistic protection (Degterev et al., 2005). Conversely, RIP3ko mice were shown to be unprotected in the MCAO model, suggesting necroptotic-independent roles for RIP1 in this setting (Newton et al., 2016a). However, pharmacological RIP3 inhibition also decreased the infarct volume following MCAO in mice, which

Table 1 Necroptosis and necroptotic-related proteins in acute neurodegenerative diseases.

Model/sample	Functional outcome/involvement of necroptosis	References
Ischemic stroke		
Transient MCAO in mice	Nec-1/Nec-1s reduced infarct volume	Degterev et al. (2005)
	Brain lesions remained unchanged between Wt and RIP3ko mice	Newton et al. (2016a)
	GSK'872 (RIP3 inhibitor) limited RIP1/3 protein expression and p-MLKL levels, attenuating brain and neurological problems	Yang et al. (2017)
	RIP1/3, MLKL, TRAF2, and inhibitory complex formation between TRAF2 and MLKL increased—all reverted by Nec-1	Li et al. (2019a)
	Nec-1 increased oligodendrocyte precursor cell number, reduced white matter injury and limited RIP1/3, MLKL and p-MLKL levels	Chen et al. (2018)
Transient MCAO in rats	Nec-1 or anti-TNF-α therapy prevented endothelial cell death, BBB disruption, and increased p-RIP1 (S166) and p-MLKL	Chen et al. (2019a)
	Nec-1 reduced RIP1/3 and MLKL protein overexpression and phosphorylation of RIP1 (S166) and MLKL, reducing infarct volume and improving neurological scores	Deng et al. (2019)
Hemorrhagic stroke		
Induction of SAH in rats	Nec-1 limited BBB disruption, while improving neurological scores; Nec-1 reduced cell death and overexpression of RIP3 and MLKL	Chen et al. (2019b)
	Overexpression and increased interaction of RIP1/3 and MLKL; increased RIP3 and MLKL phosphorylation; genetic or pharmacological inhibition of RIP3 (GSK'872) or TNF-α reduced these markers and attenuated BBB disruption	Yuan et al. (2019b)

Continued

Table 1 Necroptosis and necroptotic-related proteins in acute neurodegenerative diseases.—cont'd

Model/sample	Functional outcome/involvement of necroptosis	References
Intrastriatal hemorrhage in mice	Mice with RIP1 kinase dead mutations (K45A and D138N) were protected from acute neuronal death	Lule et al. (2017)
	RIP3 increased; Nec-1 reduced RIP1/RIP3 interaction, necrotic neuronal death and neuroinflammation	Su et al. (2015)
Traumatic brain injury		
Controlled cortical impact in mice	Nec-1 improved motor performance and reduced lesion volume, cell death and neuroinflammation, without affecting caspase-3-activation	You et al. (2008)
	Nec-1 reverted Bcl-2 overexpression, caspase-3 activation, and neuronal cell death	Wang et al. (2012a)
	RIP1/3 and p-MLKL increased; RIP3 deficiency improved cognitive function, limited inflammation and caspase-3/8 activation	Liu et al. (2018b)
Traumatic spinal cord injury		
SCI in mice	Nec-1 reduced SCI lesions, TUNEL staining and expression of RIP1/3, caspase-3 and BAX	Wang et al. (2014a)
	Nec-1 decreased necrotic cell death and improved functional recovery, without affecting caspase-3 activation and TUNEL staining	Liu et al. (2015)
	RIP3 deficiency or Nec-1 limited MLKL and/or RIP3 expression in reactive astrocytes, protecting from cell death and inflammation	Fan et al. (2016)
	RIP1/3 and MLKL accumulated in neurons	Liu et al. (2018a)

Retinal ischemia		
Intraocular pressure in rats	Nec-1 protected from acute neuronal death in the inner retinal layers without affecting caspase- or TUNEL-stained cells	Rosenbaum et al. (2010)
	RIP1/3 increased in retinal ganglion neurons; Nec-1 prevented neuronal death and limited inflammation	Dvoriantchikova et al. (2014)
	RIP1/3 overexpressed in the retinal ganglion neurons; Nec-1 decreased RIP1 expression and neuronal death	Kim et al. (2017)
Glaucoma		
Intraocular pressure in rats	RIP1-inhibitory compound protected against retinal ganglion cell death and limited RIP1 phosphorylation	Do et al. (2017)
Retinal detachment		
Retinal detachment in rats	Nec-1 protected from photoreceptor cell death in the ONL, abrogated RIP1 phosphorylation and increased functional recovery	Dong et al. (2012)
Age-related macular degeneration		
Subretinal injection of poly(I:C) in mice	Retinal RIP3 increased; RIP3ko limited neuronal death in ONL and retinal pigment epithelium, HMGB1 release and inflammation	Murakami et al. (2014)

BAX, Bcl-2 associated X; Bcl-2, B-cell lymphoma 2; BBB, Blood brain barrier; HMGB1, High mobility group box 1; MCAO, middle cerebral artery occlusion; MLKL, mixed lineage kinase domain-like, Nec, necrostatin; ONL, outer nuclear layer; RIP, receptor interacting protein; SAH, subarachnoid hemorrhage, SCI, Spinal cord injury; TNF-α, tumor necrosis factor-α; TRAF2, TNFR-associated factor 2; TUNEL, terminal deoxynucleotidyl transferase dUTP nick end labeling.

Table 2 Necroptosis and necroptotic-related proteins in chronic neurodegenerative diseases.

Model/sample	Functional outcome/involvement of necroptosis	References
Huntington's disease		
R6/2 transgenic mouse model	Full-length RIP1 increased; Nec-1 improved motor performance and increased longevity	Zhu et al. (2011)
Alzheimer's disease		
Human AD brains	RIP1, MLKL and p-MLKL protein and mRNA levels were elevated; RIP1/RIP3/MLKL were shown to interact and/or be contained in insoluble fractions/functional dimers; RIP1 and MLKL mRNA levels positively correlated with Braak stage and inversely correlated with brain weight and cognitive scores; RIP1-regulated gene network overlapped with AD transcriptomic signatures	Caccamo et al. (2017)
	Active RIP1, as evaluated by recruitment into insoluble fractions and S166 phosphorylation, was increased	Ofengeim et al. (2017)
5 × FAD mice	Nec-1s limited MLKL phosphorylation and neurodegeneration	Caccamo et al. (2017)
APP/PS1 mice	Overexpression of a constitutively active MLKL form in neurons induced higher hippocampal neuronal loss than in Wt mice	Caccamo et al. (2017)
	RIP1 inhibition (Nec-1s or $RIP1^{D138N}$ kinase dead mutant) reduced amyloidosis, Aβ levels, microglial-driven inflammation and behavioral deficits independently of necroptosis; RIP1 activity can modulate the expression of several microglial genes, such as cystatin F, which decreases lysosomal protein turnover	Ofengeim et al. (2017)

Multiple sclerosis		
Human MS samples	Increased RIP1/3 and MLKL activation, as evaluated by total protein levels, functional interaction, insoluble fraction recruitment and activating phosphorylations; defective caspase-8 activation and increased $cFLIP_L$ levels; insoluble proteome overlapped with the equivalent proteomes of AD and PD	Ofengeim et al. (2015)
Cuprizone and EAE models	Nec-1s and RIP3 deficiency protected from demyelination by preventing TNF-α-mediated necroptosis of oligodendrocytes	Ofengeim et al. (2015)
EAE mouse model	Compound 22 (RIP1 inhibitor) ameliorated disease pathology	Yoshikawa et al. (2018)
Amyotrophic lateral sclerosis		
Human ALS samples	RIP1/3, MLKL and activating phosphorylations increased	Ito et al. (2016)
OPTN-deficient mice	Axonal loss and demyelination derived from oligodendrocyte death and microglial activation; RIP1/3, MLKL and activating phosphorylations increased, while Nec-1s or RIP3 deficiency were protective	Ito et al. (2016)
$SOD1^{G93A}$ mice	RIP1/3 and p-MLKL increased; Nec-1s or RIP3 deficiency attenuated myelinating and axonal pathology and improved motor function	Ito et al. (2016)
	RIP1 mRNA and protein levels increased; no oligodendrocyte-dependent myelinating pathology detected; RIP3 deficiency was unprotective	Dermentzaki et al. (2019)
MN co-cultured with iPSC-astrocytes	Nec-1 protected from neuronal death induced by astrocytes derived from patients with familial or sporadic ALS	Re et al. (2014)

Continued

Table 2 Necroptosis and necroptotic-related proteins in chronic neurodegenerative diseases.—cont'd

Model/sample	Functional outcome/involvement of necroptosis	References
ALS/frontotemporal dementia		
$TBK1^{+/-}$ mice with $TAK1^{+/-}$ in myeloid cells	RIP1 kinase-dependent apoptosis/necroptosis, microgliosis and inflammation in CNS, resulting in the hallmarks of ALS/FTD; genetic RIP1 inhibition was protective	Xu et al. (2018)
Parkinson's disease		
Human PD samples	Increased RIP1/3 and p-MLKL mostly in nigral dopaminergic neurons	Hu et al. (2019) and Onate et al. (2019)
MPTP mouse models	Nec-1/Nec-1s/miR-425 decreased RIP1/3 and MLKL overexpression and activation, striatal dopamine loss and nigral dopaminergic neuronal death; RIP3 and MLKL deficiency were similarly protective	Hu et al. (2019), Iannielli et al. (2018), andLin et al. (2019)
6-OHDA mouse model	Nec-1s or RIP3/MLKL deficiency attenuated necroptotic-driven axonal degeneration in the striatum, as evaluated by p-MLKL/RIP1 interaction, and retrograde nigral dopaminergic neurodegeneration	Onate et al. (2019)
Gaucher's disease		
GBA-deficient mice; CBE-injected mice	RIP1/3 and $cFLIP_S$ protein levels increased; RIP3 deficiency increased longevity and motor function in the CBE mouse model	Vitner et al. (2014)
Niemann-Pick, type C1		
NPC1-deficient mice	Increased RIP1/RIP3 protein levels and interaction; Nec-1 and GSK'547 (RIP1 inhibitor) attenuated cerebellar Purkinje neuronal death, progression of motor deficits and RIP1/RIP3 interaction; Genetic RIP1 inhibition was protective, unlike RIP3 deficiency	Cougnoux et al. (2016, 2018)

6-OHDA, 6-hydroxydopamine; Aβ, amyloid-β; AD, Alzheimer's disease; ALS, amyotrophic lateral sclerosis; APP, amyloid precursor protein; CBE, conduritol B epoxide (GBA inhibitor); cFLIP, cellular FLICE/caspase-8-like inhibitory protein; CNS, central nervous system; EAE, experimental autoimmune encephalomyelitis; FAD, familial AD; GBA, glucocerebrosidase; FTD, frontotemporal dementia; iPSC, induced pluripotent stem cell; ko, knockout; MLKL, mixed lineage kinase domain-like MN, motor neurons; MS, multiple sclerosis; Nec, necrostatin; OPTN, optineurin; PD, Parkinson's disease; PS1, presenilin 1; RIP, receptor interacting protein; SOD1, superoxide dismutase 1; TAK1, TGF-β-activated kinase 1; TBK1, TANK-binding kinase 1; TNF-α, tumor necrosis factor-α; TUNEL, terminal deoxynucleotidyl transferase dUTP nick end labeling.

resulted in limited RIP1/3 protein expression and MLKL phosphorylation, while zVAD co-administration once again conferred additional protection (Yang et al., 2017). Other recent study reported the detection of RIP1 phosphorylation at S166 in neurons, along with increased p-MLKL and IL-1β in striatal tissue of rats exposed to MCAO, which were reverted by Nec-1 treatment (Deng et al., 2019). TRAF2 may also be protective in the MCAO mouse model by directly interacting with MLKL and thus preventing necroptosis (Li et al., 2019a). Nec-1 was also shown to improve recovery in a mouse model of MCAO-driven ischemia followed by reperfusion by protecting oligodendrocyte precursor cells from necroptosis (Chen et al., 2018). Microglial-dependent TNF-α production was also shown to contribute to endothelial necroptosis and blood-brain barrier (BBB) disruption in the MCAO model of ischemic stroke, which was ameliorated by Nec-1 treatment (Chen et al., 2019a). Similarly, Nec-1 administration protected from BBB disruption after subarachnoid hemorrhage induction in a rat model of hemorrhagic stroke (Chen et al., 2019b). Genetic and pharmacological inhibition of RIP3 or TNF-α were also protective in this model by downregulating several necroptotic markers (Yuan et al., 2019b). Similarly, Nec-1 and genetic inhibition of RIP1 kinase activity was protective in mouse models of hemorrhagic stroke by decreasing RIP1/RIP3 interaction in the brain and attenuating necrotic cell death and inflammation (Lule et al., 2017; Su et al., 2015). Additionally, Nec-1 treatment also attenuated RIP1-RIP3 complex formation, oxidative damage and inflammation in a model of neonatal hypoxia-ischemia, and this protection was also dependent on improved mitochondrial complex I activity and decreased mitochondrial structural abnormalities in neurons and astrocytes (Chavez-Valdez et al., 2012; Northington et al., 2011). Nec-1 also has protective effects in traumatic brain injury models, despite disparate results reporting either simultaneous protection from caspase-3 activation or lack of it (Wang et al., 2012a; You et al., 2008). Moreover, RIP3ko mice are also protected in this model, but concurrent protection from apoptosis and inflammasome activation besides necroptosis were also reported (Liu et al., 2018b). Likewise, Nec-1 is protective in models of spinal cord injury, despite controversy whether it can also attenuate apoptosis besides necroptosis (Liu et al., 2015; Wang et al., 2014b). In these models, reactive astrocytes have been shown to undergo necroptosis induced by M1-activated microglia, while RIP3 deficiency was protective and reinstated proper neurotrophic functions in astrocytes (Fan et al., 2016). Additionally, neurons can become susceptible to necroptosis following lysosomal damage and subsequent RIP1/3 accumulation (Liu et al., 2018a). Nec-1 was also extensively shown to be protective in retinal ischemia-reperfusion injury models, probably by reducing inflammation and inhibiting TNF-α-dependent necroptosis in retinal ganglion cells (Dvoriantchikova et al., 2014; Kim et al., 2017; Rosenbaum et al., 2010). A novel RIP1 inhibitor was shown to protect retinal ganglion cells from ischemic injury in a rat glaucoma model (Do et al., 2017). In a model of retinal detachment, Nec-1 was also reported to inhibit necroptosis in the outer nuclear layer (Dong et al., 2012). Finally, in a mouse model of age-related macular degeneration, RIP3 deficiency prevented necrosis of the retinal pigment epithelium and inflammation (Murakami et al., 2014).

7.2 Chronic neurodegenerative diseases

Necroptotic-related proteins have also been involved in neurodegeneration and inflammation in several chronic neurodegenerative diseases (Table 2) (Yuan et al., 2019a). Nec-1 attenuated symptoms in the R6/2 transgenic mouse model of Huntington's disease (Zhu et al., 2011). In AD, RIP1/3 and MLKL activation were detected in human AD brains, especially in neurons and microglia, and their presence positively correlated with Braak disease stage (Caccamo et al., 2017; Ofengeim et al., 2017). Moreover, Nec-1s attenuated p-MLKL recruitment and necroptosis-related neurodegeneration in transgenic mice expressing five human familial AD mutations (Caccamo et al., 2017). Inhibition of RIP1 kinase activity by either Nec-1s or $RIP1^{D138N}$ kinase dead mutation knock-in in APP/PS1 mice also reduced β-amyloid plaque burden, association of microglia with plaques and expression of pro-inflammatory cytokines. Moreover, genetic RIP1 inactivation in these mice was also shown to modulate the expression of several disease-associated genes in microglia, especially Cst7, which encodes for the endosomal/lysosomal cathepsin inhibitor cystatin F. By reducing lysosomal activity in microglia, higher cystatin F expression may decrease the intracellular degradation of phagocytosed β-amyloid and contribute to RIP1 kinase activation in a positive feedback loop (Ofengeim et al., 2017). Lesioned cortical samples from MS patients and mouse models were also extensively demonstrated to present active RIP1/3 and MLKL, as determined by activating phosphorylations and recruitment into insoluble aggregates, along with defective caspase-8 activation and increased $cFLIP_L$ expression (Ofengeim et al., 2015). Further, Nec-1s and RIP3 deficiency protected oligodendrocytes against necroptotic induction and decreased neuroinflammation in the demyelinating cuprizone mouse model of MS (Ofengeim et al., 2015). An independent study also demonstrated that a novel brain penetrating RIP1 inhibitor could ameliorate disease pathology in a mouse model of MS (Yoshikawa et al., 2018). Similarly, necroptosis of oligodendrocytes has been implicated in ALS. Genetic and pharmacological inhibition of RIP1 kinase activity, as well as RIP3 deficiency, protected from oligodendrocyte death, M1-microglial activation and subsequent axonal degeneration in the spinal cords of the $SOD1^{G93A}$ transgenic mouse model of ALS and OPTN-deficient mice, whose loss-of-function gene mutations are linked to ALS. Interestingly, specific conditional OPTN deletion from oligodendrocytes or microglia in mice were sufficient to induce deficient axonal myelination, which was linked to reduced proteasomal degradation and subsequent activation of RIP1 for necroptotic induction and pro-inflammatory gene expression. Several necroptotic markers were also detected in spinal cord samples from sporadic ALS patients, especially in oligodendrocytes and microglia contained in demyelinated white matter (Ito et al., 2016). Conversely, a recent study detected no alterations in RIP1 protein levels in spinal cords samples of ALS patients, and further reported that RIP3ko provided no neuropathological protection in the $SOD1^{G93A}$ transgenic mouse model, while OPTN-deficient mice presented normal myelination in the spinal cord (Dermentzaki et al., 2019). Moreover, a report showed that adult primary astrocytes obtained from sporadic ALS patients can induce necroptosis in co-cultured human embryonic stem cell-derived motor neurons (Re et al., 2014), while others reported the absence of motor neuron loss in OPTN-deficient mice (Ito et al., 2016). Further research is thus warranted to clarify the necroptotic and non-necroptotic roles of RIP1 in different cell types during ALS progression. Interestingly, it was recently shown that TBK1 haploinsufficiency, a genetic cause of comorbid ALS and frontotemporal dementia (FTD), combined with a specific single allelic loss of TAK1 in myeloid-lineage cells in mice resulted in significant microgliosis in the cortex and spinal cord along with increased expression of proinflammatory molecules, development of axonal pathology, aggregation of TAR DNA-binding protein 43, neuronal loss and motor deficits, which were reverted by $RIP1^{D138N}$ kinase dead heterozygosity. TBK1-deficient mice die as embryos due to excessive RIP1 kinase-dependent apoptosis and necroptosis, which is fully blocked by heterozygous $RIP1^{D138N}$ kinase dead knock-in. Interestingly, besides protecting from necroptosis, apoptosis could also be partially inhibited by concomitant RIP3 deletion in these conditions. Importantly, this study also showed that TAK1 mRNA and protein levels are greatly reduced in healthy aged human brains, giving rise to the compelling hypothesis that inefficient suppression of RIP1 kinase activity may constitute a driving force in aging human brains for the development of neurodegenerative disorders, such as PD (Xu et al., 2018).

Necroptosis has also very recently emerged as a probable contributor to dopaminergic neurodegeneration in PD. Immunohistochemical data from two independent studies showed increased RIP1, RIP3 and p-MLKL levels in the SN of PD patients, mostly co-localized with dopaminergic neurons (Hu et al., 2019; Onate et al., 2019). In the acute MPTP mouse model, Nec-1 administration reverted the decrease in striatal dopamine and number of nigral dopaminergic neurons along with the increase of RIP1/3 and MLKL protein levels 7 days post-injection, when activating phosphorylations of RIP3 and MLKL were also detectable. RIP3- and MLKL-deficient mice were similarly protected from the detrimental effects of MPTP in the nigrostriatal circuit and presented decreased expression of pro-inflammatory cytokines in the midbrain (Lin et al., 2019). Nec-1s administration also attenuated dopaminergic neurodegeneration in the sub-chronic MPTP mouse model and protected from ROS-dependent lipid peroxidation, although no necroptotic markers were evaluated in this study (Iannielli et al., 2018). Treatment with a miR-425 mimic also protected from elevated RIP1 and p-MLKL levels in TH-positive neurons in the sub-chronic MPTP model, which correlated with higher dopaminergic neuronal survival and function (Hu et al., 2019). A very recent study reported that the necroptotic machinery, as assessed by p-MLKL levels and its interaction with RIP1, becomes engaged firstly in the striatum following striatal 6-OHDA injection in mice, and then propagates retrogradely along the axons of the nigrostriatal pathway to become detectable in the SN only at later time-points. Intriguingly, Nec-1s treatment or deletion of RIP3 or MLKL conferred partial protection from 6-OHDA-driven axonal degeneration nearer to the neuronal somas and from loss of dopaminergic neurons within the SN but could not rescue striatal TH-fiber loss, which was still correlated with ameliorated motor deficits (Onate et al., 2019). A concurrent study from the same group also reported that inhibition of all necroptotic proteins attenuated axonal degeneration induced by mechanical or toxic insults, which was dependent on p-MLKL recruitment and mitochondrial fragmentation even in the absence of caspase inhibition. These observations suggest that necroptosis may not only contribute to axonal degeneration in multiple pathologies, as it may even become active first in axons and then spread to neuronal somas, although more research is needed to validate this hypothesis (Arrazola et al., 2019).

In Gaucher's disease (GD), a lysosomal storage disorder with neurological consequences, both genetic and pharmacological mouse models of GD displayed increased protein levels of $cFLIP_S$ in the brain, suggestive of decreased caspase-8 activity, and higher RIP1 and RIP3 levels, with RIP3 levels increasing the most in microglia. In the pharmacological GD model, RIP3 deficiency ameliorated clinical symptoms and increased survival (Vitner et al., 2014). Niemann-Pick disease, type C1 (NPC1) is other lysosomal storage disease that can lead to cerebellar ataxia and cognitive decline. Pharmacological or genetic inhibition of RIP1 was shown to extend the lifespan of NPC1-deficient mice, which presented increased RIP1/RIP3 interactions in cerebellar tissue, indicative of necroptosis. However, RIP3 deficiency failed to mimic the positive effects of RIP1 inhibition in the NPC1 mouse model, indicating that non-necroptotic functions of RIP1 are more relevant in this context (Cougnoux et al., 2016, 2018).

Overall, extensive research demonstrate that necroptosis is relevant in acute and chronic neurodegenerative diseases. However, non-necroptotic functions of RIP1, such as RIP1 kinase-dependent apoptosis and regulation of pro-inflammatory gene expression, are gaining increased attention in these contexts, with RIP1 kinase inhibition being more beneficial than either RIP3 or MLKL deficiency in several CNS pathologies. Therefore, RIP1 kinase inhibition, which has no detrimental effects described *in vivo*, stands as the prime therapeutic target for the treatment of several pathological mechanisms in neurodegenerative diseases (Yuan et al., 2019a). Non-necroptotic roles for RIP3 have been more seldomly studied in CNS pathologies, thus warranting further research dedicated to the phenotypic comparison between MLKL- and RIP3-deficient mice in these conditions (Liu et al., 2018b; Newton et al., 2016a).

8. Therapeutic perspectives

RIP1 is an exquisite druggable target due to the presence of a hydrophobic pocket behind the active center between the amino and the carboxy lobe of its kinase domain. Binding of Nec-1s, Nec-3 and Nec-4 to this allosteric pocket locks RIP1 in an inactive conformation, which confers high specificity to these type III RIP1 inhibitors (Degterev et al., 2019; Xie et al., 2013a). Nec-1s (also known as 7-Cl-O-Nec-1) is an analogue of Nec-1 with higher specificity for both human and murine RIP1, improved pharmacokinetic properties, excellent BBB permeability and acceptable oral availability (Ofengeim et al., 2015). However, Nec-1s still possesses a poor half-life *in vivo* (~1 h), which warrants the need for the development of improved RIP1 inhibitors (Ofengeim et al., 2015). In this regard, GSK2982772 is a non-BBB permeable potent and specific RIP1 inhibitor that also targets the same allosteric pocket as Nec-1s, being currently under phase IIa clinical trials for the treatment of inflammatory diseases (Harris et al., 2017). Recently, Denali Therapeutics advanced DNL747, a highly selective brain penetrant RIP1 inhibitor, into phase I clinical trials in AD and ALS, while DNL788, which is another BBB permeable RIP1 inhibitor, is already in preclinical development (Yuan et al., 2019a). Takeda Pharmaceuticals has also recently developed 7-Oxo-2,4,5, 7-tetrahydro-6H-pyrazolo[3,4-*c*]pyridine derivatives as novel potent, orally available and brain penetrating type III RIP1 inhibitors with excellent pharmacokinetic profiles, some of which were already shown to ameliorate disease progression in a mouse model of MS (Yoshikawa et al., 2018). Screening of clinically approved kinase inhibitors have also revealed type I RIP1 inhibitors, which target the ATP-binding pocket, such as pazopanib, and type II inhibitors, which coordinate with residues in the ATP and back pocket, such as ponatinib, which is also a RIP3 inhibitor. However, these kinase inhibitors lack specificity, which argues against their usage in the clinic (Degterev et al., 2019; Fauster et al., 2015). Recently, GSK'074 was identified as a potent dual type II inhibitor for both RIP1 and RIP3. Importantly, although GSK'074 presents structural similarity with the already established RIP3 kinase inhibitor GSK'843, which induces toxicity at high concentrations due to activation of RIP3-dependent apoptosis, GSK'074 showed no detectable cytotoxicity even at high levels (Zhou et al., 2019). Therefore, a dual inhibitor would have the advantage of blocking RIP1-dependent and -independent forms of necroptosis along with blocking additional non-necroptotic RIP1 functions that depend on its kinase activity. However, the specificity, BBB permeability and pharmacokinetic profiles of this compound are still lacking. Necrosulfonamide can inhibit necroptosis by directly blocking human MLKL activation, but not the mouse ortholog, which invalidates testing in preclinical mouse models of disease (Sun et al., 2012). In fact, RIP3 and, especially, MLKL are poorly evolutionary conserved among mammals, thus increasing the probability that inhibitors that are functional in one species are rendered inactive in others (Dondelinger et al., 2016; Sun et al., 2012). This fact, allied to the well-established roles of RIP1 in neurodegenerative processes independently of necroptosis and its structural characteristics that allow for the design of highly specific type III inhibitors, support RIP1 as a prime candidate for therapeutic intervention.

9. Concluding remarks

Several questions of outstanding interest remain to be further addressed regarding necroptosis. For example, a proper understanding of how necroptosis is recruited *in vivo* during pathological conditions is still lacking. In this regard, characterization of the balance between the expression of cFLIP isoforms and caspase-8 activity and the protein levels of necroptosis-related proteins during disease progression would be instrumental to delineate therapeutic targets. In chronic neurodegenerative diseases that present decreased Ub-proteasome system activity, such as PD, it would be interesting to determine whether the amount of RIP1/3 marked with K48 poly-Ub chains is increased, which could sensitize to necroptotic engagement.

Further work is also needed to unravel the complex interactions between necroptotic signaling and other cell death pathways. Moreover, both RIP1 and RIP3 display cell death-independent signaling functions involved in the regulation of the immune response, inflammasome activation, mitochondrial function and altered metabolic responses, which are dependent on their kinase or scaffolding activities. A proper understanding of these functions will probably expand the therapeutic strategies targeting these proteins during different pathological conditions.

References

Aaes, T.L., Kaczmarek, A., Delvaeye, T., De Craene, B., De Koker, S., Heyndrickx, L., Delrue, I., Taminau, J., Wiernicki, B., De Groote, P., Garg, A.D., Leybaert, L., Grooten, J., Bertrand, M.J., Agostinis, P., Berx, G., Declercq, W., Vandenabeele, P., Krysko, D.V., 2016. Vaccination with necroptotic cancer cells induces efficient anti-tumor immunity. Cell Rep. 15, 274–287.

Alturki, N.A., McComb, S., Ariana, A., Rijal, D., Korneluk, R.G., Sun, S.C., Alnemri, E., Sad, S., 2018. Triad3a induces the degradation of early necrosome to limit RipK1-dependent cytokine production and necroptosis. Cell Death Dis. 9, 592.

Amin, P., Florez, M., Najafov, A., Pan, H., Geng, J., Ofengeim, D., Dziedzic, S.A., Wang, H., Barrett, V.J., Ito, Y., LaVoie, M.J., Yuan, J., 2018. Regulation of a distinct activated RIPK1 intermediate bridging complex I and complex II in TNFalpha-mediated apoptosis. Proc. Natl. Acad. Sci. U. S. A. 115, E5944–E5953.

Annibaldi, A., Wicky John, S., Vanden Berghe, T., Swatek, K.N., Ruan, J., Liccardi, G., Bianchi, K., Elliott, P.R., Choi, S.M., Van Coillie, S., Bertin, J., Wu, H., Komander, D., Vandenabeele, P., Silke, J., Meier, P., 2018. Ubiquitin-mediated regulation of RIPK1 kinase activity independent of IKK and MK2. Mol. Cell 69, 566-580.e5.

Arrazola, M.S., Saquel, C., Catalan, R.J., Barrientos, S.A., Hernandez, D.E., Catenaccio, A., Court, F.A., 2019. Axonal degeneration is mediated by necroptosis activation. J. Neurosci. 39, 3832–3844.

Basit, F., van Oppen, L.M., Schockel, L., Bossenbroek, H.M., van Emst-de Vries, S.E., Hermeling, J.C., Grefte, S., Kopitz, C., Heroult, M., Hgm Willems, P., Koopman, W.J., 2017. Mitochondrial complex I inhibition triggers a mitophagy-dependent ROS increase leading to necroptosis and ferroptosis in melanoma cells. Cell Death Dis. 8, e2716.

Belizario, J., Vieira-Cordeiro, L., Enns, S., 2015. Necroptotic cell death signaling and execution pathway: lessons from knockout mice. Mediators Inflamm. 2015, 128076.

Bertrand, M.J., Milutinovic, S., Dickson, K.M., Ho, W.C., Boudreault, A., Durkin, J., Gillard, J.W., Jaquith, J.B., Morris, S.J., Barker, P.A., 2008. cIAP1 and cIAP2 facilitate cancer cell survival by functioning as E3 ligases that promote RIP1 ubiquitination. Mol. Cell 30, 689–700.

Bertrand, M.J., Lippens, S., Staes, A., Gilbert, B., Roelandt, R., De Medts, J., Gevaert, K., Declercq, W., Vandenabeele, P., 2011. cIAP1/2 are direct E3 ligases conjugating diverse types of ubiquitin chains to receptor interacting proteins kinases 1 to 4 (RIP1-4). PLoS One 6, e22356.

Bigenzahn, J.W., Fauster, A., Rebsamen, M., Kandasamy, R.K., Scorzoni, S., Vladimer, G.I., Muller, A.C., Gstaiger, M., Zuber, J., Bennett, K.L., Superti-Furga, G., 2016. An inducible retroviral expression system for tandem affinity purification mass-Spectrometry-based proteomics identifies mixed lineage kinase domain-like protein (MLKL) as an heat shock protein 90 (HSP90) client. Mol. Cell. Proteomics 15, 1139–1150.

Blonska, M., Shambharkar, P.B., Kobayashi, M., Zhang, D., Sakurai, H., Su, B., Lin, X., 2005. TAK1 is recruited to the tumor necrosis factor-alpha (TNF-alpha) receptor 1 complex in a receptor-interacting protein (RIP)-dependent manner and cooperates with MEKK3 leading to NF-kappaB activation. J. Biol. Chem. 280, 43056–43063.

Bohgaki, T., Mozo, J., Salmena, L., Matysiak-Zablocki, E., Bohgaki, M., Sanchez, O., Strasser, A., Hakem, A., Hakem, R., 2011. Caspase-8 inactivation in T cells increases necroptosis and suppresses autoimmunity in Bim$^{-/-}$ mice. J. Cell Biol. 195, 277–291.

Caccamo, A., Branca, C., Piras, I.S., Ferreira, E., Huentelman, M.J., Liang, W.S., Readhead, B., Dudley, J.T., Spangenberg, E.E., Green, K.N., Belfiore, R., Winslow, W., Oddo, S., 2017. Necroptosis activation in Alzheimer's disease. Nat. Neurosci. 20, 1236–1246.

Cai, Z., Jitkaew, S., Zhao, J., Chiang, H.C., Choksi, S., Liu, J., Ward, Y., Wu, L.G., Liu, Z.G., 2014. Plasma membrane translocation of trimerized MLKL protein is required for TNF-induced necroptosis. Nat. Cell Biol. 16, 55–65.

Chavez-Valdez, R., Martin, L.J., Flock, D.L., Northington, F.J., 2012. Necrostatin-1 attenuates mitochondrial dysfunction in neurons and astrocytes following neonatal hypoxia-ischemia. Neuroscience 219, 192–203.

Chen, W., Wu, J., Li, L., Zhang, Z., Ren, J., Liang, Y., Chen, F., Yang, C., Zhou, Z., Su, S.S., Zheng, X., Zhong, C.Q., Wan, H., Xiao, M., Lin, X., Feng, X.H., Han, J., 2015. Ppm1b negatively regulates necroptosis through dephosphorylating Rip3. Nat. Cell Biol. 17, 434–444.

Chen, Y., Zhang, L., Yu, H., Song, K., Shi, J., Chen, L., Cheng, J., 2018. Necrostatin-1 Improves long-term functional recovery through protecting oligodendrocyte precursor cells after transient focal cerebral ischemia in mice. Neuroscience 371, 229–241.

Chen, A.Q., Fang, Z., Chen, X.L., Yang, S., Zhou, Y.F., Mao, L., Xia, Y.P., Jin, H.J., Li, Y.N., You, M.F., Wang, X.X., Lei, H., He, Q.W., Hu, B., 2019a. Microglia-derived TNF-alpha mediates endothelial necroptosis aggravating blood brain-barrier disruption after ischemic stroke. Cell Death Dis. 10, 487.

Chen, J., Jin, H., Xu, H., Peng, Y., Jie, L., Xu, D., Chen, L., Li, T., Fan, L., He, P., Ying, G., Gu, C., Wang, C., Wang, L., Chen, G., 2019b. The neuroprotective effects of necrostatin-1 on subarachnoid hemorrhage in rats are possibly mediated by preventing blood-brain barrier disruption and RIP3-mediated necroptosis. Cell Transplant. 28, 1358–1372.

Cho, Y.S., Challa, S., Moquin, D., Genga, R., Ray, T.D., Guildford, M., Chan, F.K., 2009. Phosphorylation-driven assembly of the RIP1-RIP3 complex regulates programmed necrosis and virus-induced inflammation. Cell 137, 1112–1123.

Choi, S.W., Park, H.H., Kim, S., Chung, J.M., Noh, H.J., Kim, S.K., Song, H.K., Lee, C.W., Morgan, M.J., Kang, H.C., Kim, Y.S., 2018. PELI1 selectively targets kinase-active RIP3 for ubiquitylation-dependent proteasomal degradation. Mol. Cell 70, 920-935.e7.

Choi, M.E., Price, D.R., Ryter, S.W., Choi, A.M.K., 2019. Necroptosis: a crucial pathogenic mediator of human disease. JCI Insight 4, e128834.

Christofferson, D.E., Li, Y., Hitomi, J., Zhou, W., Upperman, C., Zhu, H., Gerber, S.A., Gygi, S., Yuan, J., 2012. A novel role for RIP1 kinase in mediating TNFalpha production. Cell Death Dis. 3, e320.

Conos, S.A., Chen, K.W., De Nardo, D., Hara, H., Whitehead, L., Nunez, G., Masters, S.L., Murphy, J.M., Schroder, K., Vaux, D.L., Lawlor, K.E., Lindqvist, L.M., Vince, J.E., 2017. Active MLKL triggers the NLRP3 inflammasome in a cell-intrinsic manner. Proc. Natl. Acad. Sci. U. S. A. 114, E961–E969.

Cook, W.D., Moujalled, D.M., Ralph, T.J., Lock, P., Young, S.N., Murphy, J.M., Vaux, D.L., 2014. RIPK1- and RIPK3-induced cell death mode is determined by target availability. Cell Death Differ. 21, 1600–1612.

Cougnoux, A., Cluzeau, C., Mitra, S., Li, R., Williams, I., Burkert, K., Xu, X., Wassif, C.A., Zheng, W., Porter, F.D., 2016. Necroptosis in Niemann-Pick disease, type C1: a potential therapeutic target. Cell Death Dis. 7, e2147.

Cougnoux, A., Clifford, S., Salman, A., Ng, S.L., Bertin, J., Porter, F.D., 2018. Necroptosis inhibition as a therapy for Niemann-Pick disease, type C1: inhibition of RIP kinases and combination therapy with 2-hydroxypropyl-beta-cyclodextrin. Mol. Genet. Metab. 125, 345–350.

Cusson-Hermance, N., Khurana, S., Lee, T.H., Fitzgerald, K.A., Kelliher, M.A., 2005. Rip1 mediates the Trif-dependent toll-like receptor 3- and 4-induced NF-{kappa}B activation but does not contribute to interferon regulatory factor 3 activation. J. Biol. Chem. 280, 36560–36566.

Davies, K.A., Tanzer, M.C., Griffin, M.D.W., Mok, Y.F., Young, S.N., Qin, R., Petrie, E.J., Czabotar, P.E., Silke, J., Murphy, J.M., 2018. The brace helices of MLKL mediate interdomain communication and oligomerisation to regulate cell death by necroptosis. Cell Death Differ. 25, 1567–1580.

de Almagro, M.C., Goncharov, T., Izrael-Tomasevic, A., Duttler, S., Kist, M., Varfolomeev, E., Wu, X., Lee, W.P., Murray, J., Webster, J.D., Yu, K., Kirkpatrick, D.S., Newton, K., Vucic, D., 2017. Coordinated ubiquitination and phosphorylation of RIP1 regulates necroptotic cell death. Cell Death Differ. 24, 26–37.

Degterev, A., Huang, Z., Boyce, M., Li, Y., Jagtap, P., Mizushima, N., Cuny, G.D., Mitchison, T.J., Moskowitz, M.A., Yuan, J., 2005. Chemical inhibitor of nonapoptotic cell death with therapeutic potential for ischemic brain injury. Nat. Chem. Biol. 1, 112–119.

Degterev, A., Ofengeim, D., Yuan, J., 2019. Targeting RIPK1 for the treatment of human diseases. Proc. Natl. Acad. Sci. U. S. A. 116, 9714–9722.

Deng, X.X., Li, S.S., Sun, F.Y., 2019. Necrostatin-1 prevents necroptosis in brains after ischemic stroke via inhibition of RIPK1-mediated RIPK3/MLKL signaling. Aging Dis. 10, 807–817.

Dermentzaki, G., Politi, K.A., Lu, L., Mishra, V., Perez-Torres, E.J., Sosunov, A.A., McKhann 2nd, G.M., Lotti, F., Shneider, N.A., Przedborski, S., 2019. Deletion of Ripk3 prevents motor neuron death in vitro but not in vivo. eNeuro 6, 0308-18.2018.

Dillon, C.P., Oberst, A., Weinlich, R., Janke, L.J., Kang, T.B., Ben-Moshe, T., Mak, T.W., Wallach, D., Green, D.R., 2012. Survival function of the FADD-CASPASE-8-cFLIP (L) complex. Cell Rep. 1, 401–407.

Dillon, C.P., Weinlich, R., Rodriguez, D.A., Cripps, J.G., Quarato, G., Gurung, P., Verbist, K.C., Brewer, T.L., Llambi, F., Gong, Y.N., Janke, L.J., Kelliher, M.A., Kanneganti, T.D., Green, D.R., 2014. RIPK1 blocks early postnatal lethality mediated by caspase-8 and RIPK3. Cell 157, 1189–1202.

Do, Y.J., Sul, J.W., Jang, K.H., Kang, N.S., Kim, Y.H., Kim, Y.G., Kim, E., 2017. A novel RIPK1 inhibitor that prevents retinal degeneration in a rat glaucoma model. Exp. Cell Res. 359, 30–38.

Dondelinger, Y., Aguileta, M.A., Goossens, V., Dubuisson, C., Grootjans, S., Dejardin, E., Vandenabeele, P., Bertrand, M.J., 2013. RIPK3 contributes to TNFR1-mediated RIPK1 kinase-dependent apoptosis in conditions of cIAP1/2 depletion or TAK1 kinase inhibition. Cell Death Differ. 20, 1381–1392.

Dondelinger, Y., Declercq, W., Montessuit, S., Roelandt, R., Goncalves, A., Bruggeman, I., Hulpiau, P., Weber, K., Sehon, C.A., Marquis, R.W., Bertin, J., Gough, P.J., Savvides, S., Martinou, J.C., Bertrand, M.J., Vandenabeele, P., 2014. MLKL compromises plasma membrane integrity by binding to phosphatidylinositol phosphates. Cell Rep. 7, 971–981.

Dondelinger, Y., Jouan-Lanhouet, S., Divert, T., Theatre, E., Bertin, J., Gough, P.J., Giansanti, P., Heck, A.J., Dejardin, E., Vandenabeele, P., Bertrand, M.J., 2015. NF-kappaB-independent role of IKKalpha/IKKbeta in preventing RIPK1 kinase-dependent apoptotic and necroptotic cell death during TNF signaling. Mol. Cell 60, 63–76.

Dondelinger, Y., Hulpiau, P., Saeys, Y., Bertrand, M.J.M., Vandenabeele, P., 2016. An evolutionary perspective on the necroptotic pathway. Trends Cell Biol. 26, 721–732.

Dondelinger, Y., Delanghe, T., Rojas-Rivera, D., Priem, D., Delvaeye, T., Bruggeman, I., Van Herreweghe, F., Vandenabeele, P., Bertrand, M.J.M., 2017. MK2 phosphorylation of RIPK1 regulates TNF-mediated cell death. Nat. Cell Biol. 19, 1237–1247.

Dong, K., Zhu, H., Song, Z., Gong, Y., Wang, F., Wang, W., Zheng, Z., Yu, Z., Gu, Q., Xu, X., Sun, X., 2012. Necrostatin-1 protects photoreceptors from cell death and improves functional outcome after experimental retinal detachment. Am. J. Pathol. 181, 1634–1641.

Dovey, C.M., Diep, J., Clarke, B.P., Hale, A.T., McNamara, D.E., Guo, H., Brown Jr., N.W., Cao, J.Y., Grace, C.R., Gough, P.J., Bertin, J., Dixon, S.J., Fiedler, D., Mocarski, E.S., Kaiser, W.J., Moldoveanu, T., York, J.D., Carette, J.E., 2018. MLKL requires the inositol phosphate code to execute necroptosis. Mol. Cell 70, 936-948.e7.

Draber, P., Kupka, S., Reichert, M., Draberova, H., Lafont, E., de Miguel, D., Spilgies, L., Surinova, S., Taraborrelli, L., Hartwig, T., Rieser, E., Martino, L., Rittinger, K., Walczak, H., 2015. LUBAC-recruited CYLD and A20 regulate gene activation and cell death by exerting opposing effects on linear ubiquitin in signaling complexes. Cell Rep. 13, 2258–2272.

Duong, B.H., Onizawa, M., Oses-Prieto, J.A., Advincula, R., Burlingame, A., Malynn, B.A., Ma, A., 2015. A20 restricts ubiquitination of pro-interleukin-1beta protein complexes and suppresses NLRP3 inflammasome activity. Immunity 42, 55–67.

Dvoriantchikova, G., Degterev, A., Ivanov, D., 2014. Retinal ganglion cell (RGC) programmed necrosis contributes to ischemia-reperfusion-induced retinal damage. Exp. Eye Res. 123, 1–7.

Dziedzic, S.A., Su, Z., Jean Barrett, V., Najafov, A., Mookhtiar, A.K., Amin, P., Pan, H., Sun, L., Zhu, H., Ma, A., Abbott, D.W., Yuan, J., 2018. ABIN-1 regulates RIPK1 activation by linking Met1 ubiquitylation with Lys63 deubiquitylation in TNF-RSC. Nat. Cell Biol. 20, 58–68.

Fan, H., Zhang, K., Shan, L., Kuang, F., Chen, K., Zhu, K., Ma, H., Ju, G., Wang, Y.Z., 2016. Reactive astrocytes undergo M1 microglia/macrohpages-induced necroptosis in spinal cord injury. Mol. Neurodegener. 11, 14.

Fauster, A., Rebsamen, M., Huber, K.V., Bigenzahn, J.W., Stukalov, A., Lardeau, C.H., Scorzoni, S., Bruckner, M., Gridling, M., Parapatics, K., Colinge, J., Bennett, K.L., Kubicek, S., Krautwald, S., Linkermann, A., Superti-Furga, G., 2015. A cellular screen identifies ponatinib and pazopanib as inhibitors of necroptosis. Cell Death Dis. 6, e1767.

Feng, S., Yang, Y., Mei, Y., Ma, L., Zhu, D.E., Hoti, N., Castanares, M., Wu, M., 2007. Cleavage of RIP3 inactivates its caspase-independent apoptosis pathway by removal of kinase domain. Cell. Signal. 19, 2056–2067.

Feoktistova, M., Geserick, P., Kellert, B., Dimitrova, D.P., Langlais, C., Hupe, M., Cain, K., MacFarlane, M., Hacker, G., Leverkus, M., 2011. cIAPs block Ripoptosome formation, a RIP1/caspase-8 containing intracellular cell death complex differentially regulated by cFLIP isoforms. Mol. Cell 43, 449–463.

Fricker, M., Vilalta, A., Tolkovsky, A.M., Brown, G.C., 2013. Caspase inhibitors protect neurons by enabling selective necroptosis of inflamed microglia. J. Biol. Chem. 288, 9145–9152.

Gaba, A., Xu, F., Lu, Y., Park, H.S., Liu, G., Zhou, Y., 2019. The NS1 protein of influenza A virus participates in necroptosis by interacting with MLKL and increasing its oligomerization and membrane translocation. J. Virol. 93, e01835-18.

Galluzzi, L., Vitale, I., Aaronson, S.A., Abrams, J.M., Adam, D., Agostinis, P., Alnemri, E.S., Altucci, L., Amelio, I., Andrews, D.W., Annicchiarico-Petruzzelli, M., Antonov, A.V., Arama, E., Baehrecke, E.H., Barlev, N.A., Bazan, N.G., Bernassola, F., Bertrand, M.J.M., Bianchi, K., Blagosklonny, M.V., Blomgren, K., Borner, C., Boya, P., Brenner, C., Campanella, M., Candi, E., Carmona-Gutierrez, D., Cecconi, F., Chan, F.K., Chandel, N.S., Cheng, E.H., Chipuk, J.E., Cidlowski, J.A., Ciechanover, A., Cohen, G.M., Conrad, M., Cubillos-Ruiz, J.R., Czabotar, P.E., D'Angiolella, V., Dawson, T.M., Dawson, V.L., De Laurenzi, V., De Maria, R., Debatin, K.M., DeBerardinis, R.J., Deshmukh, M., Di Daniele, N., Di Virgilio, F., Dixit, V.M., Dixon, S.J., Duckett, C.S., Dynlacht, B.D., El-Deiry, W.S., Elrod, J.W., Fimia, G.M., Fulda, S., Garcia-Saez, A.J., Garg, A.D., Garrido, C., Gavathiotis, E., Golstein, P., Gottlieb, E., Green, D.R., Greene, L.A., Gronemeyer, H., Gross, A., Hajnoczky, G., Hardwick, J.M., Harris, I.S., Hengartner, M.O., Hetz, C., Ichijo, H., Jaattela, M., Joseph, B., Jost, P.J., Juin, P.P., Kaiser, W.J., Karin, M., Kaufmann, T., Kepp, O., Kimchi, A., Kitsis, R.N., Klionsky, D.J., Knight, R.A., Kumar, S., Lee, S.W., Lemasters, J.J., Levine, B., Linkermann, A., Lipton, S.A., Lockshin, R.A., Lopez-Otin, C., Lowe, S.W., Luedde, T., Lugli, E., MacFarlane, M., Madeo, F., Malewicz, M., Malorni, W., Manic, G., Marine, J.C., Martin, S.J., Martinou, J.C., Medema, J.P., Mehlen, P., Meier, P., Melino, S., Miao, E.A., Molkentin, J.D., Moll, U.M., Munoz-Pinedo, C., Nagata, S., Nunez, G., Oberst, A., Oren, M., Overholtzer, M., Pagano, M., Panaretakis, T., Pasparakis, M., Penninger, J.M., Pereira, D.M., Pervaiz, S., Peter, M.E., Piacentini, M., Pinton, P., Prehn, J.H.M., Puthalakath, H., Rabinovich, G.A., Rehm, M., Rizzuto, R., Rodrigues, C.M.P., Rubinsztein, D.C., Rudel, T., Ryan, K.M., Sayan, E., Scorrano, L., Shao, F., Shi, Y., Silke, J., Simon, H.U., Sistigu, A., Stockwell, B.R., Strasser, A., Szabadkai, G., Tait, S.W.G., Tang, D., Tavernarakis, N., Thorburn, A., Tsujimoto, Y., Turk, B., Vanden Berghe, T., Vandenabeele, P., Vander Heiden, M.G., Villunger, A., Virgin, H.W., Vousden, K.H., Vucic, D., Wagner, E.F., Walczak, H., Wallach, D., Wang, Y., Wells, J.A., Wood, W., Yuan, J., Zakeri, Z., Zhivotovsky, B., Zitvogel, L., Melino, G., Kroemer, G., 2018. Molecular mechanisms of cell death: recommendations of the Nomenclature Committee on Cell Death 2018. Cell Death Differ. 25, 486–541.

Geng, J., Ito, Y., Shi, L., Amin, P., Chu, J., Ouchida, A.T., Mookhtiar, A.K., Zhao, H., Xu, D., Shan, B., Najafov, A., Gao, G., Akira, S., Yuan, J., 2017. Regulation of RIPK1 activation by TAK1-mediated phosphorylation dictates apoptosis and necroptosis. Nat. Commun. 8, 359.

Gerlach, B., Cordier, S.M., Schmukle, A.C., Emmerich, C.H., Rieser, E., Haas, T.L., Webb, A.I., Rickard, J.A., Anderton, H., Wong, W.W., Nachbur, U., Gangoda, L., Warnken, U., Purcell, A.W., Silke, J., Walczak, H., 2011. Linear ubiquitination prevents inflammation and regulates immune signalling. Nature 471, 591–596.

Gong, Y.N., Guy, C., Olauson, H., Becker, J.U., Yang, M., Fitzgerald, P., Linkermann, A., Green, D.R., 2017. ESCRT-III acts downstream of MLKL to regulate necroptotic cell death and its consequences. Cell 169, (286–300). e16.

Goossens, V., Stange, G., Moens, K., Pipeleers, D., Grooten, J., 1999. Regulation of tumor necrosis factor-induced, mitochondria- and reactive oxygen species-dependent cell death by the electron flux through the electron transport chain complex I. Antioxid. Redox Signal. 1, 285–295.

Gunther, C., He, G.W., Kremer, A.E., Murphy, J.M., Petrie, E.J., Amann, K., Vandenabeele, P., Linkermann, A., Poremba, C., Schleicher, U., Dewitz, C., Krautwald, S., Neurath, M.F., Becker, C., Wirtz, S., 2016. The pseudokinase MLKL mediates programmed hepatocellular necrosis independently of RIPK3 during hepatitis. J. Clin. Invest. 126, 4346–4360.

Guo, H., Omoto, S., Harris, P.A., Finger, J.N., Bertin, J., Gough, P.J., Kaiser, W.J., Mocarski, E.S., 2015. Herpes simplex virus suppresses necroptosis in human cells. Cell Host Microbe 17, 243–251.

Haas, T.L., Emmerich, C.H., Gerlach, B., Schmukle, A.C., Cordier, S.M., Rieser, E., Feltham, R., Vince, J., Warnken, U., Wenger, T., Koschny, R., Komander, D., Silke, J., Walczak, H., 2009. Recruitment of the linear ubiquitin chain assembly complex stabilizes the TNF-R1 signaling complex and is required for TNF-mediated gene induction. Mol. Cell 36, 831–844.

Hanggi, K., Vasilikos, L., Valls, A.F., Yerbes, R., Knop, J., Spilgies, L.M., Rieck, K., Misra, T., Bertin, J., Gough, P.J., Schmidt, T., de Almodovar, C.R., Wong, W.W., 2017. RIPK1/RIPK3 promotes vascular permeability to allow tumor cell extravasation independent of its necroptotic function. Cell Death Dis. 8, e2588.

Harris, P.A., Berger, S.B., Jeong, J.U., Nagilla, R., Bandyopadhyay, D., Campobasso, N., Capriotti, C.A., Cox, J.A., Dare, L., Dong, X., Eidam, P.M., Finger, J.N., Hoffman, S.J., Kang, J., Kasparcova, V., King, B.W., Lehr, R., Lan, Y., Leister, L.K., Lich, J.D., MacDonald, T.T., Miller, N.A., Ouellette, M.T., Pao, C.S., Rahman, A., Reilly, M.A., Rendina, A.R., Rivera, E.J., Schaeffer, M.C., Sehon, C.A., Singhaus, R.R., Sun, H.H., Swift, B.A., Totoritis, R.D., Vossenkamper, A., Ward, P., Wisnoski, D.D., Zhang, D., Marquis, R.W., Gough, P.J., Bertin, J., 2017. Discovery of a first-in-class receptor interacting protein 1 (RIP1) kinase specific clinical candidate (GSK2982772) for the treatment of inflammatory diseases. J. Med. Chem. 60, 1247–1261.

He, S., Wang, X., 2018. RIP kinases as modulators of inflammation and immunity. Nat. Immunol. 19, 912–922.

He, S., Wang, L., Miao, L., Wang, T., Du, F., Zhao, L., Wang, X., 2009. Receptor interacting protein kinase-3 determines cellular necrotic response to TNF-alpha. Cell 137, 1100–1111.

He, S., Liang, Y., Shao, F., Wang, X., 2011. Toll-like receptors activate programmed necrosis in macrophages through a receptor-interacting kinase-3-mediated pathway. Proc. Natl. Acad. Sci. U. S. A. 108, 20054–20059.

Hildebrand, J.M., Tanzer, M.C., Lucet, I.S., Young, S.N., Spall, S.K., Sharma, P., Pierotti, C., Garnier, J.M., Dobson, R.C., Webb, A.I., Tripaydonis, A., Babon, J.J., Mulcair, M.D., Scanlon, M.J., Alexander, W.S., Wilks, A.F., Czabotar, P.E., Lessene, G., Murphy, J.M., Silke, J., 2014. Activation of the pseudokinase MLKL unleashes the four-helix bundle domain to induce membrane localization and necroptotic cell death. Proc. Natl. Acad. Sci. U. S. A. 111, 15072–15077.

Hu, Y.B., Zhang, Y.F., Wang, H., Ren, R.J., Cui, H.L., Huang, W.Y., Cheng, Q., Chen, H.Z., Wang, G., 2019. miR-425 deficiency promotes necroptosis and dopaminergic neurodegeneration in Parkinson's disease. Cell Death Dis. 10, 589.

Huang, D., Zheng, X., Wang, Z.A., Chen, X., He, W.T., Zhang, Y., Xu, J.G., Zhao, H., Shi, W., Wang, X., Zhu, Y., Han, J., 2017. The MLKL channel in necroptosis is an octamer formed by tetramers in a dyadic process. Mol. Cell. Biol. 37, e00497-16.

Iannielli, A., Bido, S., Folladori, L., Segnali, A., Cancellieri, C., Maresca, A., Massimino, L., Rubio, A., Morabito, G., Caporali, L., Tagliavini, F., Musumeci, O., Gregato, G., Bezard, E., Carelli, V., Tiranti, V., Broccoli, V., 2018. Pharmacological inhibition of necroptosis protects from dopaminergic neuronal cell death in Parkinson's disease models. Cell Rep. 22, 2066–2079.

Ingram, J.P., Thapa, R.J., Fisher, A., Tummers, B., Zhang, T., Yin, C., Rodriguez, D.A., Guo, H., Lane, R., Williams, R., Slifker, M.J., Basagoudanavar, S.H., Rall, G.F., Dillon, C.P., Green, D.R., Kaiser, W.J., Balachandran, S., 2019. ZBP1/DAI drives RIPK3-mediated cell death induced by IFNs in the absence of RIPK1. J. Immunol. 203, 1348–1355.

Irrinki, K.M., Mallilankaraman, K., Thapa, R.J., Chandramoorthy, H.C., Smith, F.J., Jog, N.R., Gandhirajan, R.K., Kelsen, S.G., Houser, S.R., May, M.J., Balachandran, S., Madesh, M., 2011. Requirement of FADD, NEMO, and BAX/BAK for aberrant mitochondrial function in tumor necrosis factor alpha-induced necrosis. Mol. Cell. Biol. 31, 3745–3758.

Ito, Y., Ofengeim, D., Najafov, A., Das, S., Saberi, S., Li, Y., Hitomi, J., Zhu, H., Chen, H., Mayo, L., Geng, J., Amin, P., DeWitt, J.P., Mookhtiar, A.K., Florez, M., Ouchida, A.T., Fan, J.B., Pasparakis, M., Kelliher, M.A., Ravits, J., Yuan, J., 2016. RIPK1 mediates axonal degeneration by promoting inflammation and necroptosis in ALS. Science 353, 603–608.

Jaco, I., Annibaldi, A., Lalaoui, N., Wilson, R., Tenev, T., Laurien, L., Kim, C., Jamal, K., Wicky John, S., Liccardi, G., Chau, D., Murphy, J.M., Brumatti, G., Feltham, R., Pasparakis, M., Silke, J., Meier, P., 2017. MK2 phosphorylates RIPK1 to prevent TNF-induced cell death. Mol. Cell 66, 698-710.e5.

Jacobsen, A.V., Lowes, K.N., Tanzer, M.C., Lucet, I.S., Hildebrand, J.M., Petrie, E.J., van Delft, M.F., Liu, Z., Conos, S.A., Zhang, J.G., Huang, D.C., Silke, J., Lessene, G., Murphy, J.M., 2016. HSP90 activity is required for MLKL oligomerisation and membrane translocation and the induction of necroptotic cell death. Cell Death Dis. 7, e2051.

Kaiser, W.J., Offermann, M.K., 2005. Apoptosis induced by the toll-like receptor adaptor TRIF is dependent on its receptor interacting protein homotypic interaction motif. J. Immunol. 174, 4942–4952.

Kaiser, W.J., Sridharan, H., Huang, C., Mandal, P., Upton, J.W., Gough, P.J., Sehon, C.A., Marquis, R.W., Bertin, J., Mocarski, E.S., 2013. Toll-like receptor 3-mediated necrosis via TRIF, RIP3, and MLKL. J. Biol. Chem. 288, 31268–31279.

Kaiser, W.J., Daley-Bauer, L.P., Thapa, R.J., Mandal, P., Berger, S.B., Huang, C., Sundararajan, A., Guo, H., Roback, L., Speck, S.H., Bertin, J., Gough, P.J., Balachandran, S., Mocarski, E.S., 2014. RIP1 suppresses innate immune necrotic as well as apoptotic cell death during mammalian parturition. Proc. Natl. Acad. Sci. U. S. A. 111, 7753–7758.

Kang, S., Fernandes-Alnemri, T., Rogers, C., Mayes, L., Wang, Y., Dillon, C., Roback, L., Kaiser, W., Oberst, A., Sagara, J., Fitzgerald, K.A., Green, D.R., Zhang, J., Mocarski, E.S., Alnemri, E.S., 2015. Caspase-8 scaffolding function and MLKL regulate NLRP3 inflammasome activation downstream of TLR3. Nat. Commun. 6, 7515.

Karch, J., Kanisicak, O., Brody, M.J., Sargent, M.A., Michael, D.M., Molkentin, J.D., 2015. Necroptosis interfaces with MOMP and the MPTP in mediating cell death. PLoS One 10, e0130520.

Kearney, C.J., Cullen, S.P., Tynan, G.A., Henry, C.M., Clancy, D., Lavelle, E.C., Martin, S.J., 2015. Necroptosis suppresses inflammation via termination of TNF- or LPS-induced cytokine and chemokine production. Cell Death Differ. 22, 1313–1327.

Kim, S.J., Li, J., 2013. Caspase blockade induces RIP3-mediated programmed necrosis in Toll-like receptor-activated microglia. Cell Death Dis. 4, e716.

Kim, Y.S., Morgan, M.J., Choksi, S., Liu, Z.G., 2007. TNF-induced activation of the Nox1 NADPH oxidase and its role in the induction of necrotic cell death. Mol. Cell 26, 675–687.

Kim, C.R., Kim, J.H., Park, H.L., Park, C.K., 2017. Ischemia reperfusion injury triggers TNFalpha induced-necroptosis in rat retina. Curr. Eye Res. 42, 771–779.

Kitur, K., Wachtel, S., Brown, A., Wickersham, M., Paulino, F., Penaloza, H.F., Soong, G., Bueno, S., Parker, D., Prince, A., 2016. Necroptosis promotes *Staphylococcus aureus* clearance by inhibiting excessive inflammatory signaling. Cell Rep. 16, 2219–2230.

Koo, G.B., Morgan, M.J., Lee, D.G., Kim, W.J., Yoon, J.H., Koo, J.S., Kim, S.I., Kim, S.J., Son, M.K., Hong, S.S., Levy, J.M., Pollyea, D.A., Jordan, C.T., Yan, P., Frankhouser, D., Nicolet, D., Maharry, K., Marcucci, G., Choi, K.S., Cho, H., Thorburn, A., Kim, Y.S., 2015. Methylation-dependent loss of RIP3 expression in cancer represses programmed necrosis in response to chemotherapeutics. Cell Res. 25, 707–725.

Kupka, S., De Miguel, D., Draber, P., Martino, L., Surinova, S., Rittinger, K., Walczak, H., 2016. SPATA2-mediated binding of CYLD to HOIP enables CYLD recruitment to signaling complexes. Cell Rep. 16, 2271–2280.

Lafont, E., Draber, P., Rieser, E., Reichert, M., Kupka, S., de Miguel, D., Draberova, H., von Massenhausen, A., Bhamra, A., Henderson, S., Wojdyla, K., Chalk, A., Surinova, S., Linkermann, A., Walczak, H., 2018. TBK1 and IKKepsilon prevent TNF-induced cell death by RIPK1 phosphorylation. Nat. Cell Biol. 20, 1389–1399.

Lawlor, K.E., Khan, N., Mildenhall, A., Gerlic, M., Croker, B.A., D'Cruz, A.A., Hall, C., Kaur Spall, S., Anderton, H., Masters, S.L., Rashidi, M., Wicks, I.P., Alexander, W.S., Mitsuuchi, Y., Benetatos, C.A., Condon, S.M., Wong, W.W., Silke, J., Vaux, D.L., Vince, J.E., 2015. RIPK3 promotes cell death and NLRP3 inflammasome activation in the absence of MLKL. Nat. Commun. 6, 6282.

Lee, S.B., Kim, J.J., Han, S.A., Fan, Y., Guo, L.S., Aziz, K., Nowsheen, S., Kim, S.S., Park, S.Y., Luo, Q., Chung, J.O., Choi, S.I., Aziz, A., Yin, P., Tong, S.Y., Fiesel, F.C., Springer, W., Zhang, J.S., Lou, Z., 2019. The AMPK-Parkin axis negatively regulates necroptosis and tumorigenesis by inhibiting the necrosome. Nat. Cell Biol. 21, 940–951.

Li, J., McQuade, T., Siemer, A.B., Napetschnig, J., Moriwaki, K., Hsiao, Y.S., Damko, E., Moquin, D., Walz, T., McDermott, A., Chan, F.K., Wu, H., 2012. The RIP1/RIP3 necrosome forms a functional amyloid signaling complex required for programmed necrosis. Cell 150, 339–350.

Li, D., Xu, T., Cao, Y., Wang, H., Li, L., Chen, S., Wang, X., Shen, Z., 2015. A cytosolic heat shock protein 90 and cochaperone CDC37 complex is required for RIP3 activation during necroptosis. Proc. Natl. Acad. Sci. U. S. A. 112, 5017–5022.

Li, J., Zhang, J., Zhang, Y., Wang, Z., Song, Y., Wei, S., He, M., You, S., Jia, J., Cheng, J., 2019a. TRAF2 protects against cerebral ischemia-induced brain injury by suppressing necroptosis. Cell Death Dis. 10, 328.

Li, X., Yao, X., Zhu, Y., Zhang, H., Wang, H., Ma, Q., Yan, F., Yang, Y., Zhang, J., Shi, H., Ning, Z., Dai, J., Li, Z., Li, C., Su, F., Xue, Y., Meng, X., Dong, G., Xiong, H., 2019b. The caspase inhibitor Z-VAD-FMK alleviates endotoxic shock via inducing macrophages necroptosis and promoting MDSCs-mediated inhibition of macrophages activation. Front. Immunol. 10, 1824.

Lin, J., Kumari, S., Kim, C., Van, T.M., Wachsmuth, L., Polykratis, A., Pasparakis, M., 2016. RIPK1 counteracts ZBP1-mediated necroptosis to inhibit inflammation. Nature 540, 124–128.
Lin, Q.S., Chen, P., Wang, W.X., Lin, C.C., Zhou, Y., Yu, L.H., Lin, Y.X., Xu, Y.F., Kang, D.Z., 2019. RIP1/RIP3/MLKL mediates dopaminergic neuron necroptosis in a mouse model of Parkinson disease. Lab. Invest. [Epub ahead of print].
Linkermann, A., Brasen, J.H., Darding, M., Jin, M.K., Sanz, A.B., Heller, J.O., De Zen, F., Weinlich, R., Ortiz, A., Walczak, H., Weinberg, J.M., Green, D.R., Kunzendorf, U., Krautwald, S., 2013. Two independent pathways of regulated necrosis mediate ischemia-reperfusion injury. Proc. Natl. Acad. Sci. U. S. A. 110, 12024–12029.
Liu, J., van Mil, A., Vrijsen, K., Zhao, J., Gao, L., Metz, C.H., Goumans, M.J., Doevendans, P.A., Sluijter, J.P., 2011. MicroRNA-155 prevents necrotic cell death in human cardiomyocyte progenitor cells via targeting RIP1. J. Cell. Mol. Med. 15, 1474–1482.
Liu, M., Wu, W., Li, H., Li, S., Huang, L.T., Yang, Y.Q., Sun, Q., Wang, C.X., Yu, Z., Hang, C.H., 2015. Necroptosis, a novel type of programmed cell death, contributes to early neural cells damage after spinal cord injury in adult mice. J. Spinal Cord Med. 38, 745–753.
Liu, S., Li, Y., Choi, H.M.C., Sarkar, C., Koh, E.Y., Wu, J., Lipinski, M.M., 2018a. Lysosomal damage after spinal cord injury causes accumulation of RIPK1 and RIPK3 proteins and potentiation of necroptosis. Cell Death Dis. 9, 476.
Liu, Z.M., Chen, Q.X., Chen, Z.B., Tian, D.F., Li, M.C., Wang, J.M., Wang, L., Liu, B.H., Zhang, S.Q., Li, F., Ye, H., Zhou, L., 2018b. RIP3 deficiency protects against traumatic brain injury (TBI) through suppressing oxidative stress, inflammation and apoptosis: dependent on AMPK pathway. Biochem. Biophys. Res. Commun. 499, 112–119.
Lu, W., Sun, J., Yoon, J.S., Zhang, Y., Zheng, L., Murphy, E., Mattson, M.P., Lenardo, M.J., 2016. Mitochondrial protein PGAM5 regulates mitophagic protection against cell necroptosis. PLoS One 11, e0147792.
Lu, J., Sun, Z., Fang, Y., Zheng, J., Xu, S., Xu, W., Shi, L., Mei, S., Wu, H., Liang, F., Zhang, J., 2019. Melatonin suppresses microglial necroptosis by regulating deubiquitinating enzyme A20 After intracerebral hemorrhage. Front. Immunol. 10, 1360.
Lule, S., Wu, L., McAllister, L.M., Edmiston 3rd, W.J., Chung, J.Y., Levy, E., Zheng, Y., Gough, P.J., Bertin, J., Degterev, A., Lo, E.H., Whalen, M.J., 2017. Genetic inhibition of receptor interacting protein kinase-1 reduces cell death and improves functional outcome after intracerebral hemorrhage in mice. Stroke 48, 2549–2556.
Mandal, P., Berger, S.B., Pillay, S., Moriwaki, K., Huang, C., Guo, H., Lich, J.D., Finger, J., Kasparcova, V., Votta, B., Ouellette, M., King, B.W., Wisnoski, D., Lakdawala, A.S., DeMartino, M.P., Casillas, L.N., Haile, P.A., Sehon, C.A., Marquis, R.W., Upton, J., Daley-Bauer, L.P., Roback, L., Ramia, N., Dovey, C.M., Carette, J.E., Chan, F.K., Bertin, J., Gough, P.J., Mocarski, E.S., Kaiser, W.J., 2014. RIP3 induces apoptosis independent of pronecrotic kinase activity. Mol. Cell 56, 481–495.
Martin, I., Kim, J.W., Dawson, V.L., Dawson, T.M., 2014. LRRK2 pathobiology in Parkinson's disease. J. Neurochem. 131, 554–565.
McComb, S., Cessford, E., Alturki, N.A., Joseph, J., Shutinoski, B., Startek, J.B., Gamero, A.M., Mossman, K.L., Sad, S., 2014. Type-I interferon signaling through ISGF3 complex is required for sustained Rip3 activation and necroptosis in macrophages. Proc. Natl. Acad. Sci. U. S. A. 111, E3206-13.
McNamara, D.E., Dovey, C.M., Hale, A.T., Quarato, G., Grace, C.R., Guibao, C.D., Diep, J., Nourse, A., Cai, C.R., Wu, H., Kalathur, R.C., Green, D.R., York, J.D., Carette, J.E., Moldoveanu, T., 2019. Direct activation of human MLKL by a select repertoire of inositol phosphate metabolites. Cell Chem. Biol. 26, 863–877.e7.

Meng, H., Liu, Z., Li, X., Wang, H., Jin, T., Wu, G., Shan, B., Christofferson, D.E., Qi, C., Yu, Q., Li, Y., Yuan, J., 2018. Death-domain dimerization-mediated activation of RIPK1 controls necroptosis and RIPK1-dependent apoptosis. Proc. Natl. Acad. Sci. U. S. A. 115, E2001–E2009.

Menon, M.B., Gropengiesser, J., Fischer, J., Novikova, L., Deuretzbacher, A., Lafera, J., Schimmeck, H., Czymmeck, N., Ronkina, N., Kotlyarov, A., Aepfelbacher, M., Gaestel, M., Ruckdeschel, K., 2017. p38(MAPK)/MK2-dependent phosphorylation controls cytotoxic RIPK1 signalling in inflammation and infection. Nat. Cell Biol. 19, 1248–1259.

Meylan, E., Burns, K., Hofmann, K., Blancheteau, V., Martinon, F., Kelliher, M., Tschopp, J., 2004. RIP1 is an essential mediator of Toll-like receptor 3-induced NF-kappa B activation. Nat. Immunol. 5, 503–507.

Mizumura, K., Cloonan, S.M., Nakahira, K., Bhashyam, A.R., Cervo, M., Kitada, T., Glass, K., Owen, C.A., Mahmood, A., Washko, G.R., Hashimoto, S., Ryter, S.W., Choi, A.M., 2014. Mitophagy-dependent necroptosis contributes to the pathogenesis of COPD. J. Clin. Invest. 124, 3987–4003.

Mompean, M., Li, W., Li, J., Laage, S., Siemer, A.B., Bozkurt, G., Wu, H., McDermott, A.E., 2018. The structure of the necrosome RIPK1-RIPK3 core, a human hetero-amyloid signaling complex. Cell 173, (1244–1253). e10.

Moquin, D.M., McQuade, T., Chan, F.K., 2013. CYLD deubiquitinates RIP1 in the TNFalpha-induced necrosome to facilitate kinase activation and programmed necrosis. PLoS One 8, e76841.

Moriwaki, K., Chan, F.K., 2017. The inflammatory signal adaptor RIPK3: functions beyond necroptosis. Int. Rev. Cell Mol. Biol. 328, 253–275.

Moriwaki, K., Balaji, S., McQuade, T., Malhotra, N., Kang, J., Chan, F.K., 2014. The necroptosis adaptor RIPK3 promotes injury-induced cytokine expression and tissue repair. Immunity 41, 567–578.

Moriwaki, K., Farias Luz, N., Balaji, S., De Rosa, M.J., O'Donnell, C.L., Gough, P.J., Bertin, J., Welsh, R.M., Chan, F.K., 2016. The mitochondrial phosphatase PGAM5 is dispensable for necroptosis but promotes inflammasome activation in macrophages. J. Immunol. 196, 407–415.

Moriwaki, K., Balaji, S., Bertin, J., Gough, P.J., Chan, F.K., 2017. Distinct kinase-independent role of RIPK3 in CD11c(+) mononuclear phagocytes in cytokine-induced tissue repair. Cell Rep. 18, 2441–2451.

Murakami, Y., Matsumoto, H., Roh, M., Giani, A., Kataoka, K., Morizane, Y., Kayama, M., Thanos, A., Nakatake, S., Notomi, S., Hisatomi, T., Ikeda, Y., Ishibashi, T., Connor, K.M., Miller, J.W., Vavvas, D.G., 2014. Programmed necrosis, not apoptosis, is a key mediator of cell loss and DAMP-mediated inflammation in dsRNA-induced retinal degeneration. Cell Death Differ. 21, 270–277.

Murphy, J.M., Czabotar, P.E., Hildebrand, J.M., Lucet, I.S., Zhang, J.G., Alvarez-Diaz, S., Lewis, R., Lalaoui, N., Metcalf, D., Webb, A.I., Young, S.N., Varghese, L.N., Tannahill, G.M., Hatchell, E.C., Majewski, I.J., Okamoto, T., Dobson, R.C., Hilton, D.J., Babon, J.J., Nicola, N.A., Strasser, A., Silke, J., Alexander, W.S., 2013. The pseudokinase MLKL mediates necroptosis via a molecular switch mechanism. Immunity 39, 443–453.

Najjar, M., Saleh, D., Zelic, M., Nogusa, S., Shah, S., Tai, A., Finger, J.N., Polykratis, A., Gough, P.J., Bertin, J., Whalen, M., Pasparakis, M., Balachandran, S., Kelliher, M., Poltorak, A., Degterev, A., 2016. RIPK1 and RIPK3 kinases promote cell-death-independent inflammation by toll-like receptor 4. Immunity 45, 46–59.

Nakazawa, S., Oikawa, D., Ishii, R., Ayaki, T., Takahashi, H., Takeda, H., Ishitani, R., Kamei, K., Takeyoshi, I., Kawakami, H., Iwai, K., Hatada, I., Sawasaki, T., Ito, H., Nureki, O., Tokunaga, F., 2016. Linear ubiquitination is involved in the pathogenesis of optineurin-associated amyotrophic lateral sclerosis. Nat. Commun. 7, 12547.

Newton, K., Sun, X., Dixit, V.M., 2004. Kinase RIP3 is dispensable for normal NF-kappa Bs, signaling by the B-cell and T-cell receptors, tumor necrosis factor receptor 1, and Toll-like receptors 2 and 4. Mol. Cell. Biol. 24, 1464–1469.

Newton, K., Dugger, D.L., Wickliffe, K.E., Kapoor, N., de Almagro, M.C., Vucic, D., Komuves, L., Ferrando, R.E., French, D.M., Webster, J., Roose-Girma, M., Warming, S., Dixit, V.M., 2014. Activity of protein kinase RIPK3 determines whether cells die by necroptosis or apoptosis. Science 343, 1357–1360.

Newton, K., Dugger, D.L., Maltzman, A., Greve, J.M., Hedehus, M., Martin-McNulty, B., Carano, R.A., Cao, T.C., van Bruggen, N., Bernstein, L., Lee, W.P., Wu, X., DeVoss, J., Zhang, J., Jeet, S., Peng, I., McKenzie, B.S., Roose-Girma, M., Caplazi, P., Diehl, L., Webster, J.D., Vucic, D., 2016a. RIPK3 deficiency or catalytically inactive RIPK1 provides greater benefit than MLKL deficiency in mouse models of inflammation and tissue injury. Cell Death Differ. 23, 1565–1576.

Newton, K., Wickliffe, K.E., Maltzman, A., Dugger, D.L., Strasser, A., Pham, V.C., Lill, J.R., Roose-Girma, M., Warming, S., Solon, M., Ngu, H., Webster, J.D., Dixit, V.M., 2016b. RIPK1 inhibits ZBP1-driven necroptosis during development. Nature 540, 129–133.

Newton, K., Wickliffe, K.E., Dugger, D.L., Maltzman, A., Roose-Girma, M., Dohse, M., Komuves, L., Webster, J.D., Dixit, V.M., 2019. Cleavage of RIPK1 by caspase-8 is crucial for limiting apoptosis and necroptosis. Nature 574, 428–431.

Northington, F.J., Chavez-Valdez, R., Graham, E.M., Razdan, S., Gauda, E.B., Martin, L.J., 2011. Necrostatin decreases oxidative damage, inflammation, and injury after neonatal HI. J. Cereb. Blood Flow Metab. 31, 178–189.

Oberst, A., Dillon, C.P., Weinlich, R., McCormick, L.L., Fitzgerald, P., Pop, C., Hakem, R., Salvesen, G.S., Green, D.R., 2011. Catalytic activity of the caspase-8-FLIP(L) complex inhibits RIPK3-dependent necrosis. Nature 471, 363–367.

O'Donnell, M.A., Perez-Jimenez, E., Oberst, A., Ng, A., Massoumi, R., Xavier, R., Green, D.R., Ting, A.T., 2011. Caspase 8 inhibits programmed necrosis by processing CYLD. Nat. Cell Biol. 13, 1437–1442.

Ofengeim, D., Ito, Y., Najafov, A., Zhang, Y., Shan, B., DeWitt, J.P., Ye, J., Zhang, X., Chang, A., Vakifahmetoglu-Norberg, H., Geng, J., Py, B., Zhou, W., Amin, P., Berlink Lima, J., Qi, C., Yu, Q., Trapp, B., Yuan, J., 2015. Activation of necroptosis in multiple sclerosis. Cell Rep. 10, 1836–1849.

Ofengeim, D., Mazzitelli, S., Ito, Y., DeWitt, J.P., Mifflin, L., Zou, C., Das, S., Adiconis, X., Chen, H., Zhu, H., Kelliher, M.A., Levin, J.Z., Yuan, J., 2017. RIPK1 mediates a disease-associated microglial response in Alzheimer's disease. Proc. Natl. Acad. Sci. U. S. A. 114, E8788–E8797.

Onate, M., Catenaccio, A., Salvadores, N., Saquel, C., Martinez, A., Moreno-Gonzalez, I., Gamez, N., Soto, P., Soto, C., Hetz, C., Court, F.A., 2019. The necroptosis machinery mediates axonal degeneration in a model of Parkinson disease. Cell Death Differ. [Epub ahead of print].

Onizawa, M., Oshima, S., Schulze-Topphoff, U., Oses-Prieto, J.A., Lu, T., Tavares, R., Prodhomme, T., Duong, B., Whang, M.I., Advincula, R., Agelidis, A., Barrera, J., Wu, H., Burlingame, A., Malynn, B.A., Zamvil, S.S., Ma, A., 2015. The ubiquitin-modifying enzyme A20 restricts ubiquitination of the kinase RIPK3 and protects cells from necroptosis. Nat. Immunol. 16, 618–627.

Orozco, S., Yatim, N., Werner, M.R., Tran, H., Gunja, S.Y., Tait, S.W., Albert, M.L., Green, D.R., Oberst, A., 2014. RIPK1 both positively and negatively regulates RIPK3 oligomerization and necroptosis. Cell Death Differ. 21, 1511–1521.

Orozco, S.L., Daniels, B.P., Yatim, N., Messmer, M.N., Quarato, G., Chen-Harris, H., Cullen, S.P., Snyder, A.G., Ralli-Jain, P., Frase, S., Tait, S.W.G., Green, D.R., Albert, M.L., Oberst, A., 2019. RIPK3 activation leads to cytokine synthesis that continues after loss of cell membrane integrity. Cell Rep. 28, 2275–2287.e5.

Petersen, S.L., Chen, T.T., Lawrence, D.A., Marsters, S.A., Gonzalvez, F., Ashkenazi, A., 2015. TRAF2 is a biologically important necroptosis suppressor. Cell Death Differ. 22, 1846–1857.

Petrie, E.J., Sandow, J.J., Jacobsen, A.V., Smith, B.J., Griffin, M.D.W., Lucet, I.S., Dai, W., Young, S.N., Tanzer, M.C., Wardak, A., Liang, L.Y., Cowan, A.D., Hildebrand, J.M., Kersten, W.J.A., Lessene, G., Silke, J., Czabotar, P.E., Webb, A.I., Murphy, J.M., 2018. Conformational switching of the pseudokinase domain promotes human MLKL tetramerization and cell death by necroptosis. Nat. Commun. 9, 2422.

Petrie, E.J., Czabotar, P.E., Murphy, J.M., 2019a. The structural basis of necroptotic cell death signaling. Trends Biochem. Sci. 44, 53–63.

Petrie, E.J., Sandow, J.J., Lehmann, W.I.L., Liang, L.Y., Coursier, D., Young, S.N., Kersten, W.J.A., Fitzgibbon, C., Samson, A.L., Jacobsen, A.V., Lowes, K.N., Au, A.E., Jousset Sabroux, H., Lalaoui, N., Webb, A.I., Lessene, G., Manning, G., Lucet, I.S., Murphy, J.M., 2019b. Viral MLKL homologs subvert necroptotic cell death by sequestering cellular RIPK3. Cell Rep. 28, 3309–3319.e5.

Polykratis, A., Hermance, N., Zelic, M., Roderick, J., Kim, C., Van, T.M., Lee, T.H., Chan, F.K.M., Pasparakis, M., Kelliher, M.A., 2014. Cutting edge: RIPK1 Kinase inactive mice are viable and protected from TNF-induced necroptosis in vivo. J. Immunol. 193, 1539–1543.

Pop, C., Oberst, A., Drag, M., Van Raam, B.J., Riedl, S.J., Green, D.R., Salvesen, G.S., 2011. FLIP(L) induces caspase 8 activity in the absence of interdomain caspase 8 cleavage and alters substrate specificity. Biochem. J. 433, 447–457.

Quarato, G., Guy, C.S., Grace, C.R., Llambi, F., Nourse, A., Rodriguez, D.A., Wakefield, R., Frase, S., Moldoveanu, T., Green, D.R., 2016. Sequential engagement of distinct MLKL phosphatidylinositol-binding sites executes necroptosis. Mol. Cell 61, 589–601.

Raju, S., Whalen, D.M., Mengistu, M., Swanson, C., Quinn, J.G., Taylor, S.S., Webster, J.D., Newton, K., Shaw, A.S., 2018. Kinase domain dimerization drives RIPK3-dependent necroptosis. Sci. Signal. 11, eaar2188.

Re, D.B., Le Verche, V., Yu, C., Amoroso, M.W., Politi, K.A., Phani, S., Ikiz, B., Hoffmann, L., Koolen, M., Nagata, T., Papadimitriou, D., Nagy, P., Mitsumoto, H., Kariya, S., Wichterle, H., Henderson, C.E., Przedborski, S., 2014. Necroptosis drives motor neuron death in models of both sporadic and familial ALS. Neuron 81, 1001–1008.

Remijsen, Q., Goossens, V., Grootjans, S., Van den Haute, C., Vanlangenakker, N., Dondelinger, Y., Roelandt, R., Bruggeman, I., Goncalves, A., Bertrand, M.J., Baekelandt, V., Takahashi, N., Berghe, T.V., Vandenabeele, P., 2014. Depletion of RIPK3 or MLKL blocks TNF-driven necroptosis and switches towards a delayed RIPK1 kinase-dependent apoptosis. Cell Death Dis. 5, e1004.

Rodriguez, D.A., Weinlich, R., Brown, S., Guy, C., Fitzgerald, P., Dillon, C.P., Oberst, A., Quarato, G., Low, J., Cripps, J.G., Chen, T., Green, D.R., 2016. Characterization of RIPK3-mediated phosphorylation of the activation loop of MLKL during necroptosis. Cell Death Differ. 23, 76–88.

Rohde, K., Kleinesudeik, L., Roesler, S., Lowe, O., Heidler, J., Schroder, K., Wittig, I., Drose, S., Fulda, S., 2017. A Bak-dependent mitochondrial amplification step contributes to Smac mimetic/glucocorticoid-induced necroptosis. Cell Death Differ. 24, 83–97.

Ros, U., Pena-Blanco, A., Hanggi, K., Kunzendorf, U., Krautwald, S., Wong, W.W., Garcia-Saez, A.J., 2017. Necroptosis execution is mediated by plasma membrane nanopores independent of calcium. Cell Rep. 19, 175–187.

Rosenbaum, D.M., Degterev, A., David, J., Rosenbaum, P.S., Roth, S., Grotta, J.C., Cuny, G.D., Yuan, J., Savitz, S.I., 2010. Necroptosis, a novel form of caspase-independent cell death, contributes to neuronal damage in a retinal ischemia-reperfusion injury model. J. Neurosci. Res. 88, 1569–1576.

Saleh, D., Najjar, M., Zelic, M., Shah, S., Nogusa, S., Polykratis, A., Paczosa, M.K., Gough, P.J., Bertin, J., Whalen, M., Fitzgerald, K.A., Slavov, N., Pasparakis, M., Balachandran, S., Kelliher, M., Mecsas, J., Degterev, A., 2017. Kinase activities of RIPK1 and RIPK3 can direct IFN-beta synthesis induced by lipopolysaccharide. J. Immunol. 198, 4435–4447.

Sarhan, J., Liu, B.C., Muendlein, H.I., Weindel, C.G., Smirnova, I., Tang, A.Y., Ilyukha, V., Sorokin, M., Buzdin, A., Fitzgerald, K.A., Poltorak, A., 2019. Constitutive interferon signaling maintains critical threshold of MLKL expression to license necroptosis. Cell Death Differ. 26, 332–347.

Schenk, B., Fulda, S., 2015. Reactive oxygen species regulate Smac mimetic/TNFalpha-induced necroptotic signaling and cell death. Oncogene 34, 5796–5806.

Schulze-Osthoff, K., Bakker, A.C., Vanhaesebroeck, B., Beyaert, R., Jacob, W.A., Fiers, W., 1992. Cytotoxic activity of tumor necrosis factor is mediated by early damage of mitochondrial functions. Evidence for the involvement of mitochondrial radical generation. J. Biol. Chem. 267, 5317–5323.

Sedger, L.M., McDermott, M.F., 2014. TNF and TNF-receptors: from mediators of cell death and inflammation to therapeutic giants—past, present and future. Cytokine Growth Factor Rev. 25, 453–472.

Seo, J., Lee, E.W., Sung, H., Seong, D., Dondelinger, Y., Shin, J., Jeong, M., Lee, H.K., Kim, J.H., Han, S.Y., Lee, C., Seong, J.K., Vandenabeele, P., Song, J., 2016. CHIP controls necroptosis through ubiquitylation- and lysosome-dependent degradation of RIPK3. Nat. Cell Biol. 18, 291–302.

Seo, J., Kim, M.W., Bae, K.H., Lee, S.C., Song, J., Lee, E.W., 2019. The roles of ubiquitination in extrinsic cell death pathways and its implications for therapeutics. Biochem. Pharmacol. 162, 21–40.

Shim, J.H., Xiao, C., Paschal, A.E., Bailey, S.T., Rao, P., Hayden, M.S., Lee, K.Y., Bussey, C., Steckel, M., Tanaka, N., Yamada, G., Akira, S., Matsumoto, K., Ghosh, S., 2005. TAK1, but not TAB1 or TAB2, plays an essential role in multiple signaling pathways in vivo. Genes Dev. 19, 2668–2681.

Strilic, B., Yang, L., Albarran-Juarez, J., Wachsmuth, L., Han, K., Muller, U.C., Pasparakis, M., Offermanns, S., 2016. Tumour-cell-induced endothelial cell necroptosis via death receptor 6 promotes metastasis. Nature 536, 215–218.

Su, X., Wang, H., Kang, D., Zhu, J., Sun, Q., Li, T., Ding, K., 2015. Necrostatin-1 ameliorates intracerebral hemorrhage-induced brain injury in mice through inhibiting RIP1/RIP3 pathway. Neurochem. Res. 40, 643–650.

Sun, X., Lee, J., Navas, T., Baldwin, D.T., Stewart, T.A., Dixit, V.M., 1999. RIP3, a novel apoptosis-inducing kinase. J. Biol. Chem. 274, 16871–16875.

Sun, L., Wang, H., Wang, Z., He, S., Chen, S., Liao, D., Wang, L., Yan, J., Liu, W., Lei, X., Wang, X., 2012. Mixed lineage kinase domain-like protein mediates necrosis signaling downstream of RIP3 kinase. Cell 148, 213–227.

Sureshbabu, A., Patino, E., Ma, K.C., Laursen, K., Finkelsztein, E.J., Akchurin, O., Muthukumar, T., Ryter, S.W., Gudas, L., Choi, A.M.K., Choi, M.E., 2018. RIPK3 promotes sepsis-induced acute kidney injury via mitochondrial dysfunction. JCI Insight 3, e98411.

Tait, S.W., Oberst, A., Quarato, G., Milasta, S., Haller, M., Wang, R., Karvela, M., Ichim, G., Yatim, N., Albert, M.L., Kidd, G., Wakefield, R., Frase, S., Krautwald, S., Linkermann, A., Green, D.R., 2013. Widespread mitochondrial depletion via mitophagy does not compromise necroptosis. Cell Rep. 5, 878–885.

Tang, Y., Tu, H., Zhang, J., Zhao, X., Wang, Y., Qin, J., Lin, X., 2019. K63-linked ubiquitination regulates RIPK1 kinase activity to prevent cell death during embryogenesis and inflammation. Nat. Commun. 10, 4157.

Tanzer, M.C., Tripaydonis, A., Webb, A.I., Young, S.N., Varghese, L.N., Hall, C., Alexander, W.S., Hildebrand, J.M., Silke, J., Murphy, J.M., 2015. Necroptosis signalling is tuned by phosphorylation of MLKL residues outside the pseudokinase domain activation loop. Biochem. J. 471, 255–265.

Tanzer, M.C., Matti, I., Hildebrand, J.M., Young, S.N., Wardak, A., Tripaydonis, A., Petrie, E.J., Mildenhall, A.L., Vaux, D.L., Vince, J.E., Czabotar, P.E., Silke, J., Murphy, J.M., 2016. Evolutionary divergence of the necroptosis effector MLKL. Cell Death Differ. 23, 1185–1197.

Tenev, T., Bianchi, K., Darding, M., Broemer, M., Langlais, C., Wallberg, F., Zachariou, A., Lopez, J., MacFarlane, M., Cain, K., Meier, P., 2011. The Ripoptosome, a signaling platform that assembles in response to genotoxic stress and loss of IAPs. Mol. Cell 43, 432–448.

Thapa, R.J., Nogusa, S., Chen, P., Maki, J.L., Lerro, A., Andrake, M., Rall, G.F., Degterev, A., Balachandran, S., 2013. Interferon-induced RIP1/RIP3-mediated necrosis requires PKR and is licensed by FADD and caspases. Proc. Natl. Acad. Sci. U. S. A. 110, E3109–E3118.

Tischner, D., Manzl, C., Soratroi, C., Villunger, A., Krumschnabel, G., 2012. Necrosis-like death can engage multiple pro-apoptotic Bcl-2 protein family members. Apoptosis 17, 1197–1209.

Upton, J.W., Kaiser, W.J., Mocarski, E.S., 2010. Virus inhibition of RIP3-dependent necrosis. Cell Host Microbe 7, 302–313.

Upton, J.W., Kaiser, W.J., Mocarski, E.S., 2012. DAI/ZBP1/DLM-1 complexes with RIP3 to mediate virus-induced programmed necrosis that is targeted by murine cytomegalovirus vIRA. Cell Host Microbe 11, 290–297.

Vanlangenakker, N., Vanden Berghe, T., Bogaert, P., Laukens, B., Zobel, K., Deshayes, K., Vucic, D., Fulda, S., Vandenabeele, P., Bertrand, M.J., 2011. cIAP1 and TAK1 protect cells from TNF-induced necrosis by preventing RIP1/RIP3-dependent reactive oxygen species production. Cell Death Differ. 18, 656–665.

Varfolomeev, E., Goncharov, T., Fedorova, A.V., Dynek, J.N., Zobel, K., Deshayes, K., Fairbrother, W.J., Vucic, D., 2008. c-IAP1 and c-IAP2 are critical mediators of tumor necrosis factor alpha (TNFalpha)-induced NF-kappaB activation. J. Biol. Chem. 283, 24295–24299.

Vince, J.E., Wong, W.W., Gentle, I., Lawlor, K.E., Allam, R., O'Reilly, L., Mason, K., Gross, O., Ma, S., Guarda, G., Anderton, H., Castillo, R., Hacker, G., Silke, J., Tschopp, J., 2012. Inhibitor of apoptosis proteins limit RIP3 kinase-dependent interleukin-1 activation. Immunity 36, 215–227.

Vitner, E.B., Salomon, R., Farfel-Becker, T., Meshcheriakova, A., Ali, M., Klein, A.D., Platt, F.M., Cox, T.M., Futerman, A.H., 2014. RIPK3 as a potential therapeutic target for Gaucher's disease. Nat. Med. 20, 204–208.

Wang, L., Du, F., Wang, X., 2008. TNF-alpha induces two distinct caspase-8 activation pathways. Cell 133, 693–703.

Wang, Y.Q., Wang, L., Zhang, M.Y., Wang, T., Bao, H.J., Liu, W.L., Dai, D.K., Zhang, L., Chang, P., Dong, W.W., Chen, X.P., Tao, L.Y., 2012a. Necrostatin-1 suppresses autophagy and apoptosis in mice traumatic brain injury model. Neurochem. Res. 37, 1849–1858.

Wang, Z., Jiang, H., Chen, S., Du, F., Wang, X., 2012b. The mitochondrial phosphatase PGAM5 functions at the convergence point of multiple necrotic death pathways. Cell 148, 228–243.

Wang, K., Liu, F., Zhou, L.Y., Ding, S.L., Long, B., Liu, C.Y., Sun, T., Fan, Y.Y., Sun, L., Li, P.F., 2013. miR-874 regulates myocardial necrosis by targeting caspase-8. Cell Death Dis. 4, e709.

Wang, H., Sun, L., Su, L., Rizo, J., Liu, L., Wang, L.F., Wang, F.S., Wang, X., 2014a. Mixed lineage kinase domain-like protein MLKL causes necrotic membrane disruption upon phosphorylation by RIP3. Mol. Cell 54, 133–146.

Wang, Y., Wang, H., Tao, Y., Zhang, S., Wang, J., Feng, X., 2014b. Necroptosis inhibitor necrostatin-1 promotes cell protection and physiological function in traumatic spinal cord injury. Neuroscience 266, 91–101.

Wang, H., Meng, H., Li, X., Zhu, K., Dong, K., Mookhtiar, A.K., Wei, H., Li, Y., Sun, S.C., Yuan, J., 2017. PELI1 functions as a dual modulator of necroptosis and apoptosis by regulating ubiquitination of RIPK1 and mRNA levels of c-FLIP. Proc. Natl. Acad. Sci. U. S. A. 114, 11944–11949.

Wang, Y., Shan, B., Liang, Y., Wei, H., Yuan, J., 2018. Parkin regulates NF-kappaB by mediating site-specific ubiquitination of RIPK1. Cell Death Dis. 9, 732.

Wertz, I.E., O'Rourke, K.M., Zhou, H., Eby, M., Aravind, L., Seshagiri, S., Wu, P., Wiesmann, C., Baker, R., Boone, D.L., Ma, A., Koonin, E.V., Dixit, V.M., 2004. De-ubiquitination and ubiquitin ligase domains of A20 downregulate NF-kappaB signalling. Nature 430, 694–699.

Wu, J., Huang, Z., Ren, J., Zhang, Z., He, P., Li, Y., Ma, J., Chen, W., Zhang, Y., Zhou, X., Yang, Z., Wu, S.Q., Chen, L., Han, J., 2013. Mlkl knockout mice demonstrate the indispensable role of Mlkl in necroptosis. Cell Res. 23, 994–1006.

Wu, X.N., Yang, Z.H., Wang, X.K., Zhang, Y., Wan, H., Song, Y., Chen, X., Shao, J., Han, J., 2014. Distinct roles of RIP1-RIP3 hetero- and RIP3-RIP3 homo-interaction in mediating necroptosis. Cell Death Differ. 21, 1709–1720.

Xia, B., Fang, S., Chen, X., Hu, H., Chen, P., Wang, H., Gao, Z., 2016. MLKL forms cation channels. Cell Res. 26, 517–528.

Xie, T., Peng, W., Liu, Y., Yan, C., Maki, J., Degterev, A., Yuan, J., Shi, Y., 2013a. Structural basis of RIP1 inhibition by necrostatins. Structure 21, 493–499.

Xie, T., Peng, W., Yan, C., Wu, J., Gong, X., Shi, Y., 2013b. Structural insights into RIP3-mediated necroptotic signaling. Cell Rep. 5, 70–78.

Xiong, Y., Li, L., Zhang, L., Cui, Y., Wu, C., Li, H., Chen, K., Yang, Q., Xiang, R., Hu, Y., Huang, S., Wei, Y., Yang, S., 2019. The bromodomain protein BRD4 positively regulates necroptosis via modulating MLKL expression. Cell Death Differ. 26, 1929–1941.

Xu, D., Jin, T., Zhu, H., Chen, H., Ofengeim, D., Zou, C., Mifflin, L., Pan, L., Amin, P., Li, W., Shan, B., Naito, M.G., Meng, H., Li, Y., Pan, H., Aron, L., Adiconis, X., Levin, J.Z., Yankner, B.A., Yuan, J., 2018. TBK1 suppresses RIPK1-driven apoptosis and inflammation during development and in aging. Cell 174, (1477–1491). e19.

Yang, H., Ma, Y., Chen, G., Zhou, H., Yamazaki, T., Klein, C., Pietrocola, F., Vacchelli, E., Souquere, S., Sauvat, A., Zitvogel, L., Kepp, O., Kroemer, G., 2016. Contribution of RIP3 and MLKL to immunogenic cell death signaling in cancer chemotherapy. Oncoimmunology 5, e1149673.

Yang, X.S., Yi, T.L., Zhang, S., Xu, Z.W., Yu, Z.Q., Sun, H.T., Yang, C., Tu, Y., Cheng, S.X., 2017. Hypoxia-inducible factor-1 alpha is involved in RIP-induced necroptosis caused by in vitro and in vivo ischemic brain injury. Sci. Rep. 7, 5818.

Yang, Z., Wang, Y., Zhang, Y., He, X., Zhong, C.Q., Ni, H., Chen, X., Liang, Y., Wu, J., Zhao, S., Zhou, D., Han, J., 2018. RIP3 targets pyruvate dehydrogenase complex to increase aerobic respiration in TNF-induced necroptosis. Nat. Cell Biol. 20, 186–197.

Yatim, N., Jusforgues-Saklani, H., Orozco, S., Schulz, O., Barreira da Silva, R., Reis e Sousa, C., Green, D.R., Oberst, A., Albert, M.L., 2015. RIPK1 and NF-kappaB signaling in dying cells determines cross-priming of CD8(+) T cells. Science 350, 328–334.

Yeh, W.C., de la Pompa, J.L., McCurrach, M.E., Shu, H.B., Elia, A.J., Shahinian, A., Ng, M., Wakeham, A., Khoo, W., Mitchell, K., El-Deiry, W.S., Lowe, S.W., Goeddel, D.V., Mak, T.W., 1998. FADD: essential for embryo development and signaling from some, but not all, inducers of apoptosis. Science 279, 1954–1958.

Yoon, S., Kovalenko, A., Bogdanov, K., Wallach, D., 2017. MLKL, the protein that mediates necroptosis, also regulates endosomal trafficking and extracellular vesicle generation. Immunity 47, (51–65). e7.

Yoshikawa, M., Saitoh, M., Katoh, T., Seki, T., Bigi, S.V., Shimizu, Y., Ishii, T., Okai, T., Kuno, M., Hattori, H., Watanabe, E., Saikatendu, K.S., Zou, H., Nakakariya, M., Tatamiya, T., Nakada, Y., Yogo, T., 2018. Discovery of 7-oxo-2,4,5,7-tetrahydro-6

H-pyrazolo[3,4- c]pyridine derivatives as potent, orally available, and brain-penetrating receptor interacting protein 1 (RIP1) kinase inhibitors: analysis of structure-kinetic relationships. J. Med. Chem. 61, 2384–2409.

You, Z., Savitz, S.I., Yang, J., Degterev, A., Yuan, J., Cuny, G.D., Moskowitz, M.A., Whalen, M.J., 2008. Necrostatin-1 reduces histopathology and improves functional outcome after controlled cortical impact in mice. J. Cereb. Blood Flow Metab. 28, 1564–1573.

Yu, P.W., Huang, B.C., Shen, M., Quast, J., Chan, E., Xu, X., Nolan, G.P., Payan, D.G., Luo, Y., 1999. Identification of RIP3, a RIP-like kinase that activates apoptosis and NFkappaB. Curr. Biol. 9, 539–542.

Yuan, J., Amin, P., Ofengeim, D., 2019a. Necroptosis and RIPK1-mediated neuroinflammation in CNS diseases. Nat. Rev. Neurosci. 20, 19–33.

Yuan, S., Yu, Z., Zhang, Z., Zhang, J., Zhang, P., Li, X., Li, H., Shen, H., Chen, G., 2019b. RIP3 participates in early brain injury after experimental subarachnoid hemorrhage in rats by inducing necroptosis. Neurobiol. Dis. 129, 144–158.

Zandi, E., Chen, Y., Karin, M., 1998. Direct phosphorylation of IkappaB by IKKalpha and IKKbeta: discrimination between free and NF-kappaB-bound substrate. Science 281, 1360–1363.

Zargarian, S., Shlomovitz, I., Erlich, Z., Hourizadeh, A., Ofir-Birin, Y., Croker, B.A., Regev-Rudzki, N., Edry-Botzer, L., Gerlic, M., 2017. Phosphatidylserine externalization, "necroptotic bodies" release, and phagocytosis during necroptosis. PLoS Biol. 15, e2002711.

Zhang, D.W., Shao, J., Lin, J., Zhang, N., Lu, B.J., Lin, S.C., Dong, M.Q., Han, J., 2009. RIP3, an energy metabolism regulator that switches TNF-induced cell death from apoptosis to necrosis. Science 325, 332–336.

Zhang, X., Fan, C., Zhang, H., Zhao, Q., Liu, Y., Xu, C., Xie, Q., Wu, X., Yu, X., Zhang, J., 2016. MLKL and FADD are critical for suppressing progressive lymphoproliferative disease and activating the NLRP3 inflammasome. Cell Rep. 16, 3247–3259.

Zhang, Y., Su, S.S., Zhao, S., Yang, Z., Zhong, C.Q., Chen, X., Cai, Q., Yang, Z.H., Huang, D., Wu, R., Han, J., 2017. RIP1 autophosphorylation is promoted by mitochondrial ROS and is essential for RIP3 recruitment into necrosome. Nat. Commun. 8, 14329.

Zhao, J., Jitkaew, S., Cai, Z., Choksi, S., Li, Q., Luo, J., Liu, Z.G., 2012. Mixed lineage kinase domain-like is a key receptor interacting protein 3 downstream component of TNF-induced necrosis. Proc. Natl. Acad. Sci. U. S. A. 109, 5322–5327.

Zhao, X.M., Chen, Z., Zhao, J.B., Zhang, P.P., Pu, Y.F., Jiang, S.H., Hou, J.J., Cui, Y.M., Jia, X.L., Zhang, S.Q., 2016. Hsp90 modulates the stability of MLKL and is required for TNF-induced necroptosis. Cell Death Dis. 7, e2089.

Zhou, T., Wang, Q., Phan, N., Ren, J., Yang, H., Feldman, C.C., Feltenberger, J.B., Ye, Z., Wildman, S.A., Tang, W., Liu, B., 2019. Identification of a novel class of RIP1/RIP3 dual inhibitors that impede cell death and inflammation in mouse abdominal aortic aneurysm models. Cell Death Dis. 10, 226.

Zhu, S., Zhang, Y., Bai, G., Li, H., 2011. Necrostatin-1 ameliorates symptoms in R6/2 transgenic mouse model of Huntington's disease. Cell Death Dis. 2, e115.

Zhu, K., Liang, W., Ma, Z., Xu, D., Cao, S., Lu, X., Liu, N., Shan, B., Qian, L., Yuan, J., 2018. Necroptosis promotes cell-autonomous activation of proinflammatory cytokine gene expression. Cell Death Dis. 9, 500.

CHAPTER THREE

Necroptosis, ADAM proteases and intestinal (dys)function

Michelle Heib[a], Stefan Rose-John[b], Dieter Adam[a,*]

[a]Institut für Immunologie, Christian-Albrechts-Universität zu Kiel, Kiel, Germany
[b]Institut für Biochemie, Christian-Albrechts-Universität zu Kiel, Kiel, Germany
[*]Corresponding author: e-mail address: dadam@email.uni-kiel.de

Contents

Abstract

Recently, an unexpected connection between necroptosis and members of the a disintegrin and metalloproteinase (ADAM) protease family has been reported. Necroptosis represents an important cell death routine which helps to protect from viral, bacterial,

International Review of Cell and Molecular Biology, Volume 353
ISSN 1937-6448
https://doi.org/10.1016/bs.ircmb.2020.02.001

fungal and parasitic infections, maintains adult T cell homeostasis and contributes to the elimination of potentially defective organisms before parturition. Equally important for organismal homeostasis, ADAM proteases control cellular processes such as development and differentiation, immune responses or tissue regeneration. Notably, necroptosis as well as ADAM proteases have been implicated in the control of inflammatory responses in the intestine. In this review, we therefore provide an overview of the physiology and pathophysiology of necroptosis, ADAM proteases and intestinal (dys)function, discuss the contribution of necroptosis and ADAMs to intestinal (dys) function, and review the current knowledge on the role of ADAMs in necroptotic signaling.

1. Necroptosis

For many years, cell death has been divided into the categories "apoptosis" and "necrosis," where apoptosis was classified as a cysteine-dependent aspartate-directed protease (caspase)-dependent, regulated and physiologic form of cell death, which was executed by a defined set of molecular components. In contrast, necrosis was considered as a form of lytic cell death which arises in an accidental, non-regulated and non-physiological manner (Galluzzi et al., 2018). Although several groups described instances of regulated cell death that displayed a necrotic rather than an apoptotic phenotype (Kawahara et al., 1998; Laster et al., 1988; Strelow et al., 2000; Vercammen et al., 1998a,b), these reports were initially met with skepticism. This changed in 2000 when Holler et al. published a seminal study showing that death receptor (DR) ligands could trigger a necrotic form of cell death in a receptor-interacting serine-threonine protein kinase (RIPK)1-dependent manner (Holler et al., 2000). This specific modality of cell death was termed "necroptosis" by Degterev et al. in 2005 (Degterev et al., 2005). Today, necroptosis is defined by the Nomenclature Committee on Cell Death as "a type of regulated cell death triggered by perturbations of extracellular or intracellular homeostasis that critically depends on MLKL, RIPK3, and (at least in some settings) on the kinase activity of RIPK1" (Galluzzi et al., 2018).

1.1 (Patho)physiology

Necroptosis ultimately causes the rupture of the cell membrane, leading to the release of immune-stimulatory intracellular danger-associated molecular patterns (DAMPs) (Kaczmarek et al., 2013). Therefore, necroptosis represents a strongly pro-inflammatory mode of cell death (Pasparakis and Vandenabeele, 2015; Sarhan et al., 2018) and has been implicated in an increasing number of physiological and pathophysiological conditions. Inflammatory processes and diseases associated with necroptosis include acute and chronic liver injury (Afonso et al., 2015; Deutsch et al., 2015; Gautheron et al., 2014; Roychowdhury et al., 2013; Wang et al., 2016), inflammatory skin diseases (Bonnet et al., 2011; Dannappel et al., 2014; Takahashi et al., 2014), acute pancreatitis (He et al., 2009), atherosclerosis (Lin et al., 2013; Meng et al., 2015), myocardial infarction (Lim et al., 2007; Luedde et al., 2014; Oerlemans et al., 2012; Smith et al., 2007), ischemia-reperfusion injury (Rosenbaum et al., 2010; Zhang et al., 2016a), acute kidney injury (Linkermann et al., 2013), systemic inflammatory syndrome (Meng et al., 2015), sepsis (Duprez et al., 2011; Linkermann et al., 2012), stroke (Degterev et al., 2005; Xu et al., 2010; Zhu et al., 2012), multiple sclerosis (Ofengeim et al., 2015), retinal disorders (Dong et al., 2012; Hanus et al., 2013; Murakami et al., 2012, 2014; Sato et al., 2013; Tamura et al., 2014; Trichonas et al., 2010), neuron degeneration (Re et al., 2014), traumatic brain (King et al., 2014; You et al., 2008) and spinal cord injuries (Fan et al., 2016; Liu et al., 2015b; Wang et al., 2014b), Morbus Alzheimer (Qinli et al., 2013), Morbus Huntington (Zhu et al., 2011), cancer (Aaes et al., 2016; He et al., 2013; Yatim et al., 2015), and, last not least, inflammatory bowel disease (IBD) (Aden et al., 2018; Caprioli et al., 2008; Dannappel et al., 2014; Günther et al., 2011; Matsuzawa-Ishimoto et al., 2017; Negroni et al., 2017; Pierdomenico et al., 2014; Takahashi et al., 2014; Weinlich et al., 2013; Welz et al., 2011; Zhang et al., 2019a,b). Necroptosis represents an integral component of autoimmune and autoinflammatory diseases (Alvarez-Diaz et al., 2016; Zhang et al., 2016b) and reportedly also contributes to diseases such as lysosomal storage disorders (Vitner et al., 2014) and Duchenne's muscular dystrophy (Morgan et al., 2018).

Regarding its physiological relevance, necroptosis was initially considered as a backup mechanism which can permit a cell to commit suicide if the caspase machinery fails or is inactivated (e.g., after virus infection or in apoptosis-resistant tumor cells) (Conrad et al., 2016; Thon et al., 2006). Adding complexity to this initial picture, several groups have shown that necroptosis can be triggered in parallel to apoptosis, with a cellular predisposition for a specific pathway (Rickard et al., 2014a; Takahashi et al., 2014). Therefore, necroptosis may not merely represent a backup pathway for apoptosis but rather a cell death routine that evolved to counter malignant and infectious threats such as viral (Cho et al., 2009; Pan et al., 2014; Polykratis et al., 2014; Upton et al., 2012), bacterial (González-Juarbe et al., 2015; McComb et al., 2012; Pearson et al., 2017; Robinson et al., 2012; Roca and Ramakrishnan, 2013), fungal (Cao et al., 2019) and parasitic infections (Osborn et al., 2010). Representing a strongly pro-inflammatory form of cell death, necroptosis not only eliminates pathogen-infected cells but also contributes to physiological immunosurveillance by activating innate and adaptive immune responses (Cho et al., 2009; Huang et al., 2015; Pearson and Murphy, 2017; Polykratis et al., 2014; Weng et al., 2014; Yatim et al., 2015). As a consequence of this role of necroptosis in immunity, manipulation of necroptosis is in turn employed by ubiquitous pathogens as an immunomodulatory strategy. Alpha herpesviruses such as human herpes simplex virus 1 and -2 have specific mediators (i.e., the proteins ICP6 and ICP10) to prevent RIPK3-dependent necroptosis (Guo et al., 2015). The beta herpes virus murine cytomegalovirus expresses the protein viral inhibitor of caspase-8-induced apoptosis to inhibit RIPK3 (Upton et al., 2008, 2010). Human cytomegalovirus infection prevents necroptosis, although the specific factor has not been identified (Omoto et al., 2015). Influenza A virus activates the necroptotic pathway (Kuriakose et al., 2016; Thapa et al., 2016), and vaccinia virus expresses the protein E3L to inhibit necroptosis (Koehler et al., 2017). In addition, necroptosis is important for the maintenance of adult T cell homeostasis (Ch'en et al., 2011; Kaiser et al., 2011; Lu et al., 2011; Osborn et al., 2010; Zhang et al., 2016b) and participates in developmental safeguard programs that ensure the elimination of potentially defective organisms before parturition (Kaiser et al., 2014).

However, when assessing the physiologic significance of necroptosis, it is important to keep in mind that key components of the necroptotic signaling machinery are poorly conserved across the animal kingdom (Dondelinger et al., 2016). Therefore, the role of necroptosis in animals may not be as universal as previously assumed, challenging the concept that necroptosis has a general, conserved function for immune defense (Pearson and Murphy, 2017). As an alternative explanation, other, yet unidentified proteins may substitute for the non-conserved components in the respective species (Dondelinger et al., 2016).

1.2 Triggers and core machinery

Necroptosis can be triggered by DRs (tumor necrosis factor (TNF) receptor (TNF-R)1 (Degterev et al., 2005; Holler et al., 2000; Vercammen et al., 1997), TNF-related apoptosis-inducing ligand (TRAIL) receptor

(TRAIL-R)1 and 2 (Holler et al., 2000), FS-7-associated surface antigen (Fas) (Degterev et al., 2005; Holler et al., 2000; Vercammen et al., 1998b), DR3 (Bittner et al., 2017) and, reportedly, DR6 (Strilic et al., 2016)), the T cell receptor (TCR) (Ch'en et al., 2008), Toll-like receptor (TLR)3 (Kaiser et al., 2013) and 4 (Kim and Li, 2013), fibroblast growth factor-inducible 14 (Fn14) (Wilson and Browning, 2002), TNF-R2 (Chan et al., 2003), interferons (IFNs) (He et al., 2011; Kalai et al., 2002; McComb et al., 2014a; Robinson et al., 2012; Thapa et al., 2013), sensors for intracellular RNA (retinoic acid inducible gene I) (Schock et al., 2017) and DNA (Z-DNA-binding protein (ZBP)1/DNA-dependent activator of IFN-regulatory factors (DAI) (Upton et al., 2012), cyclic GMP-AMP synthase (cGAS)-stimulator of IFN genes (STING) (Brault et al., 2018)), by inhibition of the proteasome (Moriwaki and Chan, 2016), by stressors such as calcium overload (Nomura et al., 2014), reactive oxygen species (ROS) (Zhang et al., 2017), ischemia-reperfusion injury (Wen et al., 2017), photodynamic therapy (Coupienne et al., 2011), and ionizing radiation (Schildkopf et al., 2010).

At the molecular level, all these triggers ultimately induce the activation of two molecules that encompass the core machinery of necroptosis: RIPK3 (Cho et al., 2009; He et al., 2009; Zhang et al., 2009) and the pseudokinase mixed-lineage kinase domain-like (MLKL) (Sun et al., 2012; Wang et al., 2014a). Necroptotic stimuli activate RIPK3 by triggering physical interactions of the RIP homotypic interaction motif (RHIM; Sun et al., 2002) of RIPK3 with other RHIM-containing proteins (e.g., RIPK1; Cho et al., 2009; He et al., 2009; Upton et al., 2008; Zhang et al., 2009), ZBP1/DAI (Lin et al., 2016; Newton et al., 2016b; Upton et al., 2012) (which was recently shown to also interact with RNA (Maelfait et al., 2017; Sridharan et al., 2017; Thapa et al., 2016)), or the TLR3 and TLR4 adaptor protein Toll/interleukin-1 receptor homology domain-containing adapter-inducing interferon-β (TRIF) (Kaiser et al., 2013; Kim and Li, 2013). This results in oligomerization of RIPK3 into β-amyloid-like structures (Li et al., 2012) (termed the "canonical necrosome" when RIPK3 is in complex with RIPK1 or the "non-canonical necrosome" when complexed with other RHIM-containing proteins (Mompeán et al., 2018)) and activation of its kinase activity. The phosphorylation of human RIPK3 at Ser227 (Sun et al., 2012) or mouse RIPK3 at Ser232 (Chen et al., 2013) is essential for the subsequent recruitment of MLKL into the necrosome. In the next step, RIPK3 phosphorylates the pseudokinase domain of MLKL at Thr357/Ser358 in the human (Sun et al., 2012) or S345 in the mouse system (Rodriguez et al., 2016), uncovering a latent N-terminal 4-helix bundle (4HB) domain that is required for cell lytic activity and for the oligomerization of MLKL (Hildebrand et al., 2014). To date, the exact stoichiometry of the MLKL oligomer is contentious, with trimers (Davies et al., 2018; Hildebrand et al., 2014), tetramers (Petrie et al., 2018), hexamers (Wang et al., 2014a), octamers (Huang et al., 2017), and polymers (Liu et al., 2017) being reported. In vitro, biophysical analyses indicate that human MLKL forms tetramers (Petrie et al., 2018) while mouse MLKL seems to assemble into trimers (Davies et al., 2018; Hildebrand et al., 2014), although the existence of higher order complexes in vivo cannot be excluded. Regardless of their exact composition, MLKL oligomers translocate and insert into the plasma membrane and into intracellular membranes where they induce the loss of membrane integrity (Cai et al., 2014; Chen et al., 2014; Dondelinger et al., 2014; Wang et al., 2014a). The exact mechanism by which MLKL causes the necroptotic death of the cell is still controversially discussed (i.e., influx of calcium (Cai et al., 2014) or sodium (Chen et al., 2014) through ion channels vs. direct binding to membrane phosphatidylinositol phosphates (Dondelinger et al., 2014; Wang et al., 2014a)).

1.3 Regulators

Both initiation and execution of necroptosis are subject to stringent regulation. In the best understood model system, i.e., necroptosis induction by TNF-R1, binding of TNF to TNF-R1 induces the recruitment of the proteins RIPK1 (Ermolaeva et al., 2008; Pobezinskaya et al., 2008), TNF-R1-associated death domain protein (TRADD) (Hsu et al., 1995), cellular inhibitor of apoptosis protein (cIAP)1 and 2 (Rothe et al., 1995; Shu et al., 1996; Varfolomeev et al., 2008), TNF-R-associated factor (TRAF)2 (Hsu et al., 1995), and the linear ubiquitin chain assembly complex (LUBAC) (Haas et al., 2009) to the receptor, forming a complex which has been termed complex I (Micheau and Tschopp, 2003). Within complex I, RIPK1 is rapidly modified with K11- (Dynek et al., 2010), K48- (Bertrand et al., 2008), K63-linked (Ea et al., 2006; Li et al., 2006) and linear (Gerlach et al., 2011) polyubiquitin chains. In this constellation, complex I promotes the canonical activation of the pro-inflammatory and pro-survival transcription factor nuclear factor kappa-light-chain-enhancer of activated B-cells (NF-κB) (Micheau and Tschopp, 2003), favoring cell survival over apoptotic or necroptotic cell death. Thus, the ubiquitination status of RIPK1 constitutes the first major regulatory checkpoint, which by default tonically suppresses necroptosis.

When RIPK1 ubiquitination is prevented by inhibition or depletion of the cIAP proteins (Dondelinger et al., 2013; Geserick et al., 2009; Varfolomeev et al., 2007; Wang et al., 2008; Yang et al., 2013) or upon deubiquitination of RIPK1 by cylindromatosis (CYLD) (Moquin et al., 2013), RIPK1 is released from TNF-R1 to form a cytosolic "complex IIb" (Declercq et al., 2009) with RIPK3, Fas-associated death domain protein (FADD), caspase-8 and the caspase-8-regulatory protein FADD-like interleukin-1β converting enzyme-like inhibitory protein long form ($FLIP_L$). Within this complex, induction of necroptosis by RIPK1 remains tonically inhibited by a second checkpoint: the enzymatic activity of caspase-8 (Dillon et al., 2014; Kaiser et al., 2011; Oberst et al., 2011; Weinlich et al., 2013). This is accomplished by caspase-8-mediated cleavage and inactivation of RIPK1 (Lin et al., 1999; Newton et al., 2019), and also RIPK3 (Feng et al., 2007) and CYLD (O'Donnell et al., 2011), although this remains to be formally demonstrated for the latter two (Pasparakis and Vandenabeele, 2015). Thus, in the presence of active caspase-8, the default response elicited by complex IIb is apoptosis rather than necroptosis. However, under conditions where caspase-8 activity within complex IIb is low or inhibited (e.g., by the pancaspase inhibitor carbobenzoxy-valyl-alanyl-aspartyl-[*O*-methyl]-fluoromethylketone or by viral proteins such as cytokine response modifier A (Vercammen et al., 1998a), viral inhibitor of caspase-8-induced apoptosis (McCormick et al., 2003) or the Herpes simplex virus proteins ICP6 and ICP10 (Guo et al., 2015)), and provided that sufficiently high levels of $FLIP_L$ and/or RIPK3 and MLKL are present (Feoktistova et al., 2011; Moujalled et al., 2013; Oberst et al., 2011; Pasparakis and Vandenabeele, 2015; Tenev et al., 2011), RHIM-RHIM interactions of RIPK1 and RIPK3 will permit an efficient formation of the necrosome and subsequent necroptosis (Pasparakis and Vandenabeele, 2015).

In addition to RIPK1 ubiquitination and caspase-8 activity as the two major checkpoints, necroptosis is modulated by multiple auxiliary factors and mechanisms. At the level of RIPK1, mitochondrial ROS are reportedly sensed by cysteine residues in RIPK1, activate RIPK1 autophosphorylation on Ser161, and facilitate the recruitment of RIPK3 (Zhang et al., 2017). As negative regulators, the kinases NF-κB inhibitor kinase (IKK)α, IKKβ (Dondelinger et al., 2015, 2019), mitogen-activated protein kinase-activated protein kinase 2 (Dondelinger et al., 2017; Jaco et al., 2017; Menon et al., 2017), TRAF family member-associated NF-κB activator-binding kinase 1 and IKKε (Lafont et al., 2018) have been reported to catalyze an inhibitory phosphorylation of RIPK1. Likewise, caspase-6 and cathepsins

B and S can cleave RIPK1 and thus interfere with necroptosis (McComb et al., 2014b; van Raam et al., 2013). Notably, RIPK1 itself has a more complex role in the regulation of necroptosis than previously anticipated, i.e., by acting not only as an activator, but also as an inhibitor of necroptosis (Dannappel et al., 2014; Dillon et al., 2014; Kaiser et al., 2014; Rickard et al., 2014b; Takahashi et al., 2014). Mechanistically, RIPK1 can suppress ZBP1/DAI-mediated necroptosis in a non-enzymatic manner via its RHIM domain which prevents ZBP1/DAI from binding and activating RIPK3 (Lin et al., 2016; Newton et al., 2016b). Moreover, Oroczo et al. have reported that RIPK1 intrinsically suppresses spontaneous RIPK3 activation in the cytosol by controlling RIPK3 oligomerization (Orozco et al., 2014). Therefore, RIPK1 can promote the assembly of RIPK3-containing pro-necroptotic complexes via its kinase activity, but it also seems to regulate or even attenuate the assembly of such complexes in a kinase-independent manner via its RHIM domain, depending on yet unknown cues (Petrie et al., 2019).

As another target of regulation, RIPK3 is deubiquitinated at Lys5 by TNF-induced protein 3/A20 (Onizawa et al., 2015), dephosphorylated at T231/S232 by protein phosphatase 1B (Ppm1b) (Chen et al., 2015), or modified at Thr467 by O-linked β-*N*-acetylglucosamine transferase (Li et al., 2019), which all restricts the formation of necrosome complexes. Very recently, Lee et al. have reported that the ubiquitin ligase Parkin ubiquitinates RIPK3 via non-degradative K33-ubiquitin linkages, which likewise blocks necrosome assembly (Lee et al., 2019).

As for MLKL, its necroptotic function can be promoted by the inositol kinases inositol polyphosphate multikinase (IPMK) and inositol-tetrakisphosphate 1-kinase (ITPK1) (Dovey et al., 2018) or inositol-pentakisphosphate 2-kinase (IPPK) (McNamara et al., 2019), which generate inositol hexaphosphate (IP_6) as a ligand and activator of MLKL. As a negative regulator of MLKL, TRAF2, in addition to its regular functions in the activation of NF-κB, attenuates DR-induced necroptosis by constitutive interaction with MLKL (Karl et al., 2014; Petersen et al., 2015). Similarly, thioredoxin-1 (Trx1) maintains MLKL in a reduced, inactivated monomeric form (Reynoso et al., 2017) and repulsive guidance molecule b (RGMb) counters MLKL membrane association (Liu et al., 2018). Furthermore, membrane damage caused by MLKL during necroptosis is counteracted by components of the endosomal sorting complexes required for transport (ESCRT)-I and -III complex via shedding of extracellular vesicles (necroptotic bodies) from cell membranes which contain activated MLKL (Gong et al., 2017; Yoon et al., 2017; Zargarian et al., 2017).

As a regulatory mechanism that does not directly target RIPK1, RIPK3 or MLKL, removal of ubiquitin chains from TRAF2 by CYLD can suppress the anti-necroptotic activity of TRAF2 and additionally promote necroptosis (Petersen et al., 2015).

Several regulatory mechanisms affect more than one component of the necrosome. The proteins heat shock protein (HSP)90 and its co-chaperone cell division cycle 37 promote necroptosis by facilitating RIPK1 (Chen et al., 2012; Fearns et al., 2006; Gentle et al., 2011; Lewis et al., 2000; Thon et al., 2005; Vanden Berghe et al., 2003) and RIPK3 activation (Cho et al., 2009; Li et al., 2015b; Park et al., 2014), as well as the oligomerization and translocation of MLKL (Bigenzahn et al., 2016; Jacobsen et al., 2016; Zhao et al., 2016). Similarly, a prolonged or hyperactivation of the kinase transforming growth factor (TGF-)β-activated kinase 1 stimulates the pro-necroptotic phosphorylation of RIPK1 and RIPK3 (Geng et al., 2017; Morioka et al., 2014). As negative regulators of necroptosis, the LUBAC complex limits the formation of the RIPK1-RIPK3-MLKL complex (Lafont et al., 2017), aurora kinase A inhibits necrosome formation by directly binding to RIPK1 and RIPK3 (Xie et al., 2017), and the E3 ubiquitin ligase heat shock cognate 71 kDa protein-interacting protein triggers the ubiquitination and lysosomal degradation of RIPK1 and RIPK3 (Seo et al., 2016). The function of another ubiquitin ligase, pellino E3 ubiquitin protein ligase (PELI)1, is presently ambiguous. Originally reported as an enhancer of necroptosis by ubiquitinating RIPK1 at Lys115 and thus promoting the formation of the necrosome (Wang et al., 2017), PELI1 can also suppress necroptosis by ubiquitinating RIPK3 at Lys363, targeting it for proteasomal degradation (Choi et al., 2018). Similarly, proteolytic cleavage of RIPK1, RIPK3, TRIF and ZBP1/DAI by the cysteine protease EspL is exploited by enteropathogenic *Escherichia coli* to suppress necroptosis during infection (Pearson et al., 2017). As another proteolytic regulator, we have previously identified the protease high-temperature requirement A2 (HtrA2)/Omi as a key component of TNF- (Sosna et al., 2013) and TRAIL- (Sosna et al., 2016) induced necroptosis. Currently, the exact molecular mechanism how HtrA2/Omi interacts with the necroptotic signaling cascade is not clear, although Zhang et al. have recently suggested that HtrA2/Omi enhances necroptosis by degrading RIPK1 (Zhang et al., 2019a). Finally, proteases of the a disintegrin and metalloproteinase

(ADAM) family have been implicated in necroptosis by us (Fuchslocher Chico et al., 2018) and others (Cai et al., 2016), which will be discussed in further detail below. For a comprehensive discussion how necroptosis is regulated, we refer the reader to recent excellent reviews, e.g., (Galluzzi et al., 2017, 2018; Pasparakis and Vandenabeele, 2015; Petrie et al., 2019). A schematic overview of necroptosis, its effectors and regulators is given in Fig. 1.

2. ADAM proteases

The ADAM family consists of cell surface proteases which control important functions such as development and differentiation, immune responses or regenerative processes (Jones et al., 2016; Scheller et al., 2011a). By shedding the ectodomains of membrane-associated proteins, ADAMs can regulate intra- or intercellular signaling cascades and thus initiate rapid cellular responses to an altered extracellular environment, i.e., during cellular stress or inflammation. Thus, ADAM proteases are fundamental for cellular and organismal homeostasis and, unsurprisingly, their dysregulation has been linked to multiple pathological states, including Alzheimer's disease, asthma, cardiovascular disease, cancer (Edwards et al., 2008; Mullooly et al., 2016) and intestinal inflammation (Chalaris et al., 2010).

2.1 Structure of ADAM proteases

In humans, 22 members of the ADAM family have been described, 12 of which have proteolytic activity (ADAM8, 9, 10, 12, 15, 17, 19, 20, 21, 28, 30, and 33) (Edwards et al., 2008; Jones et al., 2016). ADAMs are type I transmembrane proteins which share a common modular structure. As outlined in Fig. 2, mature ADAMs consist of a prodomain, a metalloproteinase domain, a disintegrin domain, a cysteine-rich domain, an epidermal growth factor (EGF)-like repeat-containing domain, a transmembrane region, and a cytoplasmic tail (White, 2003). ADAM10 and its closest relative ADAM17 differ from the other members of the family in that they lack the cysteine-rich domain and the EGF-like repeat-containing domain, but rather contain a distinct membrane-proximal domain (Düsterhöft et al., 2013; Janes et al., 2005; Takeda, 2009). The prodomain acts as an inhibitor of proteinase activity and is cleaved during ADAM activation by furin proteinases (Gonzales et al., 2004; Leonard et al., 2005; Moss et al., 2007). The metalloprotease domain contains the signature motif HEXGHXXGXXHD in enzymatically active ADAMs (Stöcker et al., 1995), but not in protease-inactive family members such as ADAM2, 7, 11, 18, 22, 23, 29 and

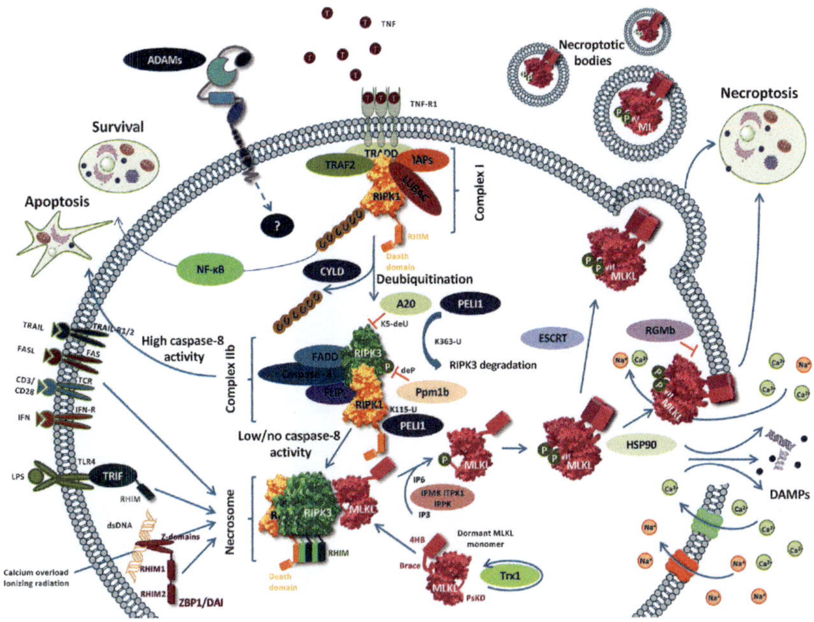

Fig. 1 Schematic overview of necroptosis, its effectors and regulators. Upon binding of TNF (T) to its receptor TNF-R1, RIPK1 is recruited into complex I which also consists of TRADD, TRAF2, IAPs and LUBAC, by default resulting in RIPK1 polyubiquitination (U) and induction of NF-κB-dependent survival signals. Under conditions that prevent *de novo* ubiquitination of RIPK1 by IAPs, deubiquitination of RIPK1 by CYLD enables its association with RIPK3, FADD, procaspase-8 and $FLIP_L$ into complex IIb. Within this complex, high caspase-8 activity results in cleavage and inactivation of RIPK1 and RIPK3, by default leading to apoptosis. If caspase-8 activity is low or compromised, RHIM-RHIM interactions of RIPK1 and RIPK3 promote the phosphorylation (P) and activation of RIPK3, which causes the recruitment of MLKL into the necrosome. Necrosome formation can also be triggered by other DRs such as TRAIL-R1/2 and Fas, the TCR or IFN receptors, or by calcium overload and ionizing radiation. In case of TLR-induced necroptosis, RIPK1 is substituted for by TRIF, while intracellular DNA triggers necrosome formation via ZBP1/DAI. As part of the necrosome, MLKL is phosphorylated and activated by RIPK3, resulting in the oligomerization and membrane translocation of MLKL, where its exposed 4HB domain induces membrane disruption, DAMP release and the necroptotic burst of the cell, possibly with contribution of calcium or sodium influx and ion channels. Necroptosis is enhanced by IPMK, ITPK1 and IPPK (MLKL activation via IP_6 generation) and by HSP90 (facilitating MLKL oligomerization and membrane translocation), but restricted by A20 (Lys5 deubiquitination), Ppm1b in mice (RIPK3 T231/S232 dephosphorylation), Trx1 (keeping MLKL in a dormant reduced monomeric state), RGMb (preventing MLKL membrane association), and proteins of the ESCRT machinery (elimination of activated MLKL from cell membranes by shedding of necroptotic bodies). PELI1 reportedly serves both activating (Lys115 ubiquitination of RIPK1, promoting its incorporation into the necrosome) and inhibitory (induction of proteasomal degradation of RIPK3 by Lys363 ubiquitination) functions in necroptosis. ADAM proteases have recently been reported as additional players in necroptosis, yet their exact role still needs to be determined. For clarity of presentation, not all known effectors and regulators of necroptosis are shown. For further details, see the main text. IAPs, inhibitor of apoptosis proteins; PsKD, pseudokinase domain.

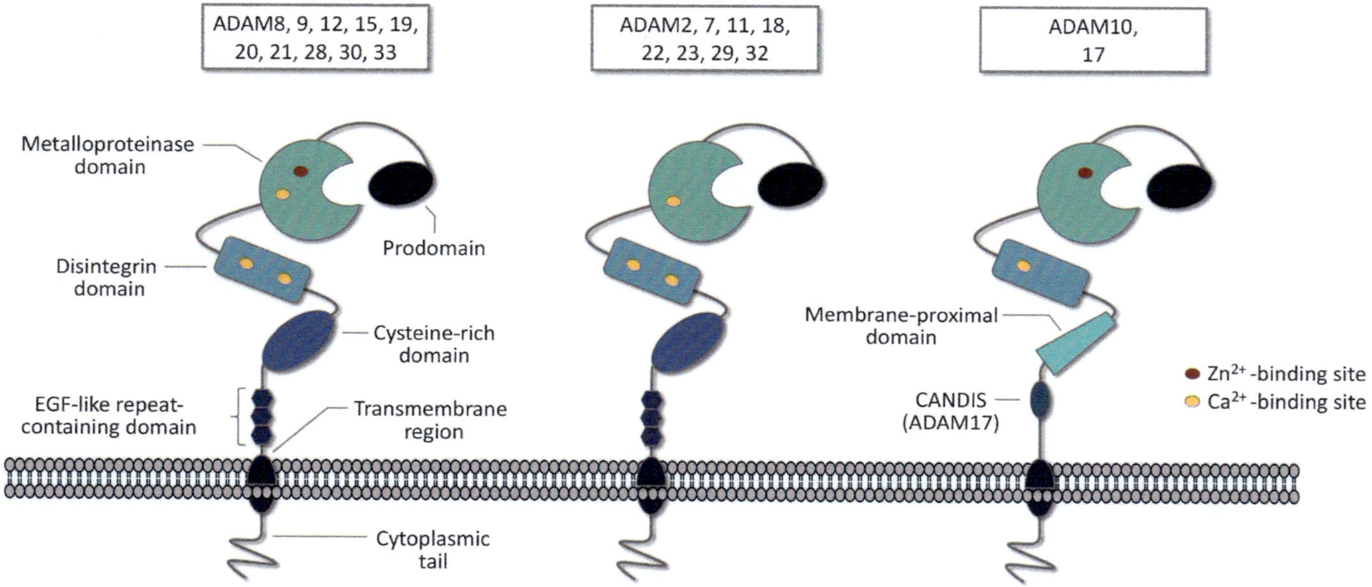

Fig. 2 Structure of ADAM proteases. ADAMs consist of an extracellular portion which encompasses a prodomain, a metalloproteinase domain, a disintegrin domain, a cysteine-rich domain and an EGF-like repeat-containing domain, followed by a transmembrane region and a cytoplasmic tail. The protease-inactive family members ADAM2, 7, 11, 18, 22, 23, 29 and 32 lack the Zn^{2+}-binding site in the metalloproteinase domain. In ADAM10 and ADAM17, only one Ca^{2+}-binding site is preserved. Moreover, ADAM10 and ADAM17 lack the cysteine-rich domain and the EGF-like repeat-containing domain, but contain a distinct membrane-proximal domain. ADAM17 additionally contains the CANDIS, which mediates the interaction of ADAM17 with the IL-6R.

32 (Takeda, 2016), which instead function as scaffolding proteins or via direct cell-cell signaling (Edwards et al., 2008; Hsia et al., 2019). The disintegrin-domain of ADAMs can influence cell adhesion and cell-cell interactions in certain settings by interacting with integrins (Edwards et al., 2008; White, 2003), and has also been implicated in substrate recognition and/or (hetero)-dimerization in concert with the EGF-like repeat-containing domains (Scheller et al., 2011a). The cysteine-rich domain contains a hypervariable region, which is important for substrate recognition and the regulation of catalytic activity (Jones et al., 2016). This region may be also be involved in regulating protein-protein interactions important for ADAM activity, such as interaction with syndecans (Iba et al., 2000). Finally, the cytoplasmic tail of ADAMs has been described as an interaction partner for a multitude of intracellular signal transducing proteins (Edwards et al., 2008). However, this region is dispensable for the sheddase activity of ADAM17 or ADAM19 induced by phorbol 12-myristate 13-acetate or other stimuli (Reddy et al., 2000; Schwarz et al., 2013; Wakatsuki et al., 2004) and thus may not be universally required for ADAMs to exert their functions (see below).

2.2 Regulation of ADAM activity

ADAM proteases are activated by a multitude of stimuli, such as ligands for receptor tyrosine kinases, TLR or G protein-coupled receptors, bacteria, lipopolysaccharide, bacterial toxins, Ca^{2+}-flux, phorbol esters, oxidative stress, ROS, physical or osmotic stress, cholesterol depletion, retinoic acid, sirtuins or interleukins (Edwards et al., 2008; Jones et al., 2016; Weber and Saftig, 2012). To prevent damage by inadvertent activation, the sheddase activity of ADAM proteins is strictly regulated at multiple levels. Even before their appearance at the cell surface, the activity of ADAMs can be restricted by intracellular retention. The endoplasmic reticulum (ER)-resident proteins iRhom1 and iRhom2 have been found to control the trafficking of immature ADAM17 from the ER to the Golgi network. Loss of iRhom function causes retention of ADAM17 in the ER, thereby preventing any extracellular proteolytic activity (Adrain et al., 2012; Christova et al., 2013; Maretzky et al., 2013; McIlwain et al., 2012). Furthermore, phosphorylation of iRhoms seems to be involved in the regulation of ADAM17 activation (Cavadas et al., 2017; Grieve et al., 2017). Recently, 4.1 protein ezrin radixin moesin domain-containing protein 8/iRhom tail-associated protein was identified as a protein that binds to iRhoms and prevents endolysosomal degradation of iRhoms and

ADAM17 (Künzel et al., 2018; Oikonomidi et al., 2018). In case of ADAM10, ER retention is mediated by a yet unidentified mechanism via an arginine-rich protein motif within the cytoplasmic tail, and counteracted by the association of ADAM10 with tetraspanins (a family of scaffolding proteins that group specific cell surface proteins into one area of the cell membrane, thereby modulating the formation and stability of signaling complexes; Hemler, 2005) and synapse-associated protein 97 (Marcello et al., 2010; Prox et al., 2012; Saraceno et al., 2014). Notably, different members of the tetraspanin family seem to differentially regulate the localization, surface export, activity and possibly even the substrate preference of ADAM10. Based on these observations, it was provocatively suggested that ADAM10 actually represents six different proteases, depending on its association with the respective tetraspanin family member (Reyat et al., 2017).

Once properly exposed at the cell surface, ADAM protease activity can be modulated by additional mechanisms such as reinternalization and degradation (Doedens and Black, 2000; Lorenzen et al., 2016), shedding of ADAMs by other ADAMs (Scharfenberg et al., 2020) or unrelated proteases (Tousseyn et al., 2009), phosphorylation of the cytoplasmic tail, protein disulfide isomerase (PDI)-mediated alterations in the ADAM extracellular domain, interactions with endogenous tissue inhibitors of metalloproteinases, regulators of ADAM di- or oligomerization, and also by substrate recognition/interaction (Jones et al., 2016).

For ADAM17, it has recently been shown that its positively charged, membrane-proximal domain interacts with negatively charged phosphatidylserine (PS) components of the cell membrane to trigger its sheddase activity (Sommer et al., 2016). Although PS is normally localized at the inner leaflet of the cell membrane, activation of protein kinase C (PKC) isoforms and a rise in intracellular Ca_2^+ concentrations can induce PS-externalization. In consequence, PKC activators may stimulate the proteolytic activity of ADAM17 via PS exposure (Fischer et al., 2006; Lambrecht et al., 2018; Sommer et al., 2016), and it has been speculated that translocation of PS to the outer membrane leaflet might even represent the penultimate event common to all routes of ADAM17 activation (Sommer et al., 2016).

2.3 Substrate cleavage by ADAMs

ADAMs cleave cell surface molecules in close proximity to their transmembrane region, shedding the ectodomains into the extracellular environment. A residual extracellular stalk, the transmembrane region and the cytoplasmic portion of the ADAM substrate remain anchored to the plasma membrane. This residual fragment can be further cleaved by regulated intramembrane proteolysis and the released intracellular domains may act as transcription factors in the nucleus. As a prototypic example, ADAM10-mediated cleavage and subsequent regulated intramembrane proteolysis of Notch controls important aspects of cellular and organismal development (Edwards et al., 2008; Jones et al., 2016; Lambrecht et al., 2018; Weber and Saftig, 2012). Conversely, the shed ectodomains of ADAM substrates can modulate cell function and signaling through juxtacrine, autocrine, and paracrine pathways, as they are often derived from cytokines, chemokines, growth factors, cell adhesion molecules and cell surface receptors (Jones et al., 2016). The shed ectodomains of cell surface receptors may also act as inhibitory decoy receptors or, as in the case of the shed interleukin (IL)-6 receptor (IL-6R), exhibit agonistic properties via a process termed "trans-signaling" (Rose-John and Heinrich, 1994). Briefly, IL-6 trans-signaling describes a signaling pathway where cells that lack the IL-6R at their surface (and which therefore cannot respond to IL-6 directly) can nevertheless be stimulated by a complex of IL-6 and the soluble, shed IL-6R. This complex binds to the signaling receptor subunit gp130 which is ubiquitously expressed on every cell, induces the subsequent dimerization of gp130 and triggers intracellular signaling via the signal transducer and activator of transcription (STAT)3 and extracellular-signal regulated kinase (ERK) pathways (Garbers et al., 2015; Scheller et al., 2011b). ADAM-mediated ectodomain shedding can also attenuate signaling by depletion of receptors from the cell surface, and regulate cell-cell or cell-matrix interactions by cleavage of cell adhesion molecules. Interestingly, Scharfenberg et al. have recently reported ectodomain shedding of ADAM17 by ADAM8 and demonstrated that the soluble ectodomains of ADAM10 and ADAM17 retain proteolytic activity, however, with an altered substrate spectrum compared to their membrane-bound counterparts (Scharfenberg et al., 2020). Finally, ADAMs and their substrates have been found in exosomes and may thereby contribute to long distance signaling (Jones et al., 2016; Shimoda et al., 2014; Stoeck et al., 2006).

In contrast to, e.g., caspase substrates, ADAM substrates show no specific cleavage motif, which complicates predictions whether a protein will be cleaved by ADAMs or not (Jones et al., 2016; Lambrecht et al., 2018). This has resulted in a complex list of reported ADAM substrates (Camodeca et al., 2019; Edwards et al., 2008; Reiss and Saftig, 2009; Zunke and Rose-John, 2017). Predictions are further complicated by redundancy between different ADAM family members as well as altered substrate specificities of ADAMs in disease vs. normal physiological states or in constitutive vs. induced shedding (Jones et al., 2016). In consequence, efforts to develop specific, clinically suitable ADAM inhibitors have so far shown only limited success (Fridman et al., 2007; Murumkar et al., 2010; Zhou et al., 2006). Nevertheless, cleavage site profiling of the proteases ADAM10 and ADAM17 using peptide libraries has recently revealed preferences for certain amino acid residues and may provide a first basis for a better substrate identification in vivo and the generation of activity-based probes or specific inhibitors (Tucher et al., 2014).

2.4 ADAM17

ADAM17 represents the prototype member of the family of ADAM proteases. ADAM17 was originally cloned as TNF-α-converting enzyme (Black et al., 1997; Moss et al., 1997). Through its ability to liberate membrane-bound TNF as well as TNF-R1/2 from the cell surface, ADAM17 constitutes an important regulator of TNF-driven inflammatory responses (Edwards et al., 2008). Notably, it was found that deletion of ADAM17 leads to embryonic lethality (Peschon et al., 1998) whereas mice lacking TNF do not show any overt defects (Marino et al., 1997; Pasparakis et al., 1996). This provided a first indication that the functions of ADAM17 must extend beyond the shedding of TNF. Indeed, more than 80 proteins have been identified to date that are cleaved by ADAM17 (Arribas and Esselens, 2009; Scheller et al., 2011a; Zunke and Rose-John, 2017). These substrates encompass cytokines, chemokines, growth factors and their receptors as well as adhesion molecules (Arribas and Esselens, 2009; Scheller et al., 2011a; Zunke and Rose-John, 2017), explaining the involvement of ADAM17 in physiological and pathophysiological processes such as development, regeneration, immunity, inflammation or tumorigenesis (Bandsma et al., 2015; Black et al., 1997; Blaydon et al., 2011; Chalaris et al., 2010; Moss et al., 1997; Peschon et al., 1998). As prominent examples, the IL-6R, L-selectin and amyloid precursor protein (APP) have been identified as ADAM17 substrates (Bell et al., 2007; Le Gall et al., 2009). By cleaving IL-6R and generating the soluble IL-6R, ADAM17 regulates the IL-6 trans-signaling pathway (Yan et al., 2016), which is responsible for most of the pro-inflammatory activities of the cytokine IL-6 (Rose-John et al., 2017). Shedding of L-selectin, intercellular adhesion molecule-1 and vascular cell adhesion molecule-1 by ADAM17 enhances the vascular permeability and transmigration of leukocytes during inflammatory responses (Garton et al., 2006). In Alzheimer's disease patients, ADAM17 colocalizes with amyloid plaques (Skovronsky et al., 2001), where it may modulate the progress of the disease by producing a soluble and non-amyloidogenic fragment of APP (Allinson et al., 2003). It was noted already in the initial studies that the embryonic lethality of ADAM17-deficient mice resulted from defects in the epithelial maturation of multiple organs, accompanied by eye, hair and skin abnormalities, and that these defects resembled those seen in mice lacking the EGF receptor (EGFR) or its ligands heparin-binding-EGF, amphiregulin and TGF-α (Horiuchi et al., 2005; Peschon et al., 1998). Indeed, ADAM17 turned out to be the major sheddase for many ligands of the EGFR (Peschon et al., 1998; Sahin et al., 2004), thus critically contributing to EGFR-activation and controlling key processes such as cell growth and survival (Blobel, 2005).

As shown in Fig. 2, ADAM17 shares a common structure with other ADAM family members but lacks the cysteine-rich domain and the EGF-like repeat-containing domain (Düsterhöft et al., 2013; Janes et al., 2005; Takeda, 2009). Instead, ADAM17 contains a distinct membrane-proximal domain and a "conserved ADAM17 dynamic interaction sequence" (CANDIS), which mediates the interaction of ADAM17 with the IL-6R, but not TNF, revealing differences in the substrate recognition mechanism of ADAM17 (Düsterhöft et al., 2014). Moreover, the adjacent membrane-proximal domain, which interacts with PS (Sommer et al., 2016), is negatively regulated by the enzyme PDI. PDI changes the structure of the membrane-proximal domain from an open to a closed conformation, thereby restricting the access of PS and thus the activation of ADAM17 (Düsterhöft et al., 2013). The cytoplasmic tail of ADAM17 is subject to phosphorylation by mitogen-activated protein kinases (MAPK) (Killock and Ivetić, 2010; Soond et al., 2005; Xu et al., 2012; Xu and Derynck, 2010) and polo-like kinase 2 (Schwarz et al., 2014). However, since the cytoplasmic tail of ADAM17 is dispensable for its shedding activity (Reddy et al., 2000; Schwarz et al., 2013), the relevance of these phosphorylation events for the regulation of ADAM17 is not fully understood. Importantly, phosphorylation of iRhom proteins enhances the response of ADAM17 to phosphorylation-inducing stimuli (Cavadas et al., 2017; Grieve et al., 2017), which might explain the dispensability of the ADAM17 cytoplasmic tail for the inducibility of shedding activity.

Remarkably, loss of ADAM17 function is compatible with life in humans. To date, three independent studies have reported patients with

ADAM17 deficiency (Bandsma et al., 2015; Blaydon et al., 2011; Tsukerman et al., 2015). In the first study, two siblings were described that suffered from a deletion of 4 base pairs in exon 5 of ADAM17 (c.603-606delCAGA). This deletion was predicted to introduce a frame shift and a premature stop codon (p.Asp201GlufsX11), leading to a severely truncated, enzymatically inactive protein, which consists only of the signal peptide and the prodomain of ADAM17, but lacks all functional domains, including the catalytic domain required for the sheddase function of ADAM17 (Blaydon et al., 2011). In the second study, a patient carried a distinct mutation in ADAM17 (c.308dupA) predicted to cause a frameshift and a premature stop codon (p.Asn103LysfsTer20) and expected to likewise result in the expression of a severely truncated, inactive ADAM17 protein, or, as a consequence of nonsense-mediated mRNA decay, in a complete lack of ADAM17 (Bandsma et al., 2015). The third study identified a patient with a deletion covering exon 1 of ADAM17, resulting in the complete absence of ADAM17 protein in Western blots (Tsukerman et al., 2015).

Notably, although not embryonic lethal, loss of ADAM17 has nevertheless severe consequences for the afflicted patients. In the study of Blaydon et al., both siblings developed skin lesions on the second day of life. When they were young, their skin was prone to infection with *Staphylococcus aureus*, leading to recurrent blepharitis and otitis externa. Their eyelashes and eyebrows were wiry and disorganized, they displayed short or broken hair and thickened nails, with frequent paronychia caused by *Candida* and *Pseudomonas* infections (Blaydon et al., 2011). Moreover, the siblings developed diarrhea within the first week of life, which worsened in parallel with increases in the severity of the skin disease and which was exacerbated by intercurrent gastrointestinal infections (Blaydon et al., 2011). At the molecular level, both children displayed a marked reduction in the expression of truncated, non-functional ADAM17 in the skin and small intestine. As a likely consequence of the ADAM17 mutation, TNF production in peripheral blood mononuclear cells (PBMCs) of the affected boy was impaired, presumably being responsible for the enhanced susceptibility of the skin to opportunistic infections (Blaydon et al., 2011). The affected girl died at 12 years of age from parvovirus-associated myocarditis, and the affected boy was found to have left ventricular dilatation. The authors suggested that an impaired regulation of desmoglein 2 by ADAM17 in the two siblings may have been responsible for these cardiac manifestations, and that the death of the girl may have been partly caused by a lack of cardioprotective TNF during acute myocarditis (Blaydon et al., 2011). In the study by Bandsma et al., the affected girl also presented with episodes of severe diarrhea, skin rash, and recurrent sepsis, which eventually resulted in her death at the age of 10 months (Bandsma et al., 2015). In this patient, production of TNF and IL-2 by T cells was impaired, and skin biopsies revealed focal neutrophilic infiltrates and spongiotic dermatitis. Moreover, the patient developed unexplained systolic hypertension and non-specific hepatitis with apoptosis (Bandsma et al., 2015). In their study, the authors suggested that the observed immunological defects were a likely cause for the increased susceptibility to severe bacterial infections and for the enhanced inflammatory state of this patient (Bandsma et al., 2015). The third study by Tsukerman et al. reported a patient who died of respiratory failure and presumed sepsis at 2 years of age (Tsukerman et al., 2015). Similar to the first two studies, the patient displayed inflammatory skin lesions and diarrhea and suffered from recurrent infections beginning at 1 month of age. Accordingly, PBMCs from the patient showed an impaired secretion of most of the cytokines tested, and TNF was barely secreted by the patient (Tsukerman et al., 2015). Notably, the natural killer cells of the patient were more potently activated by antibodies against CD16, an activating receptor on the surface of natural killer cells, which is cleaved and negatively regulated by ADAM17. Nevertheless, the overall antibody-dependent cell-mediated cytotoxicity of the natural killer cells from the patient was not different from that of natural killer cells from healthy controls (Tsukerman et al., 2015).

When comparing the phenotypes of the above patients to the phenotypes of ADAM17-deficient mouse models, the most striking difference is the embryonic lethality in the mouse, whereas humans can survive well beyond birth. Blaydon et al. suggested that humans may have mechanisms that compensate for the lack of ADAM17 (Blaydon et al., 2011). However, facing the early death of their patient, Bandsma et al. alternatively proposed that disease severity in humans is variable and can nevertheless lead to very early mortality (Bandsma et al., 2015). As another point to take note, even in the initial mouse studies, not all mice with a full knockout of ADAM17 died before or soon after birth, but a few animals survived for several weeks (Peschon et al., 1998). Therefore, it may well be possible that loss of ADAM17 in humans is likewise embryonic lethal in most cases, but simply not detected due to lack of a corresponding screening, and that the few ADAM17-deficient patients described so far represent the surviving minority of cases.

In the initial mouse studies, the surviving ADAM17-deficient animals had eye degeneration, altered hair, and impaired epithelial-cell maturation in multiple organs, including the intestine (Peschon et al., 1998). Particularly with regard to the hair, similarities exist to the patients described by Bandsma et al. (2015) and Blaydon et al. (2011). Furthermore, although the three existing studies vary in the extent and scope of the performed investigations, they all consistently show that ADAM17-deficient patients suffer from episodes of severe inflammatory skin and bowel disease and from recurrent infections (Bandsma et al., 2015; Blaydon et al., 2011; Tsukerman et al., 2015). With regard to the intestine, Blaydon et al. reported that both patients had early-onset diarrhea and speculated that this might have a similar origin as the diarrhea in mice lacking ADAM17 (Blaydon et al., 2011). Nevertheless, the intestines of the two patients appeared substantially normal later in life, different from the hypoplastic crypts in the small bowel and the very low rate of epithelial production in mice lacking ADAM17 (Blaydon et al., 2011). In their study, Blaydon et al. attributed these differences to the presence of compensatory mechanisms in humans or to tethered ligand sufficient to maintain the intestinal epithelial renewal in the absence of ADAM17 (Blaydon et al., 2011). Moreover, and consistent with the initial reports from ADAM17-deficient mice (Peschon et al., 1998), all three studies showed a diminished extracellular production of TNF upon stimulation of patient cells (Bandsma et al., 2015; Blaydon et al., 2011; Tsukerman et al., 2015), and Blaydon et al. postulated that this scant production of TNF may have been a determinant of the increased susceptibility of the patients to opportunistic infections of the skin (Blaydon et al., 2011). Overall, the phenotype of ADAM17-deficient humans shows significant overlaps with the defects observed in the initial studies using mice carrying a full ADAM17 knockout (Peschon et al., 1998). As outlined below, this was later confirmed in more refined mouse models that do not cause embryonic lethality, underscoring the important role of ADAM17 for epithelial cell homeostasis in both mice and humans.

To avoid the lethality of the full ADAM17 knockout, conditional knockout mice were generated which revealed functions of ADAM17 in bone formation (Horiuchi et al., 2009), TNF release (Horiuchi et al., 2007), in the regulation of skin inflammation (Franzke et al., 2012) and in the protection of hepatocytes from apoptosis (Murthy et al., 2010). In an alternative approach, hypomorphic mice (ADAM17$^{ex/ex}$ mice) were produced which expressed only ~5% of the original levels of ADAM17 in all tissues and which displayed severely reduced shedding of ADAM17 substrates (Chalaris et al., 2010). Due to the residual ADAM17 activity, these animals were viable and helped to uncover protective functions of ADAM17 in *Listeria* infection (Yan et al., 2016) as well as in atherosclerosis

(Nicolaou et al., 2017). Notably, ADAM17$^{ex/ex}$ mice showed eye, hair, and skin defects comparable to those in TGF-α-deficient mice, and milk duct formation in the breast of female mice was severely impaired. Since milk duct formation depends on the EGFR ligand amphiregulin, this pointed to a general impairment of EGFR signaling in ADAM17$^{ex/ex}$ mice (Chalaris et al., 2010), in line with the findings in the full ADAM17 knockout models (Peschon et al., 1998; Sahin et al., 2004). Moreover, the compromised shedding of EGFR ligands in ADAM17$^{ex/ex}$ mice causes a greatly enhanced susceptibility in IBD models (Chalaris et al., 2010). The role of ADAM17 in IBD will be discussed in further detail below.

2.5 Other ADAM family members

Here, we present a brief overview of the other human ADAM family members. For more comprehensive discussions of their properties and functions, please see (Edwards et al., 2008; Hsia et al., 2019).

ADAM2, 7, 18, 20, 21, 29 and 30 are important for spermatogenesis and sperm function, consistent with their expression primarily in the testis (Edwards et al., 2008).

ADAM8 was first identified in hematopoietic cell types (Yoshida et al., 1990). This protease may have a central role in the brain, as it modulates neuroinflammation (Bartsch et al., 2010) and is expressed in cell types such as neurons, microglia and astrocytes (Hsia et al., 2019). As an exception among ADAM proteases, ADAM8 does not depend on furin cleavage for activation, but rather removes its prodomain autocatalytically (Schlomann et al., 2002).

ADAM9 and ADAM15 are expressed broadly in somatic tissues (Edwards et al., 2008) and have been linked to heart development (Horiuchi et al., 2005), bone formation (Marzia et al., 2011), retinopathy (Parry et al., 2009) and inflammation (Dreymueller et al., 2015). Notably, ADAM9 and ADAM15 may act as regulators of ADAM10 activity by shedding its ectodomain and thereby inactivating ADAM10 (Tousseyn et al., 2009). Furthermore, an alternatively spliced variant of ADAM9 can promote tumor invasion by cleavage of the extracellular matrix protein laminin (Mazzocca et al., 2005). A possible role of ADAM9 in necroptosis (Cai et al., 2016) will be discussed further below.

ADAM10 is—together with ADAM17—the most intensively studied ADAM protease. As the major Notch-cleaving enzyme (Hartmann et al., 2002), ADAM10 controls central developmental processes, e.g., in adult crypt homeostasis in the intestine (Tsai et al., 2014). ADAM10 has major functions in the nervous system, where it is important for proper axon targeting and neurite growth as well as synapse formation and maintenance (Hsia et al., 2019; Kuhn et al., 2016). ADAM10 is particularly important in the brain, where it cleaves APP (Lammich et al., 1999; Postina et al., 2004) and the prion protein (Altmeppen et al., 2015), thereby affecting the course of Alzheimer's and prion diseases (Hsia et al., 2019). ADAM10 has also been implicated in the progression of several other human diseases such as chronic liver inflammation (Müller et al., 2016), systemic lupus erythematosus (Hoffmann et al., 2015; Orme et al., 2016), rheumatoid arthritis (Li et al., 2015a), atherosclerosis (van der Vorst et al., 2015) as well as cancer entities such as glioblastoma (Kohutek et al., 2009; Yang et al., 2011), T cell acute lymphoblastic leukemia (Sulis et al., 2011), squamous cell carcinoma (Jones et al., 2013) and breast cancer (Mullooly et al., 2015). Like ADAM9, ADAM10 has recently been linked to necroptosis (Cai et al., 2016), see below.

ADAM11, ADAM22 and ADAM23 are protease-inactive members of the ADAM family, which exert their functions through protein-protein interactions (Hsia et al., 2019). They are primarily expressed in the brain (Sagane et al., 1999), where they have been implicated in myelination, synapse maturation and synaptic transmission as well as in epilepsy (Hsia et al., 2019).

ADAM12 is a less well-explored ADAM family member, which has originally been linked to muscle formation (Yagami-Hiromasa et al., 1995). However, results from overexpression studies (Moghadaszadeh et al., 2003) and knockout mice (Edwards et al., 2008; Sahin et al., 2004) have not corroborated this assumption. ADAM12 may play a role in adipocyte proliferation and differentiation (Masaki et al., 2005) and may have relevance for the pathology of Alzheimer's disease (Hsia et al., 2019). However, given that even ADAM9/ADAM12/ADAM15 triple-deficient mice reveal no major evident deficiencies (Sahin et al., 2004), the precise physiological role of ADAM12 has so far remained elusive.

ADAM19 is essential for cardiovascular morphogenesis (Zhou et al., 2004) and important for regeneration processes in the peripheral nerve system (Wakatsuki et al., 2009). Although shedding of neuregulins has been observed upon overexpression of ADAM19 (Shirakabe et al., 2001), this was not confirmed in vivo (Zhou et al., 2004), and it remains unclear to which extent ADAM19 participates in neuregulin-regulated biological processes. ADAM19 is able to cleave canonical ADAM substrates such as TNF (Chesneau et al., 2003) and APP (Tanabe et al., 2007), however, much less efficiently as compared to other ADAMs (Tanabe et al., 2007). Therefore, cleavage of these substrates by ADAM19 may become relevant only under conditions when ADAM19 is highly expressed (Hsia et al., 2019). Notably, ADAM19 expression is elevated in IBD (Franzè et al., 2013).

ADAM21 is expressed in brain (Yang et al., 2005), in addition to its above-mentioned expression in testis. In consequence, roles for this protease have been postulated in axonal growth and neurogenesis (Yang et al., 2005), and independently, in hepatocellular carcinoma (Honda et al., 2017).

ADAM22 lacks protease activity and acts as a scaffolding protein. In mice, deletion of ADAM22 causes hypomyelination of peripheral nerves and ataxia (Sagane et al., 2005). Mutants of leucine-rich glioma-inactivated protein 1 that cannot interact with ADAM22 anymore cause autosomal dominant lateral temporal epilepsy (Fukata et al., 2006), suggesting a potential link of ADAM22 to this form of epilepsy (Snyder, 2006). However, this was questioned in another study (Chabrol et al., 2007), so that the exact role of ADAM22 in neuronal functions still awaits clarification.

ADAM28 has been implicated in the proliferation and metastasis of diverse carcinomas (Mitsui et al., 2006; Ohtsuka et al., 2006; Rudnicka et al., 2016; Zhang et al., 2019d). Yet, a protective function of ADAM28 against metastasis was also reported recently (Gérard et al., 2018), suggesting that the role of ADAM28 in cancer is more complex than previously anticipated. Independently, ADAM28 was recently ascribed a role in metabolic dysfunction (Herat et al., 2017).

ADAM30 was initially found to be expressed in testis (Cerretti et al., 1999) but later also in neurons and the human brain. There, ADAM30 may counteract the pathogenesis of Alzheimer's disease by activating the protease cathepsin D and thus decreasing the amount of Aβ peptide (Letronne et al., 2016).

ADAM32 is again predominantly expressed in testes in mice (Choi et al., 2003) and humans (Fagerberg et al., 2014). A recent study has identified ADAM32 as a marker for cancer-associated fibroblasts which contribute to the progression of hepatocellular carcinoma (Zou et al., 2018).

Finally, ADAM33 has been associated with asthma by genetic linkage analyses (Tripathi et al., 2014; Van Eerdewegh et al., 2002).

3. Intestinal (dys)function

As one of the main barrier organs, the intestine is permanently exposed to a wealth of antigens and potential pathogens that are present in the food or the microbial flora (O'Hara and Shanahan, 2006). To provide an efficient defense against these pathogens while sustaining tolerance to diet and commensal microbes, the intestine must prevent an excessive activation of immune cells. This task is mainly accomplished by a layer of intestinal epithelial cells (IECs), which—aside from mediating electrolyte passage and nutrient absorption—provides a tight physical barrier and thus helps to keep the cells of the immune system separate from the microbiota (Negroni et al., 2015; van der Flier and Clevers, 2009). This layer of IECs is the second largest mucosal surface in the body, covering an area of $32\,m^2$ (Helander and Fändriks, 2014). As depicted in Fig. 3, the intestinal epithelium is organized into villi, which are present in the small intestine, but not in the colon, and into crypts (van der Flier and Clevers, 2009). At the bottom of the crypts, proliferating stem cells constantly generate new IECs (Crosnier et al., 2006) which differentiate into absorptive enterocytes or into secretory IECs (Dagenais et al., 2014) while moving upward to the intestinal surface (van der Flier and Clevers, 2009). IECs have a lifetime of 4–5 days, making the intestinal epithelium the most vigorously self-renewing tissue of adult individuals (Crosnier et al., 2006; van der Flier and Clevers, 2009). Once having reached the luminal surface, the IECs die and are shed from the epithelial cell layer (Bullen et al., 2006).

Several subtypes of secretory IECs provide highly specialized innate immune functions to sustain intestinal homeostasis. Paneth cells secrete antimicrobial peptides (Bevins and Salzman, 2011) to spatially regulate the colonization and penetration of the epithelium by luminal microbes and to actively shape the composition of the intestinal microbiota (Günther et al., 2014; Jones et al., 2016; Maloy and Powrie, 2011). Unlike other secretory IECs, Paneth cells have a longer lifespan (up to 57 days) (Ireland et al., 2005) and remigrate downward to their residence at the crypt base (Bjerknes and Cheng, 1981). Goblet cells secrete mucins which form a viscous mucous layer on the gut wall (Specian and Oliver, 1991), thereby restricting the access of microbes to the epithelial surface (Bevins and Salzman, 2011). Here, cleavage of mucin 2 by the protease meprin β is required for proper mucus detachment and barrier function (Schütte et al., 2014; Wichert et al., 2017). Enteroendocrine cells coordinate the function of the gut by secreting specific peptide hormones in a paracrine manner or into the bloodstream (Gribble and Reimann, 2016). Tuft cells secrete an uncommon set of biological effector molecules such as allergy-associated eicosanoids and IL-25, and seem to be a major component of the immune defense of the small intestine against parasitic helminths and protists (Gerbe et al., 2016;

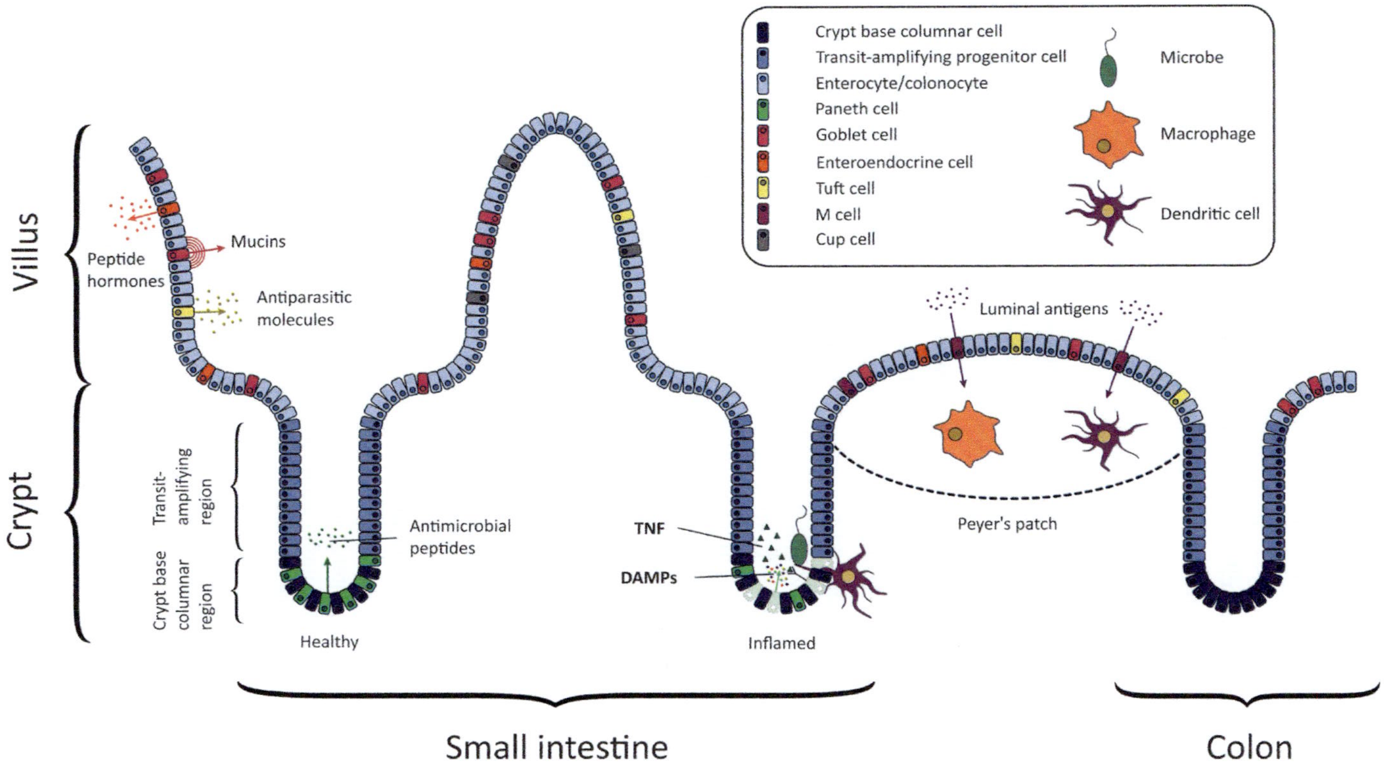

Fig. 3 See figure legend on next page

Howitt et al., 2016; von Moltke et al., 2016). As a distinct IEC subtype, M (microfold) cells are found in epithelial areas that overlie mucosal lymphoid tissues such as Peyer's patches. Their function in mucosal immune surveillance consists in sampling luminal contents and in presenting antigens to resident immune cells (Dillon and Lo, 2019). Finally, cup cells constitute another, not well understood type of IEC in the small intestine (Dagenais et al., 2014). In the colon, IECs are mainly restricted to goblet cells and enterocytes, which there are termed colonocytes (Dagenais et al., 2014; de Santa Barbara et al., 2003).

When considering the remarkable self-renewing potential of the intestine, it is not surprising that a strict balance between cell proliferation and cell death must be kept to maintain the intestinal barrier (Negroni et al., 2015). Physiologically, cell death is rarely observed in the intestinal epithelium (Edelblum et al., 2006). Aged epithelial cells are replaced at the tip of the villus by neighboring cells (Watson et al., 2009; Yuan and Kroemer, 2010) in a process termed homeostatic cell shedding (Becker et al., 2013). Under steady-state conditions, it appears that apoptosis is not essential for the death of IECs after homeostatic cell shedding and that the turnover of IECs may also involve necroptotic cell death (Blander, 2018). Regardless of the exact mode, dysregulation of IEC death can result in severe pathologies. The excessive death of IECs is thought to cause barrier disruption, and the ensuing aberrant stimulation of the intestinal immune system by the entering microbial flora and noxious agents has been held responsible for

Fig. 3 Overview of intestinal (dys)function. In the small intestine, multipotent crypt base columnar stem cells are located at the bottom of the crypts. They give rise to transit-amplifying progenitor cells which rapidly migrate upward from the crypt to the villus junction. Before exiting the crypts, these cells differentiate into absorptive enterocytes and secretory IECs, i.e., Paneth cells, which secrete antimicrobial peptides; goblet cells, which secrete mucins; enteroendocrine cells, which secrete peptide hormones; and tuft cells which secrete antiparasitic molecules such as eicosanoids and IL-25. Unlike other secretory IECs, Paneth cells remigrate downward to the crypt base where they reside. M cells are generally only found in epithelial areas that overlie Peyer's patches or other mucosal lymphoid tissues and present luminal antigens to resident immune cells such as dendritic cells or macrophages. The colon differs from the small intestine in that it lacks villi and that it consists mainly of two cell types: absorptive enterocytes (termed colonocytes) and goblet cells. In the inflamed intestine, mediators such as TNF induce an excessive death of IECs. The subsequent loss of barrier integrity, together with an entering microbial flora and the release of pro-inflammatory DAMPs from necroptotic cells promotes an aberrant stimulation of the immune system and intestinal inflammation.

the development of IBD (Maloy and Powrie, 2011; Negroni et al., 2015). Notably, TNF plays a major role in this process, and the successful application of anti-TNF therapy constituted a major breakthrough in the treatment of IBD (Pouillon et al., 2016).

3.1 Inflammatory bowel disease

The term IBD is most often used for two separate chronic relapsing inflammatory disorders of the gastrointestinal tract, Crohn's disease (Harris et al., 1933) and ulcerative colitis (Wilks, 1859). Crohn's disease can affect all segments of the gastrointestinal tract (Torres et al., 2017), whereas ulcerative colitis is restricted to the colonic mucosa (Ungaro et al., 2017). The exact etiology of IBD is unknown, although it is generally assumed that the co-operative action of environmental factors, an altered gut microbiota and genetic susceptibility culminates in a compromised epithelial barrier function and an abnormal mucosal immune response (Torres et al., 2017; Ungaro et al., 2017). To date, more than 200 genetic loci have been associated with IBD (Jostins et al., 2012; Liu et al., 2015a), which affect epithelial barrier function and integrity (Dagenais et al., 2014), bacterial sensing and innate immunity, autophagy and the homeostasis of the mucus layer (Torres et al., 2017). Yet, it should be kept in mind that genetic variations in these loci explain less than 25% of disease heritability, which underscores the importance of epigenetic and non-genetic environmental factors (Vedamurthy and Ananthakrishnan, 2019).

3.2 Colorectal cancer

Patients with IBD, and especially those with ulcerative colitis, have a greatly increased risk of developing colorectal cancer (CRC) (Waldner and Neurath, 2009). CRC is one of the most common cancer types and a leading cause of cancer-related deaths worldwide (Chen et al., 2016; Siegel et al., 2017). In addition to chronic intestinal inflammation (Waldner and Neurath, 2009), the overexpression and increased activity of the EGFR has been recognized as a risk factor for CRC (Sibilia et al., 2007). The development of CRC is initiated in intestinal epithelial stem cells at the bottom of the crypt by the loss of tumor suppressor genes such as adenomatous-polyposis-coli (APC), first leading to the formation of long-lived, benign adenomas. Subsequently, the progressive accumulation of mutations in oncogenes such as *Kirsten rat sarcoma, p53* and *Smad4* ultimately results in metastatic CRC (Fearon and Vogelstein, 1990; Kinzler and Vogelstein, 1996).

4. Necroptosis and intestinal (dys)function

4.1 Necroptosis and inflammatory bowel disease

In the healthy intestine, necroptosis has been linked to physiologic IEC death in the course of homeostatic cell shedding (Blander, 2018). In the pathologic situation, several studies in mouse models have implicated IEC necroptosis as a mechanism that contributes to intestinal inflammation and thus to intestinal dysfunction and IBD (Dannappel et al., 2014; Günther et al., 2011; Takahashi et al., 2014; Weinlich et al., 2013; Welz et al., 2011). An IEC-specific knockout of caspase-8 as a negative regulator of necroptosis resulted in spontaneous inflammatory lesions in the terminal ileum and was characterized by extensive TNF-dependent epithelial necroptosis, Paneth cell loss, and high susceptibility of mice to dextran sulfate sodium (DSS)-induced colitis (Günther et al., 2011). In an independent study, mice with an IEC-specific deletion of caspase-8 showed inflammation that extended to both the small intestine and the colon (Weinlich et al., 2013). Essentially the same results were observed in mice carrying an IEC-specific deletion of FADD (Welz et al., 2011), RIPK1 (Takahashi et al., 2014) or of FADD together with RIPK1 (Dannappel et al., 2014). The additional deletion of the necroptotic core protein RIPK3 or of the necroptosis regulator CYLD in mice with an IEC-specific deletion of FADD (Welz et al., 2011) or deletion of RIPK3 in mice with an IEC-specific deletion of caspase-8 (Weinlich et al., 2013) prevented IEC death and inflammation, arguing in favor of necroptosis as the underlying cause.

As another line of evidence, we and others have recently shown that phosphorylation of RIPK3 and MLKL as an indicator for necroptosis occurs in the colons of mice during DSS-induced colitis (Fuchslocher Chico et al., 2018; Zhang et al., 2019b). Along these lines, necrostatin-1, an inhibitor of RIPK1, attenuated DSS-induced colitis in mice (Liu et al., 2015c; Zhang et al., 2019a), and necrosulfonamide, an inhibitor of MLKL, protected in a murine colitis model triggered by 2,4,6-trinitrobenzenesulfonic acid (TNBS) (Dong et al., 2017). However, these latter two studies should be interpreted with some caution, since necrostatin-1 also non-specifically inhibits indoleamine-2,3-dioxygenase, an enzyme which modulates inflammation (Vandenabeele et al., 2013), and since the in vivo occurrence of necroptosis was not conclusively demonstrated in the TNBS model.

As an additional point to consider, the knockout models used in the above studies lead to an artificial overactivation of necroptosis and may not accurately reflect the conditions that occur in IBD patients. Also, at least in DSS-induced colitis models, we and others have found that defective necroptotic signaling in ADAM17$^{ex/ex}$ (Fuchslocher Chico et al., 2018) or RIPK3-deficient mice (Newton et al., 2016a) has no protective impact on the intestinal damage caused by DSS, questioning a critical contribution of necroptosis. In two additional studies, deletion of RIPK3 even exacerbated DSS-induced colitis (Moriwaki et al., 2014; Xu et al., 2017). Notably, this was attributed to novel, necroptosis-independent functions by which RIPK3 controls inflammatory responses (Moriwaki and Chan, 2017; Newton and Manning, 2016; Xu et al., 2017). Therefore, knockout or inhibition of RIPK3 should no longer be considered as sufficient to conclusively demonstrate an involvement of necroptosis.

Alternatively employing MLKL-deficient mice, Zhang et al. have reported that deletion of MLKL leads to a significant improvement with regard to survival rate, clinical score, intestinal damage and intestinal mucosal barrier integrity in DSS-induced colitis models (Zhang et al., 2019c), consistent with an independent study (Zhang et al., 2019a). However, IEC death after DSS administration was not prevented in MLKL-deficient mice. Rather, the authors suggested that the protective effects of MLKL deficiency are due to an inhibition of pro-inflammatory signaling by MAPK (Zhang et al., 2019c), which would argue against a role of necroptosis in IBD. Adding further complexity to this picture, a third study has recently reported that MLKL knockout mice are not resistant, but rather highly susceptible to DSS-induced colitis, showing increased rather than reduced inflammatory responses (Zhao et al., 2019). This study identified hyperactivation of MAPK kinase/ERK as the underlying cause while Zhang et al. had rather observed a reduced phosphorylation/activation of ERK and other MAPKs in their experiments (Zhang et al., 2019c). The reasons for these contradicting results are currently unclear, especially since both groups used similar experimental conditions to induce colitis. As a possible explanation, microbiota differences are known to influence the response of mouse models of intestinal inflammation and may account for the observed differences.

In an independent approach, necroptosis was recently described to cause the death of IECs in mice carrying a specific deletion of the protein autophagy related 16 like 1 (ATG16L1) by triggering IL-22-mediated signaling through the cGAS-STING pathway (Aden et al., 2018). In concordance, a related study demonstrated that the presence of ATG16L1 prevents necroptosis in organoids of the murine intestinal epithelium (Matsuzawa-Ishimoto et al., 2017). Additional evidence in favor of a role of necroptosis in IBD was recently obtained using LY3009120, a pan-rat fibrosarcoma

(RAF) inhibitor. LY3009120 inhibited necroptosis in vitro and significantly ameliorated DSS-induced colitis in vivo by protecting IECs from necroptosis in a RAF-independent manner, decreasing DSS-induced colonic inflammation and preventing intestinal barrier function loss (Zhang et al., 2019b). Independently, HtrA2/Omi (a protease that we have previously identified as a regulator of necroptosis; Sosna et al., 2013, 2016) was isolated in a proteomic screen for proteins that are dysregulated in the inflamed colon. Subsequent pharmacologic inhibition of HtrA2/Omi led to a decreased necroptotic phosphorylation of MLKL and ameliorated DSS-induced necroptotic IEC death, intestinal barrier disruption and colonic inflammation (Zhang et al., 2019a). In summary, the above studies have provided some evidence for an in vivo role of necroptosis in IBD in the mouse; however, the results are not yet fully coherent and require further validation.

Similarly, definite evidence that necroptosis indeed participates in human IBD remains limited. A reduced expression of caspase-8 and a concomitantly increased expression of RIPK3 and MLKL was observed in the inflamed tissues of pediatric patients with Crohn's disease and ulcerative colitis (Pierdomenico et al., 2014). In biopsies from adult patients with Crohn's disease, elevated RIPK3 (Günther et al., 2011) and $FLIP_L$ (Caprioli et al., 2008) expression was observed, and bioptic inflamed ileal and colonic tissues from patients with Crohn's disease and ulcerative colitis showed a significant increase in phosphorylated MLKL (Negroni et al., 2017), collectively leading to the suggestion that necroptosis is associated with intestinal inflammation and thereby contributes to the pathogenesis of human IBD. However, the evidence underlying this conclusion is mostly correlative. Moreover, it is not clear yet whether necroptosis actually promotes inflammation in human IBD or whether it is only a secondary consequence of the inflammatory reaction. Clearly, more research will be needed to unequivocally establish the role of necroptosis in the pathogenesis of IBD.

4.2 Necroptosis and colorectal cancer

RIPK3 has been assigned a tumor suppressive role in CRC and its expression levels have been reported as a predictor of patient survival (Bozec et al., 2016; Feng et al., 2015a). Accordingly, RIPK3-deficient mice were highly susceptible to colitis-associated CRC, and this was attributed to an enhanced ability of IECs to aberrantly proliferate in a sustained inflammatory microenvironment (Bozec et al., 2016). However, another study reported that

RIPK3 promotes rather than inhibits colitis-associated CRC by controlling tumor cell proliferation and (C-X-C motif) ligand 1-induced immune suppression (Liu et al., 2019). Clearly, the potential tumor-suppressive or tumor-promoting functions of RIPK3, the relevance of RIPK3 in the pathogenesis of CRC in more physiological settings, and especially whether the effects of RIPK3 are mediated by necroptotic or non-necroptotic mechanisms require further exploration.

Similar to RIPK3, a reduced expression of MLKL in colon tissues from CRC patients has been linked to a poor prognosis (Li et al., 2017). Consistent with these findings, MLKL knockout mice were highly sensitive to colitis-associated tumorigenesis, suggesting a role of MLKL in maintaining intestinal homeostasis and protecting against CRC. However, this was attributed to necroptosis-independent functions of MLKL (Zhao et al., 2019). Therefore, it currently remains open whether necroptosis is an important protective mechanism which prevents CRC under physiologic conditions.

Notwithstanding, necroptosis can be potentially employed as a therapeutic approach to combat CRC. Human CRCs frequently acquire mutations in caspase-8 (Cancer Genome Atlas Network, 2012), which protects against apoptosis but simultaneously renders cells highly susceptible to necroptosis (Günther et al., 2011; He et al., 2009; Kaiser et al., 2011; Oberst et al., 2011). He et al. have analyzed mouse models of caspase-8-deficient CRC. In both a hereditary and a xenograft model, induction of necroptosis by second mitochondrial derived activator of caspases mimetics triggered massive cell death and led to the regression of tumors (He et al., 2017). Moreover, in mouse syngeneic CRC models, the administration of MLKL mRNA caused necroptosis of tumor cells and even induced anti-tumoral immunity (Van Hoecke et al., 2018). A similar induction of anti-tumor immunity was seen in mice after prophylactic vaccination with necroptotic colon carcinoma cells (Aaes et al., 2016).

5. ADAM proteases and intestinal (dys)function

5.1 ADAM function in intestinal homeostasis

In the intestine, ADAMs control important cellular processes associated with intestinal homeostasis, i.e., the maintenance of intestinal stem cell and progenitor cell populations, cell fate specification, and the development of the intestine (Jones et al., 2016). By shedding the ectodomains of Notch and ligands/receptors from the EGF and TNF families, ADAMs regulate the intercellular communication during normal intestinal homeostasis, but also during injury, inflammation and regeneration of the intestine. In consequence, aberrant ADAM expression or activity can result in IBD or CRC (Jones et al., 2016). Also, deregulation of ADAM proteases has been implicated in the pathophysiology and pathogenesis mediated by intestinal pathogens such as *Helicobacter* and *Lactobacillus* where the infections drive ADAM-dependent EGFR signaling that leads to epithelial hyperplasia (Jones et al., 2016; Posselt et al., 2017).

ADAM10 and ADAM17 represent the best investigated ADAM family members in the gastrointestinal tract and are ubiquitously expressed in all cells of the intestine (Jones et al., 2016; Tsai et al., 2014). By cleaving Notch, ADAM10 activates Notch signaling in IECs, which is pivotal for the maintenance and survival of crypt base columnar stem cells and for specifying the fate of transit-amplifying progenitor cells (Tsai et al., 2014). In contrast to this essential role of ADAM10, ADAM17 signaling in IECs is apparently not required for the maintenance of intestinal homeostasis under physiologic conditions. Unchallenged ADAM17 hypomorphic mice lack an overt intestinal phenotype and display a regular crypt-villus architecture as well as normal crypt proliferation and differentiation (Brandl et al., 2010; Chalaris et al., 2010). Similarly, no intestinal phenotype is detectable in mice with an IEC-specific deletion of ADAM17 (Feng et al., 2015b; Yan et al., 2013).

5.2 ADAMs and inflammatory bowel disease

Although dispensable for normal intestinal homeostasis, ADAM17 has essential functions in the inflamed intestine by mediating IEC cytoprotection and intestinal regeneration after inflammatory damage. Evidence for this comes from the study of ADAM17 hypomorphic mice. ADAM17$^{ex/ex}$ mice display a substantially increased susceptibility to DSS-induced colitis. This susceptibility is caused by a defect in the STAT3-mediated IEC proliferation in these mice, which strongly impairs the regeneration of the injured intestinal epithelium. Notably, this defect could be partially corrected by administration of recombinant EGFR ligands (Chalaris et al., 2010). Since ADAM17 is the main sheddase for many ligands of the EGFR (Peschon et al., 1998; Sahin et al., 2004), it was concluded that an impaired shedding of EGFR ligands in ADAM17$^{ex/ex}$ mice resulted in a failure of the EGFR to phosphorylate STAT3 and, in consequence, in a strongly reduced regeneration of IECs of the basal crypts and in the breakdown of the intestinal barrier (Chalaris et al., 2010). These findings were independently confirmed by studies in a different ADAM17 hypomorphic mouse model (Brandl et al., 2010) and in IEC-specific ADAM17 knockout mice (Tsai et al., 2012). Concordantly, mouse models deficient or hypomorphic for EGFR ligands or members of the EGFR family likewise show an exacerbated DSS-induced intestinal inflammation (Egger et al., 1997; Zhang et al., 2012).

With regard to the pathogenesis of IBD in humans, an increase of ADAM17 expression and activity was found in the mucosa of IBD patients (Brynskov et al., 2002; Cesaro et al., 2009). Similar to ADAM17, Franzè et al. have reported increased expression levels of ADAM19 in IBD patients (Franzè et al., 2013). Since ADAM19 also cleaves TNF (Chesneau et al., 2003) and members of the EGFR ligand family (Horiuchi et al., 2005; Shirakabe et al., 2001), ADAM19 might be able to complement the activity of ADAM17 in IBD to some extent. However, the hypersensitivity of ADAM17 hypomorphic mice in DSS colitis models argues against a significant contribution of ADAM19 to intestinal regeneration. Independently, ADAM10 is potently activated by ROS (Sanderson et al., 2005; Tsai et al., 2014), and patients with ulcerative colitis (and thus elevated colonic ROS levels) show a decreased goblet cell differentiation indicative of an increased Notch signaling (Gersemann et al., 2011; Zheng et al., 2011). Thus, ADAM10 may be an important regulator of Notch signaling not only during intestinal homeostasis, but also during intestinal inflammation (Jones et al., 2016).

5.3 ADAMs and colorectal cancer

Given that ADAMs regulate signaling through Notch, EGFR ligands, IL-6 and STAT3, which all can contribute to the development of CRC, it is not surprising that an activation of ADAM proteases in human CRCs has been correlated with the progression and invasiveness of tumors and with poor clinical outcomes (Duffy et al., 2009; Murphy, 2008).

In CRC, activation of the Notch signaling pathway is pro-tumorigenic (Noah and Shroyer, 2013), and its components Notch, Jagged, and hairy and enhancer of split-1 are highly expressed in human CRCs and CRC cell lines (Reedijk et al., 2008; Zagouras et al., 1995). In APC mutant mice, which are predisposed to develop benign intestinal adenomas as CRC precursors (Moser et al., 1995; Yamada and Mori, 2007), pharmacological inhibition of Notch blocked adenoma proliferation and induced their differentiation into goblet cells (van Gijn et al., 2005). Conversely, activation of Notch led to enhanced adenoma formation and reduced survival

(Fre et al., 2009). Consistent with the central role of ADAM10 as an activator of Notch, deletion of the *Adam10* gene in a different APC mutant mouse model led to a pronounced reduction of adenoma formation and enhanced survival (Tsai et al., 2013). Therefore, by controlling Notch signaling, ADAM10 may hold a key role not only in the development of benign adenomas, but also in the subsequent emergence of CRC.

For the treatment of CRC patients, targeting EGFR signaling is an established therapeutic approach (Martins et al., 2018). As the main sheddase of EGFR ligands, ADAM17 has a critical function in the development of CRC. Accordingly, a combined inhibition of the EGFR and of ADAM17 led to growth inhibition and enhanced apoptosis in CRC cell lines (Dong et al., 1999; Merchant et al., 2008). In vivo, APC mutant mice, which expressed ADAM17, developed high-grade dysplasia and invasive carcinomas, whereas CRC formation was largely abrogated in APC mutant $ADAM17^{ex/ex}$ mice, and the few remaining tumors were only of low-grade dysplasia. Notably, they did not differ in size from those seen in APC mutant ADAM17 wildtype mice, indicating that ADAM17 is required in the tumor-initiating stage rather than for the outgrowth of the tumors (Schmidt et al., 2018). Since it had been shown that CRC formation requires EGFR signaling in myeloid cells rather than in IECs (Srivatsa et al., 2017), and since blockade of IL-6 trans-signaling was sufficient for inhibition of tumor formation (Schmidt et al., 2018), a model was suggested in which CRC formation depends on the shedding of both the EGFR ligand amphiregulin and the membrane-bound IL-6R by ADAM17. Amphiregulin-dependent activation of the EFGR on myeloid cells then triggers the synthesis of IL-6, which, after binding to the shed, now soluble IL-6R, acts on IECs via trans-signaling to induce CRC formation (Schumacher and Rose-John, 2019). As a side note, the expression levels of ADAM17 in biopsies from CRC patients were found to be no prognostic factor for the course of disease and survival (Schmidt et al., 2018), different from expression of the EGFR (Srivatsa et al., 2017).

Altogether, these findings further underline the significance of ADAM17 in CRC pathology. Moreover, they indicate that blockade of ADAM17-triggered IL-6 trans-signaling could represent a novel therapeutic strategy in the treatment of CRC. Currently, most patients suffering from CRC display an intrinsic resistance to EGFR blockade as a result of mutations in signaling pathways downstream of the EGFR (Van Emburgh et al., 2014). Even patients that initially respond to anti-EFGR antibody treatment usually acquire resistance mechanisms within 3–12 months

(Van Emburgh et al., 2014). Blockade of IL-6 trans-signaling would circumvent these resistance mechanisms, and thus offer another line of treatment for CRC patients not amenable to established anti-EGFR therapies anymore.

The above study also found that the role of ADAM17 in CRC is apparently more complex than anticipated. ADAM17$^{ex/ex}$ mice show alterations in their intestinal microflora which resemble those of CRC patients (Schmidt et al., 2018). In particular, the abundance of butyrate-producing bacterial taxa was reduced. Since butyrate production by these taxa protects from colonic inflammation and carcinogenesis (Byndloss et al., 2017), ADAM17 not only contributes to the development of CRC via IL-6 trans-signaling under pathological conditions, but also seems to protect the organism from CRC by regulating the constitution of the intestinal microbiota under physiological conditions.

5.4 ADAMs and mucosal atrophy

Mucosal atrophy is morphologically characterized by a decrease in villous height, crypt depth, surface area, and epithelial cell numbers (Shaw et al., 2012), as well as by impaired crypt proliferation, increased death of IECs and loss of epithelial barrier function at the functional level (Demehri et al., 2013). It is most commonly observed in the absence of enteral nutrition, e.g., in patients that have to receive total parenteral nutrition (TPN, a method to provide nutrients and calories intravenously; Abunnaja et al., 2013). Deprivation of enteral nutrition shifts the intestinal microbiota to predominantly Gram-negative Proteobacteria, concomitant with an increased expression of pro-inflammatory cytokines (Demehri et al., 2013). As demonstrated in TPN-treated mice that lack TNF-R1, TNF-R2, or both, TNF and its receptors play a central role in TPN-induced epithelial barrier dysfunction (Feng and Teitelbaum, 2013). Independently, the analysis of TPN-treated mice carrying an IEC-specific deletion of ADAM17 has revealed a deleterious function of ADAM17 in the development of mucosal atrophy. In these mice, a loss of ADAM17-mediated IEC-specific TNF signaling resulted in decreased IEC death, improved crypt proliferation, and reduced expression of pro-inflammatory cytokines (Feng et al., 2012, 2015b; Tsai et al., 2012).

Overall, ADAM17 signaling in IECs has essential but, dependent on the respective model, sometimes opposing functions in intestinal inflammation. On the one hand, ADAM17 can exert a cytoprotective role as a sheddase of EGFR ligands. On the other hand, ADAM17 can exacerbate mucosal atrophy by the shedding of TNF and thus by enhancing subsequent TNF-dependent inflammatory signaling. Nevertheless, these seemingly contradictory functions of ADAM17 may allow for flexible, context-dependent reactions to maintain intestinal homeostasis when faced with specific triggers of stress and inflammation (Jones et al., 2016).

6. Necroptosis and ADAM proteases

As discussed above, both necroptosis and ADAM proteases have been implicated in intestinal (dys)function. Therefore, it is not completely surprising that there are also indications for functional connections between ADAMs and necroptosis. Cai et al. have recently reported that ADAM10 and ADAM17 in murine cells or ADAM9 and ADAM10 in human cells are activated at an early stage of necroptosis, i.e., when MLKL is already phosphorylated and translocated to the cell membrane but when cells are still intact (Cai et al., 2016). For ADAM9 and ADAM10, the authors showed that activation required their binding to phosphorylated MLKL and speculated that this induces a conformational change and a transition of ADAM homodimers to monomers as a possible activation mechanism.

Subsequently, ADAM activation seems to play a role as a new mechanism by which necroptosis directly induces inflammation. A macrophage cell line treated with supernatants from necroptotic ADAM10/ADAM17 double-deficient cells showed much less upregulation of IFN-γ, IL-6, TNF and vascular endothelial growth factor than macrophage cells after treatment with supernatants from necroptotic wildtype cells (Cai et al., 2016). This was confirmed in vivo by injecting necroptotic ADAM10/ADAM17 double-deficient or wildtype cells into the peritoneal cavity of mice. While mice injected with wildtype cells showed peritoneal infiltration of neutrophils, this was much less pronounced in mice injected with ADAM10/ADAM17 double-deficient cells (Cai et al., 2016). In a model of cerulein-induced pancreatitis, mice treated with the ADAM metalloproteinase inhibitor GW280264X showed reduced levels of serum amylase and were significantly protected from loss of pancreatic acinar cells and necrosis (Cai et al., 2016). Independently, treatment with GW280264X, double knockout of ADAM10/ADAM17 in murine cells or double knockdown of ADAM9/ADAM10 in human cells delayed necroptosis. Since the effect was limited to adherent, not suspension cells, the authors suggested disruption of cell adhesion by ADAMs as the underlying mechanism. Consistently, cells treated with supernatants from necroptotic wildtype cells but not from ADAM9/ADAM10 knockdown cells showed enhanced migration and invasion, which was prevented by depleting the ADAM substrate E-cadherin.

In summary, the authors proposed a model in which activation of ADAMs during necroptosis leads to the ectodomain release of adhesion molecules, receptors, growth factors and cytokines. Shedding of adhesion molecules such as E-cadherin impairs cell adhesion and further promotes necroptosis as well as cell migration and invasion. Simultaneously, the necroptotic activation of ADAMs triggers inflammatory responses via the cleaved ectodomains of other substrates such as cytokines. By concurrently affecting cell adhesion and inflammation, necroptotic activation of ADAMs may contribute to the disruption of the epithelial barrier and exacerbate inflammation in the intestine in vivo.

In the model of Cai et al., a currently unresolved question arises from the observation that ADAM10 and ADAM17 are the relevant ADAMs in the murine system, whereas ADAM10 and ADAM9 are important in the human system. Given the high degree of similarity between human and mouse ADAM17, it is somewhat surprising that ADAM17 plays a role in murine, but not in human cells, where ADAM9 is required instead. It will certainly be interesting to further unravel the underlying mechanisms for this difference.

We have recently obtained independent evidence for a functional connection between necroptosis and ADAMs (Fuchslocher Chico et al., 2018). To evaluate the contribution of necroptosis to intestinal inflammation in vivo, we employed ADAM17$^{ex/ex}$ mice as a particularly sensitive and suitable model system for DSS-induced colitis (Chalaris et al., 2010). When comparing their responses to ADAM17$^{ex/ex}$ mice that additionally lacked RIPK3 and were thus defective for necroptosis, we surprisingly found that mice of both genotypes showed the same elevated susceptibility in acute and chronic models of DSS-induced colitis. ADAM17$^{ex/ex}$ and ADAM17$^{ex/ex}$/RIPK3-deficient mice displayed comparable weight loss, disease activity index and colitis-associated changes of inner organs. We detected similar tissue destruction, loss of barrier integrity, immune cell infiltration and cell death in the intestine, and similar levels of the pro-inflammatory chemokine keratinocyte chemoattractant in the sera from mice of both genotypes (Fuchslocher Chico et al., 2018). As an explanation to these unexpected results, necroptotic phosphorylation of RIPK3 or MLKL was not detectable in ADAM17$^{ex/ex}$ mice during DSS-induced colitis, although both proteins were clearly expressed. Furthermore, murine embryonic fibroblasts derived from ADAM17$^{ex/ex}$ mice were protected from TNF-induced necroptosis and failed to show phosphorylation of MLKL in vitro. In summary, these results indicate that necroptosis is compromised in ADAM17$^{ex/ex}$ mice even when RIPK3 is present, and lend further support to a role of ADAM17 as a novel regulator of necroptosis (Fuchslocher Chico et al., 2018). It is currently unclear how exactly ADAM17 couples to the necroptotic signaling machinery and how necroptosis is suppressed in hypomorphic ADAM17 cells. Nevertheless, the results obtained in the studies by us and by Cai et al. provide evidence for previously unknown interconnections between ADAMs and necroptosis.

7. Concluding remarks

Necroptosis as well as ADAM proteases have important functions in maintaining the integrity and homeostasis of higher organisms. Both are integral components of the immune system, contribute to an effective immune defense against invading pathogens, and have been implicated in the control of inflammatory responses in the intestine. Nevertheless, up to now, necroptosis and ADAM proteases have been mostly studied as independent entities, and only recently, direct functional connections between necroptosis and ADAMs have been uncovered. While the importance of ADAMs in the healthy as well as in the diseased intestine is strongly established, conclusive evidence for an in vivo role of necroptosis in intestinal (dys)function is currently still limited, highlighting the need for additional research. Along these lines, the connection of ADAM proteases to the necroptotic signaling network has only recently been discovered, and it will be of outstanding interest to further characterize the underlying mechanisms. An overview of our current knowledge on the interrelationship between necroptosis and ADAM family action in intestinal (dys)function is given in Fig. 4. In summary, deeper insights into the interplay of necroptosis, ADAMs and intestinal dysfunction may prove invaluable for a better treatment of patients suffering from Crohn's disease, ulcerative colitis, colon cancer or mucosal atrophy. Clearly, many questions are still unanswered, but many exciting discoveries and answers lie also ahead.

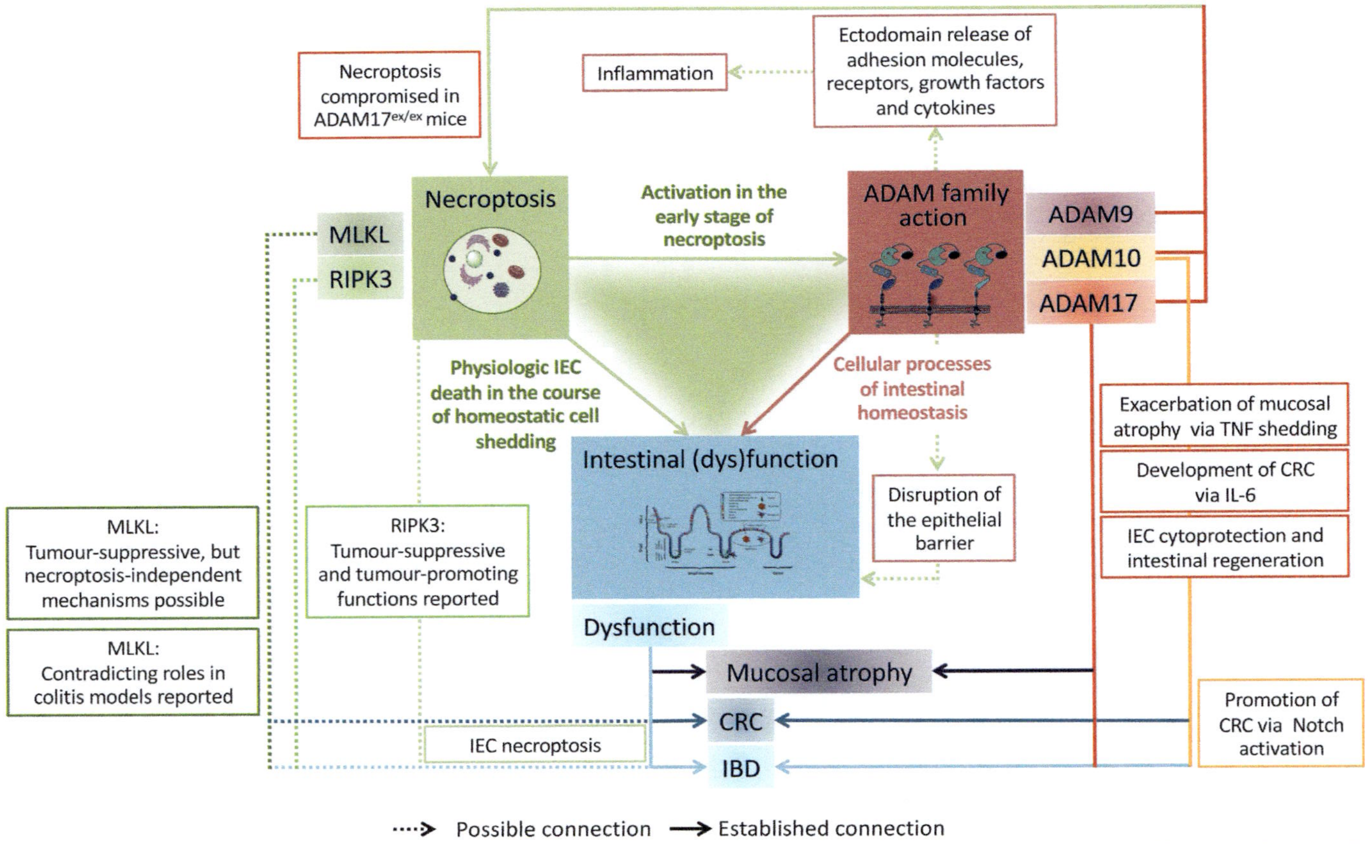

Fig. 4 Schematic overview of the interrelationship between necroptosis and ADAM family action in intestinal (dys)function. See text for details.

Funding

Work in the laboratories of S.R.-J. and D.A. is supported by grants from the Deutsche Forschungsgemeinschaft (DFG, German Research Foundation)—Projektnummer 125440785—SFB 877 (projects A1 and A14 to S.R.-J. and project B2 to D.A.), SFB 841 (project C1 to S.R.-J.), by the Deutsche Krebshilfe and by the German-Israeli Foundation for Scientific Research and Development.

Conflict of interest

M.H and D.A. declare no competing interests. S.R.-J. has acted as a consultant and speaker for AbbVie, Chugai, Genentech Roche, Pfizer, Sanofi, Eli Lilly and Boehringer Ingelheim. He also declares that he is an inventor on patents owned by CONARIS Research Institute, which develops the sgp130Fc protein Olamkicept together with the company I-Mab. S.R.-J. has stock ownership in CONARIS.

References

Aaes, T.L., Kaczmarek, A., Delvaeye, T., De Craene, B., De Koker, S., Heyndrickx, L., Delrue, I., Taminau, J., Wiernicki, B., De Groote, P., Garg, A.D., Leybaert, L., Grooten, J., Bertrand, M.J., Agostinis, P., Berx, G., Declercq, W., Vandenabeele, P., Krysko, D.V., 2016. Vaccination with necroptotic cancer cells induces efficient anti-tumor immunity. Cell Rep. 15, 274–287.

Abunnaja, S., Cuviello, A., Sanchez, J.A., 2013. Enteral and parenteral nutrition in the perioperative period: state of the art. Nutrients 5, 608–623.

Aden, K., Tran, F., Ito, G., Sheibani-Tezerji, R., Lipinski, S., Kuiper, J.W., Tschurtschenthaler, M., Saveljeva, S., Bhattacharyya, J., Häsler, R., Bartsch, K., Luzius, A., Jentzsch, M., Falk-Paulsen, M., Stengel, S.T., Welz, L., Schwarzer, R., Rabe, B., Barchet, W., Krautwald, S., Hartmann, G., Pasparakis, M., Blumberg, R.S., Schreiber, S., Kaser, A., Rosenstiel, P., 2018. ATG16L1 orchestrates interleukin-22 signaling in the intestinal epithelium via cGAS-STING. J. Exp. Med. 215, 2868–2886.

Adrain, C., Zettl, M., Christova, Y., Taylor, N., Freeman, M., 2012. Tumor necrosis factor signaling requires iRhom2 to promote trafficking and activation of TACE. Science 335, 225–228.

Afonso, M.B., Rodrigues, P.M., Carvalho, T., Caridade, M., Borralho, P., Cortez-Pinto, H., Castro, R.E., Rodrigues, C.M., 2015. Necroptosis is a key pathogenic event in human and experimental murine models of non-alcoholic steatohepatitis. Clin. Sci. (Lond.) 129, 721–739.

Allinson, T.M., Parkin, E.T., Turner, A.J., Hooper, N.M., 2003. ADAMs family members as amyloid precursor protein alpha-secretases. J. Neurosci. Res. 74, 342–352.

Altmeppen, H.C., Prox, J., Krasemann, S., Puig, B., Kruszewski, K., Dohler, F., Bernreuther, C., Hoxha, A., Linsenmeier, L., Sikorska, B., Liberski, P.P., Bartsch, U., Saftig, P., Glatzel, M., 2015. The sheddase ADAM10 is a potent modulator of prion disease. Elife 4, e04260.

Alvarez-Diaz, S., Dillon, C.P., Lalaoui, N., Tanzer, M.C., Rodriguez, D.A., Lin, A., Lebois, M., Hakem, R., Josefsson, E.C., O'Reilly, L.A., Silke, J., Alexander, W.S., Green, D.R., Strasser, A., 2016. The pseudokinase MLKL and the kinase RIPK3 have distinct roles in autoimmune disease caused by loss of death-receptor-induced apoptosis. Immunity 45, 513–526.

Arribas, J., Esselens, C., 2009. ADAM17 as a therapeutic target in multiple diseases. Curr. Pharm. Des. 15, 2319–2335.

Bandsma, R.H., van Goor, H., Yourshaw, M., Horlings, R.K., Jonkman, M.F., Schölvinck, E.H., Karrenbeld, A., Scheenstra, R., Kömhoff, M., Rump, P., Koopman-Keemink, Y., Nelson, S.F., Escher, J.C., Cutz, E., Martín, M.G., 2015. Loss of ADAM17 is associated with severe multiorgan dysfunction. Hum. Pathol. 46, 923–928.

Bartsch, J.W., Wildeboer, D., Koller, G., Naus, S., Rittger, A., Moss, M.L., Minai, Y., Jockusch, H., 2010. Tumor necrosis factor-alpha (TNF-alpha) regulates shedding of TNF-alpha receptor 1 by the metalloprotease-disintegrin ADAM8: evidence for a protease-regulated feedback loop in neuroprotection. J. Neurosci. 30, 12210–12218.

Becker, C., Watson, A.J., Neurath, M.F., 2013. Complex roles of caspases in the pathogenesis of inflammatory bowel disease. Gastroenterology 144, 283–293.

Bell, J.H., Herrera, A.H., Li, Y., Walcheck, B., 2007. Role of ADAM17 in the ectodomain shedding of TNF-alpha and its receptors by neutrophils and macrophages. J. Leukoc. Biol. 82, 173–176.

Bertrand, M.J., Milutinovic, S., Dickson, K.M., Ho, W.C., Boudreault, A., Durkin, J., Gillard, J.W., Jaquith, J.B., Morris, S.J., Barker, P.A., 2008. cIAP1 and cIAP2 facilitate cancer cell survival by functioning as E3 ligases that promote RIP1 ubiquitination. Mol. Cell 30, 689–700.

Bevins, C.L., Salzman, N.H., 2011. Paneth cells, antimicrobial peptides and maintenance of intestinal homeostasis. Nat. Rev. Microbiol. 9, 356–368.

Bigenzahn, J.W., Fauster, A., Rebsamen, M., Kandasamy, R.K., Scorzoni, S., Vladimer, G.I., Müller, A.C., Gstaiger, M., Zuber, J., Bennett, K.L., Superti-Furga, G., 2016. An inducible retroviral expression system for tandem affinity purification mass-spectrometry-based proteomics identifies mixed lineage kinase domain-like protein (MLKL) as an heat shock protein 90 (HSP90) client. Mol. Cell. Proteomics 15, 1139–1150.

Bittner, S., Knoll, G., Ehrenschwender, M., 2017. Death receptor 3 mediates necroptotic cell death. Cell. Mol. Life Sci. 74, 543–554.

Bjerknes, M., Cheng, H., 1981. The stem-cell zone of the small intestinal epithelium. I. Evidence from Paneth cells in the adult mouse. Am. J. Anat. 160, 51–63.

Black, R.A., Rauch, C.T., Kozlosky, C.J., Peschon, J.J., Slack, J.L., Wolfson, M.F., Castner, B.J., Stocking, K.L., Reddy, P., Srinivasan, S., Nelson, N., Boiani, N., Schooley, K.A., Gerhart, M., Davis, R., Fitzner, J.N., Johnson, R.S., Paxton, R.J., March, C.J., Cerretti, D.P., 1997. A metalloproteinase disintegrin that releases tumour-necrosis factor-alpha from cells. Nature 385, 729–733.

Blander, J.M., 2018. On cell death in the intestinal epithelium and its impact on gut homeostasis. Curr. Opin. Gastroenterol. 34, 413–419.

Blaydon, D.C., Biancheri, P., Di, W.L., Plagnol, V., Cabral, R.M., Brooke, M.A., van Heel, D.A., Ruschendorf, F., Toynbee, M., Walne, A., O'Toole, E.A., Martin, J.E., Lindley, K., Vulliamy, T., Abrams, D.J., MacDonald, T.T., Harper, J.I., Kelsell, D.P., 2011. Inflammatory skin and bowel disease linked to ADAM17 deletion. N. Engl. J. Med. 365, 1502–1508.

Blobel, C.P., 2005. ADAMs: key components in EGFR signalling and development. Nat. Rev. Mol. Cell Biol. 6, 32–43.

Bonnet, M.C., Preukschat, D., Welz, P.S., van Loo, G., Ermolaeva, M.A., Bloch, W., Haase, I., Pasparakis, M., 2011. The adaptor protein FADD protects epidermal keratinocytes from necroptosis in vivo and prevents skin inflammation. Immunity 35, 572–582.

Bozec, D., Iuga, A.C., Roda, G., Dahan, S., Yeretssian, G., 2016. Critical function of the necroptosis adaptor RIPK3 in protecting from intestinal tumorigenesis. Oncotarget 7, 46384–46400.

Brandl, K., Sun, L., Neppl, C., Siggs, O.M., Le Gall, S.M., Tomisato, W., Li, X., Du, X., Maennel, D.N., Blobel, C.P., Beutler, B., 2010. MyD88 signaling in nonhematopoietic cells protects mice against induced colitis by regulating specific EGF receptor ligands. Proc. Natl. Acad. Sci. U. S. A. 107, 19967–19972.

Brault, M., Olsen, T.M., Martinez, J., Stetson, D.B., Oberst, A., 2018. Intracellular nucleic acid sensing triggers necroptosis through synergistic type I IFN and TNF signaling. J. Immunol. 200, 2748–2756.

Brynskov, J., Foegh, P., Pedersen, G., Ellervik, C., Kirkegaard, T., Bingham, A., Saermark, T., 2002. Tumour necrosis factor alpha converting enzyme (TACE) activity in the colonic mucosa of patients with inflammatory bowel disease. Gut 51, 37–43.

Bullen, T.F., Forrest, S., Campbell, F., Dodson, A.R., Hershman, M.J., Pritchard, D.M., Turner, J.R., Montrose, M.H., Watson, A.J., 2006. Characterization of epithelial cell shedding from human small intestine. Lab. Invest. 86, 1052–1063.

Byndloss, M.X., Olsan, E.E., Rivera-Chávez, F., Tiffany, C.R., Cevallos, S.A., Lokken, K.L., Torres, T.P., Byndloss, A.J., Faber, F., Gao, Y., Litvak, Y., Lopez, C.A., Xu, G., Napoli, E., Giulivi, C., Tsolis, R.M., Revzin, A., Lebrilla, C.B., Bäumler, A.J., 2017. Microbiota-activated PPAR-gamma signaling inhibits dysbiotic Enterobacteriaceae expansion. Science 357, 570–575.

Cai, Z., Jitkaew, S., Zhao, J., Chiang, H.C., Choksi, S., Liu, J., Ward, Y., Wu, L.G., Liu, Z.G., 2014. Plasma membrane translocation of trimerized MLKL protein is required for TNF-induced necroptosis. Nat. Cell Biol. 16, 55–65.

Cai, Z., Zhang, A., Choksi, S., Li, W., Li, T., Zhang, X.M., Liu, Z.G., 2016. Activation of cell-surface proteases promotes necroptosis, inflammation and cell migration. Cell Res. 26, 886–900.

Camodeca, C., Cuffaro, D., Nuti, E., Rossello, A., 2019. ADAM metalloproteinases as potential drug targets. Curr. Med. Chem. 26, 2661–2689.

Cancer Genome Atlas Network, 2012. Comprehensive molecular characterization of human colon and rectal cancer. Nature 487, 330–337.

Cao, M., Wu, Z., Lou, Q., Lu, W., Zhang, J., Li, Q., Zhang, Y., Yao, Y., Zhao, Q., Li, M., Zhang, H., Qian, Y., 2019. Dectin-1-induced RIPK1 and RIPK3 activation protects host against Candida albicans infection. Cell Death Differ. 26, 2622–2636.

Caprioli, F., Stolfi, C., Caruso, R., Fina, D., Sica, G., Biancone, L., Pallone, F., Monteleone, G., 2008. Transcriptional and post-translational regulation of Flip, an inhibitor of Fas-mediated apoptosis, in human gut inflammation. Gut 57, 1674–1680.

Cavadas, M., Oikonomidi, I., Gaspar, C.J., Burbridge, E., Badenes, M., Félix, I., Bolado, A., Hu, T., Bileck, A., Gerner, C., Domingos, P.M., von Kriegsheim, A., Adrain, C., 2017. Phosphorylation of iRhom2 controls stimulated proteolytic shedding by the metalloprotease ADAM17/TACE. Cell Rep. 21, 745–757.

Cerretti, D.P., DuBose, R.F., Black, R.A., Nelson, N., 1999. Isolation of two novel metalloproteinase-disintegrin (ADAM) cDNAs that show testis-specific gene expression. Biochem. Biophys. Res. Commun. 263, 810–815.

Cesaro, A., Abakar-Mahamat, A., Brest, P., Lassalle, S., Selva, E., Filippi, J., Hébuterne, X., Hugot, J.P., Doglio, A., Galland, F., Naquet, P., Vouret-Craviari, V., Mograbi, B., Hofman, P.M., 2009. Differential expression and regulation of ADAM17 and TIMP3 in acute inflamed intestinal epithelia. Am. J. Physiol. Gastrointest. Liver Physiol. 296, G1332–G1343.

Chabrol, E., Gourfinkel-An, I., Scheffer, I.E., Picard, F., Couarch, P., Berkovic, S.F., McMahon, J.M., Bajaj, N., Mota-Vieira, L., Mota, R., Trouillard, O., Depienne, C., Baulac, M., LeGuern, E., Baulac, S., 2007. Absence of mutations in the LGI1 receptor ADAM22 gene in autosomal dominant lateral temporal epilepsy. Epilepsy Res. 76, 41–48.

Chalaris, A., Adam, N., Sina, C., Rosenstiel, P., Lehmann-Koch, J., Schirmacher, P., Hartmann, D., Cichy, J., Gavrilova, O., Schreiber, S., Jostock, T., Matthews, V.,

Häsler, R., Becker, C., Neurath, M.F., Reiss, K., Saftig, P., Scheller, J., Rose-John, S., 2010. Critical role of the disintegrin metalloprotease ADAM17 for intestinal inflammation and regeneration in mice. J. Exp. Med. 207, 1617–1624.

Chan, F.K.M., Shisler, J., Bixby, J.G., Felices, M., Zheng, L., Appel, M., Orenstein, J., Moss, B., Lenardo, M.J., 2003. A role for tumor necrosis factor receptor-2 and receptor-interacting protein in programmed necrosis and antiviral responses. J. Biol. Chem. 278, 51613–51621.

Chen, W.W., Yu, H., Fan, H.B., Zhang, C.C., Zhang, M., Zhang, C., Cheng, Y., Kong, J., Liu, C.F., Geng, D., Xu, X., 2012. RIP1 mediates the protection of geldanamycin on neuronal injury induced by oxygen-glucose deprivation combined with zVAD in primary cortical neurons. J. Neurochem. 120, 70–77.

Chen, W., Zhou, Z., Li, L., Zhong, C.Q., Zheng, X., Wu, X., Zhang, Y., Ma, H., Huang, D., Li, W., Xia, Z., Han, J., 2013. Diverse sequence determinants control human and mouse receptor interacting protein 3 (RIP3) and mixed lineage kinase domain-like (MLKL) interaction in necroptotic signaling. J. Biol. Chem. 288, 16247–16261.

Chen, X., Li, W., Ren, J., Huang, D., He, W.T., Song, Y., Yang, C., Zheng, X., Chen, P., Han, J., 2014. Translocation of mixed lineage kinase domain-like protein to plasma membrane leads to necrotic cell death. Cell Res. 24, 105–121.

Chen, W., Wu, J., Li, L., Zhang, Z., Ren, J., Liang, Y., Chen, F., Yang, C., Zhou, Z., Su, S.S., Zheng, X., Zhang, Z., Zhong, C.Q., Wan, H., Xiao, M., Lin, X., Feng, X.H., Han, J., 2015. Ppm1b negatively regulates necroptosis through dephosphorylating Rip3. Nat. Cell Biol. 17, 434–444.

Chen, W., Zheng, R., Baade, P.D., Zhang, S., Zeng, H., Bray, F., Jemal, A., Yu, X.Q., He, J., 2016. Cancer statistics in China, 2015. CA-Cancer J. Clin. 66, 115–132.

Ch'en, I.L., Beisner, D.R., Degterev, A., Lynch, C., Yuan, J., Hoffmann, A., Hedrick, S.M., 2008. Antigen-mediated T cell expansion regulated by parallel pathways of death. Proc. Natl. Acad. Sci. U. S. A. 105, 17463–17468.

Ch'en, I.L., Tsau, J.S., Molkentin, J.D., Komatsu, M., Hedrick, S.M., 2011. Mechanisms of necroptosis in T cells. J. Exp. Med. 208, 633–641.

Chesneau, V., Becherer, J.D., Zheng, Y., Erdjument-Bromage, H., Tempst, P., Blobel, C.P., 2003. Catalytic properties of ADAM19. J. Biol. Chem. 278, 22331–22340.

Cho, Y.S., Challa, S., Moquin, D., Genga, R., Ray, T.D., Guildford, M., Chan, F.K., 2009. Phosphorylation-driven assembly of the RIP1-RIP3 complex regulates programmed necrosis and virus-induced inflammation. Cell 137, 1112–1123.

Choi, I., Woo, J.M., Hong, S., Jung, Y.K., Kim, D.H., Cho, C., 2003. Identification and characterization of ADAM32 with testis-predominant gene expression. Gene 304, 151–162.

Choi, S.W., Park, H.H., Kim, S., Chung, J.M., Noh, H.J., Kim, S.K., Song, H.K., Lee, C.W., Morgan, M.J., Kang, H.C., Kim, Y.S., 2018. PELI1 selectively targets kinase-active RIP3 for ubiquitylation-dependent proteasomal degradation. Mol. Cell 70, 920–935.

Christova, Y., Adrain, C., Bambrough, P., Ibrahim, A., Freeman, M., 2013. Mammalian iRhoms have distinct physiological functions including an essential role in TACE regulation. EMBO Rep. 14, 884–890.

Conrad, M., Angeli, J.P., Vandenabeele, P., Stockwell, B.R., 2016. Regulated necrosis: disease relevance and therapeutic opportunities. Nat. Rev. Drug Discov. 15, 348–366.

Coupienne, I., Fettweis, G., Rubio, N., Agostinis, P., Piette, J., 2011. 5-ALA-PDT induces RIP3-dependent necrosis in glioblastoma. Photochem. Photobiol. Sci. 10, 1868–1878.

Crosnier, C., Stamataki, D., Lewis, J., 2006. Organizing cell renewal in the intestine: stem cells, signals and combinatorial control. Nat. Rev. Genet. 7, 349–359.

Dagenais, M., Douglas, T., Saleh, M., 2014. Role of programmed necrosis and cell death in intestinal inflammation. Curr. Opin. Gastroenterol. 30, 566–575.

Dannappel, M., Vlantis, K., Kumari, S., Polykratis, A., Kim, C., Wachsmuth, L., Eftychi, C., Lin, J., Corona, T., Hermance, N., Zelic, M., Kirsch, P., Basic, M., Bleich, A., Kelliher, M., Pasparakis, M., 2014. RIPK1 maintains epithelial homeostasis by inhibiting apoptosis and necroptosis. Nature 513, 90–94.

Davies, K.A., Tanzer, M.C., Griffin, M.D.W., Mok, Y.F., Young, S.N., Qin, R., Petrie, E.J., Czabotar, P.E., Silke, J., Murphy, J.M., 2018. The brace helices of MLKL mediate interdomain communication and oligomerisation to regulate cell death by necroptosis. Cell Death Differ. 25, 1567–1580.

de Santa Barbara, P., van den Brink, G.R., Roberts, D.J., 2003. Development and differentiation of the intestinal epithelium. Cell. Mol. Life Sci. 60, 1322–1332.

Declercq, W., Vanden Berghe, T., Vandenabeele, P., 2009. RIP kinases at the crossroads of cell death and survival. Cell 138, 229–232.

Degterev, A., Huang, Z., Boyce, M., Li, Y., Jagtap, P., Mizushima, N., Cuny, G.D., Mitchison, T.J., Moskowitz, M.A., Yuan, J., 2005. Chemical inhibitor of nonapoptotic cell death with therapeutic potential for ischemic brain injury. Nat. Chem. Biol. 1, 112–119.

Demehri, F.R., Barrett, M., Ralls, M.W., Miyasaka, E.A., Feng, Y., Teitelbaum, D.H., 2013. Intestinal epithelial cell apoptosis and loss of barrier function in the setting of altered microbiota with enteral nutrient deprivation. Front. Cell. Infect. Microbiol. 3, 105.

Deutsch, M., Graffeo, C.S., Rokosh, R., Pansari, M., Ochi, A., Levie, E.M., Van Heerden, E., Tippens, D.M., Greco, S., Barilla, R., Tomkötter, L., Zambirinis, C.P., Avanzi, N., Gulati, R., Pachter, H.L., Torres-Hernandez, A., Eisenthal, A., Daley, D., Miller, G., 2015. Divergent effects of RIP1 or RIP3 blockade in murine models of acute liver injury. Cell Death Dis. 6, e1759.

Dillon, A., Lo, D.D., 2019. M cells: intelligent engineering of mucosal immune surveillance. Front. Immunol. 10, 1499.

Dillon, C.P., Weinlich, R., Rodriguez, D.A., Cripps, J.G., Quarato, G., Gurung, P., Verbist, K.C., Brewer, T.L., Llambi, F., Gong, Y.N., Janke, L.J., Kelliher, M.A., Kanneganti, T.D., Green, D.R., 2014. RIPK1 blocks early postnatal lethality mediated by caspase-8 and RIPK3. Cell 157, 1189–1202.

Doedens, J.R., Black, R.A., 2000. Stimulation-induced down-regulation of tumor necrosis factor-alpha converting enzyme. J. Biol. Chem. 275, 14598–14607.

Dondelinger, Y., Aguileta, M.A., Goossens, V., Dubuisson, C., Grootjans, S., Dejardin, E., Vandenabeele, P., Bertrand, M.J., 2013. RIPK3 contributes to TNFR1-mediated RIPK1 kinase-dependent apoptosis in conditions of cIAP1/2 depletion or TAK1 kinase inhibition. Cell Death Differ. 20, 1381–1392.

Dondelinger, Y., Declercq, W., Montessuit, S., Roelandt, R., Goncalves, A., Bruggeman, I., Hulpiau, P., Weber, K., Sehon, C.A., Marquis, R.W., Bertin, J., Gough, P.J., Savvides, S., Martinou, J.C., Bertrand, M.J., Vandenabeele, P., 2014. MLKL compromises plasma membrane integrity by binding to phosphatidylinositol phosphates. Cell Rep. 7, 971–981.

Dondelinger, Y., Jouan-Lanhouet, S., Divert, T., Theatre, E., Bertin, J., Gough, P.J., Giansanti, P., Heck, A.J., Dejardin, E., Vandenabeele, P., Bertrand, M.J., 2015. NF-kappaB-independent role of IKKalpha/IKKbeta in preventing RIPK1 kinase-dependent apoptotic and necroptotic cell death during TNF signaling. Mol. Cell 60, 63–76.

Dondelinger, Y., Hulpiau, P., Saeys, Y., Bertrand, M.J., Vandenabeele, P., 2016. An evolutionary perspective on the necroptotic pathway. Trends Cell Biol. 26, 721–732.

Dondelinger, Y., Delanghe, T., Rojas-Rivera, D., Priem, D., Delvaeye, T., Bruggeman, I., Van Herreweghe, F., Vandenabeele, P., Bertrand, M.J.M., 2017. MK2 phosphorylation of RIPK1 regulates TNF-mediated cell death. Nat. Cell Biol. 19, 1237–1247.

Dondelinger, Y., Delanghe, T., Priem, D., Wynosky-Dolfi, M.A., Sorobetea, D., Rojas-Rivera, D., Giansanti, P., Roelandt, R., Gropengiesser, J., Ruckdeschel, K., Savvides, S.N., Heck, A.J.R., Vandenabeele, P., Brodsky, I.E., Bertrand, M.J.M., 2019. Serine 25 phosphorylation inhibits RIPK1 kinase-dependent cell death in models of infection and inflammation. Nat. Commun. 10, 1729.

Dong, J., Opresko, L.K., Dempsey, P.J., Lauffenburger, D.A., Coffey, R.J., Wiley, H.S., 1999. Metalloprotease-mediated ligand release regulates autocrine signaling through the epidermal growth factor receptor. Proc. Natl. Acad. Sci. U. S. A. 96, 6235–6240.

Dong, K., Zhu, H., Song, Z., Gong, Y., Wang, F., Wang, W., Zheng, Z., Yu, Z., Gu, Q., Xu, X., Sun, X., 2012. Necrostatin-1 protects photoreceptors from cell death and improves functional outcome after experimental retinal detachment. Am. J. Pathol. 181, 1634–1641.

Dong, W., Zhang, M., Zhu, Y., Chen, Y., Zhao, X., Li, R., Zhang, L., Ye, Z., Liang, X., 2017. Protective effect of NSA on intestinal epithelial cells in a necroptosis model. Oncotarget 8, 86726–86735.

Dovey, C.M., Diep, J., Clarke, B.P., Hale, A.T., McNamara, D.E., Guo, H., Brown Jr., N.W., Cao, J.Y., Grace, C.R., Gough, P.J., Bertin, J., Dixon, S.J., Fiedler, D., Mocarski, E.S., Kaiser, W.J., Moldoveanu, T., York, J.D., Carette, J.E., 2018. MLKL requires the inositol phosphate code to execute necroptosis. Mol. Cell 70, 936–948.

Dreymueller, D., Uhlig, S., Ludwig, A., 2015. ADAM-family metalloproteinases in lung inflammation: potential therapeutic targets. Am. J. Physiol. Lung Cell. Mol. Physiol. 308, L325–L343.

Duffy, M.J., McKiernan, E., O'Donovan, N., McGowan, P.M., 2009. Role of ADAMs in cancer formation and progression. Clin. Cancer Res. 15, 1140–1144.

Duprez, L., Takahashi, N., Van Hauwermeiren, F., Vandendriessche, B., Goossens, V., Vanden Berghe, T., Declercq, W., Libert, C., Cauwels, A., Vandenabeele, P., 2011. RIP kinase-dependent necrosis drives lethal systemic inflammatory response syndrome. Immunity 35, 908–918.

Düsterhöft, S., Jung, S., Hung, C.W., Tholey, A., Sönnichsen, F.D., Grötzinger, J., Lorenzen, I., 2013. Membrane-proximal domain of a disintegrin and metalloprotease-17 represents the putative molecular switch of its shedding activity operated by protein-disulfide isomerase. J. Am. Chem. Soc. 135, 5776–5781.

Düsterhöft, S., Höbel, K., Oldefest, M., Lokau, J., Waetzig, G.H., Chalaris, A., Garbers, C., Scheller, J., Rose-John, S., Lorenzen, I., Grötzinger, J., 2014. A disintegrin and metalloprotease 17 dynamic interaction sequence, the sweet tooth for the human interleukin 6 receptor. J. Biol. Chem. 289, 16336–16348.

Dynek, J.N., Goncharov, T., Dueber, E.C., Fedorova, A.V., Izrael-Tomasevic, A., Phu, L., Helgason, E., Fairbrother, W.J., Deshayes, K., Kirkpatrick, D.S., Vucic, D., 2010. c-IAP1 and UbcH5 promote K11-linked polyubiquitination of RIP1 in TNF signalling. EMBO J. 29, 4198–4209.

Ea, C.K., Deng, L., Xia, Z.P., Pineda, G., Chen, Z.J., 2006. Activation of IKK by TNFalpha requires site-specific ubiquitination of RIP1 and polyubiquitin binding by NEMO. Mol. Cell 22, 245–257.

Edelblum, K.L., Yan, F., Yamaoka, T., Polk, D.B., 2006. Regulation of apoptosis during homeostasis and disease in the intestinal epithelium. Inflamm. Bowel Dis. 12, 413–424.

Edwards, D.R., Handsley, M.M., Pennington, C.J., 2008. The ADAM metalloproteinases. Mol. Aspects Med. 29, 258–289.

Egger, B., Procaccino, F., Lakshmanan, J., Reinshagen, M., Hoffmann, P., Patel, A., Reuben, W., Gnanakkan, S., Liu, L., Barajas, L., Eysselein, V.E., 1997. Mice lacking transforming growth factor alpha have an increased susceptibility to dextran sulfate-induced colitis. Gastroenterology 113, 825–832.

Ermolaeva, M.A., Michallet, M.C., Papadopoulou, N., Utermöhlen, O., Kranidioti, K., Kollias, G., Tschopp, J., Pasparakis, M., 2008. Function of TRADD in tumor necrosis factor receptor 1 signaling and in TRIF-dependent inflammatory responses. Nat. Immunol. 9, 1037–1046.

Fagerberg, L., Hallström, B.M., Oksvold, P., Kampf, C., Djureinovic, D., Odeberg, J., Habuka, M., Tahmasebpoor, S., Danielsson, A., Edlund, K., Asplund, A., Sjöstedt, E., Lundberg, E., Szigyarto, C.A., Skogs, M., Takanen, J.O., Berling, H., Tegel, H., Mulder, J., Nilsson, P., Schwenk, J.M., Lindskog, C., Danielsson, F., Mardinoglu, A., Sivertsson, A., von Feilitzen, K., Forsberg, M., Zwahlen, M., Olsson, I., Navani, S., Huss, M., Nielsen, J., Ponten, F., Uhlén, M., 2014. Analysis of the human tissue-specific expression by genome-wide integration of transcriptomics and antibody-based proteomics. Mol. Cell. Proteomics 13, 397–406.

Fan, H., Zhang, K., Shan, L., Kuang, F., Chen, K., Zhu, K., Ma, H., Ju, G., Wang, Y.Z., 2016. Reactive astrocytes undergo M1 microglia/macrohpages-induced necroptosis in spinal cord injury. Mol. Neurodegener. 11, 14.

Fearns, C., Pan, Q., Mathison, J.C., Chuang, T.H., 2006. Triad3A regulates ubiquitination and proteasomal degradation of RIP1 following disruption of Hsp90 binding. J. Biol. Chem. 281, 34592–34600.

Fearon, E.R., Vogelstein, B., 1990. A genetic model for colorectal tumorigenesis. Cell 61, 759–767.

Feng, Y., Teitelbaum, D.H., 2013. Tumour necrosis factor—induced loss of intestinal barrier function requires TNFR1 and TNFR2 signalling in a mouse model of total parenteral nutrition. J. Physiol. 591, 3709–3723.

Feng, S., Yang, Y., Mei, Y., Ma, L., Zhu, D.E., Hoti, N., Castanares, M., Wu, M., 2007. Cleavage of RIP3 inactivates its caspase-independent apoptosis pathway by removal of kinase domain. Cell. Signal. 19, 2056–2067.

Feng, Y., Tsai, Y.H., Teitelbaum, D.H., Dempsey, P.J., 2012. Intestinal epithelial cell responses to total parenteral nutrition (TPN) are under complex regulation by ADAM17. Gastroenterology 142, S167–S168.

Feng, X., Song, Q., Yu, A., Tang, H., Peng, Z., Wang, X., 2015a. Receptor-interacting protein kinase 3 is a predictor of survival and plays a tumor suppressive role in colorectal cancer. Neoplasma 62, 592–601.

Feng, Y., Tsai, Y.H., Xiao, W., Ralls, M.W., Stoeck, A., Wilson, C.L., Raines, E.W., Teitelbaum, D.H., Dempsey, P.J., 2015b. Loss of ADAM17-mediated tumor necrosis factor alpha signaling in intestinal cells attenuates mucosal atrophy in a mouse model of parenteral nutrition. Mol. Cell. Biol. 35, 3604–3621.

Feoktistova, M., Geserick, P., Kellert, B., Dimitrova, D.P., Langlais, C., Hupe, M., Cain, K., Macfarlane, M., Häcker, G., Leverkus, M., 2011. cIAPs block ripoptosome formation, a RIP1/caspase-8 containing intracellular cell death complex differentially regulated by cFLIP isoforms. Mol. Cell 43, 449–463.

Fischer, K., Voelkl, S., Berger, J., Andreesen, R., Pomorski, T., Mackensen, A., 2006. Antigen recognition induces phosphatidylserine exposure on the cell surface of human CD8+ T cells. Blood 108, 4094–4101.

Franzè, E., Caruso, R., Stolfi, C., Sarra, M., Cupi, M.L., Ascolani, M., Sedda, S., Antenucci, C., Ruffa, A., Caprioli, F., MacDonald, T.T., Pallone, F., Monteleone, G., 2013. High expression of the "a disintegrin and metalloprotease" 19 (ADAM19), a sheddase for TNF-alpha in the mucosa of patients with inflammatory bowel diseases. Inflamm. Bowel Dis. 19, 501–511.

Franzke, C.W., Cobzaru, C., Triantafyllopoulou, A., Löffek, S., Horiuchi, K., Threadgill, D.W., Kurz, T., van Rooijen, N., Bruckner-Tuderman, L., Blobel, C.P., 2012. Epidermal ADAM17 maintains the skin barrier by regulating EGFR ligand-dependent terminal keratinocyte differentiation. J. Exp. Med. 209, 1105–1119.

Fre, S., Pallavi, S.K., Huyghe, M., Laé, M., Janssen, K.P., Robine, S., Artavanis-Tsakonas, S., Louvard, D., 2009. Notch and Wnt signals cooperatively control cell proliferation and tumorigenesis in the intestine. Proc. Natl. Acad. Sci. U. S. A. 106, 6309–6314.

Fridman, J.S., Caulder, E., Hansbury, M., Liu, X., Yang, G., Wang, Q., Lo, Y., Zhou, B.B., Pan, M., Thomas, S.M., Grandis, J.R., Zhuo, J., Yao, W., Newton, R.C., Friedman, S.M., Scherle, P.A., Vaddi, K., 2007. Selective inhibition of ADAM metalloproteases as a novel approach for modulating ErbB pathways in cancer. Clin. Cancer Res. 13, 1892–1902.

Fuchslocher Chico, J., Falk-Paulsen, M., Luzius, A., Ruder, B., Bolik, J., Schmidt-Arras, D., Linkermann, A., Becker, C., Rosenstiel, P., Rose-John, S., Adam, D., 2018. The enhanced susceptibility of ADAM-17 hypomorphic mice to DSS-induced colitis is not ameliorated by loss of RIPK3, revealing an unexpected function of ADAM-17 in necroptosis. Oncotarget 9, 12941–12958.

Fukata, Y., Adesnik, H., Iwanaga, T., Bredt, D.S., Nicoll, R.A., Fukata, M., 2006. Epilepsy-related ligand/receptor complex LGI1 and ADAM22 regulate synaptic transmission. Science 313, 1792–1795.

Galluzzi, L., Kepp, O., Chan, F.K., Kroemer, G., 2017. Necroptosis: mechanisms and relevance to disease. Annu. Rev. Pathol. Mech. Dis. 12, 103–130.

Galluzzi, L., Vitale, I., Aaronson, S.A., Abrams, J.M., Adam, D., Agostinis, P., Alnemri, E.S., Altucci, L., Amelio, I., Andrews, D.W., Annicchiarico-Petruzzelli, M., Antonov, A.V., Arama, E., Baehrecke, E.H., Barlev, N.A., Bazan, N.G., Bernassola, F., Bertrand, M.J.M., Bianchi, K., Blagosklonny, M.V., Blomgren, K., Borner, C., Boya, P., Brenner, C., Campanella, M., Candi, E., Carmona-Gutierrez, D., Cecconi, F., Chan, F.K., Chandel, N.S., Cheng, E.H., Chipuk, J.E., Cidlowski, J.A., Ciechanover, A., Cohen, G.M., Conrad, M., Cubillos-Ruiz, J.R., Czabotar, P.E., D'Angiolella, V., Dawson, T.M., Dawson, V.L., De Laurenzi, V., De Maria, R., Debatin, K.M., DeBerardinis, R.J., Deshmukh, M., Di Daniele, N., Di Virgilio, F., Dixit, V.M., Dixon, S.J., Duckett, C.S., Dynlacht, B.D., El-Deiry, W.S., Elrod, J.W., Fimia, G.M., Fulda, S., García-Sáez, A.J., Garg, A.D., Garrido, C., Gavathiotis, E., Golstein, P., Gottlieb, E., Green, D.R., Greene, L.A., Gronemeyer, H., Gross, A., Hajnoczky, G., Hardwick, J.M., Harris, I.S., Hengartner, M.O., Hetz, C., Ichijo, H., Jäättelä, M., Joseph, B., Jost, P.J., Juin, P.P., Kaiser, W.J., Karin, M., Kaufmann, T., Kepp, O., Kimchi, A., Kitsis, R.N., Klionsky, D.J., Knight, R.A., Kumar, S., Lee, S.W., Lemasters, J.J., Levine, B., Linkermann, A., Lipton, S.A., Lockshin, R.A., López-Otín, C., Lowe, S.W., Luedde, T., Lugli, E., MacFarlane, M., Madeo, F., Malewicz, M., Malorni, W., Manic, G., Marine, J.C., Martin, S.J., Martinou, J.C., Medema, J.P., Mehlen, P., Meier, P., Melino, S., Miao, E.A., Molkentin, J.D., Moll, U.M., Muñoz-Pinedo, C., Nagata, S., Nuñez, G., Oberst, A., Oren, M., Overholtzer, M., Pagano, M., Panaretakis, T., Pasparakis, M., Penninger, J.M., Pereira, D.M., Pervaiz, S., Peter, M.E., Piacentini, M., Pinton, P., Prehn, J.H.M., Puthalakath, H., Rabinovich, G.A., Rehm, M., Rizzuto, R., Rodrigues, C.M.P., Rubinsztein, D.C., Rudel, T., Ryan, K.M., Sayan, E., Scorrano, L., Shao, F., Shi, Y., Silke, J., Simon, H.U., Sistigu, A., Stockwell, B.R., Strasser, A., Szabadkai, G., Tait, S.W.G., Tang, D., Tavernarakis, N., Thorburn, A., Tsujimoto, Y., Turk, B., Vanden Berghe, T., Vandenabeele, P., Vander Heiden, M.G., Villunger, A., Virgin, H.W., Vousden, K.H., Vucic, D., Wagner, E.F., Walczak, H., Wallach, D., Wang, Y., Wells, J.A., Wood, W., Yuan, J., Zakeri, Z., Zhivotovsky, B., Zitvogel, L., Melino, G., Kroemer, G., 2018. Molecular mechanisms of cell death: recommendations of the Nomenclature Committee on Cell Death 2018. Cell Death Differ. 25, 486–541.

Garbers, C., Aparicio-Siegmund, S., Rose-John, S., 2015. The IL-6/gp130/STAT3 signaling axis: recent advances towards specific inhibition. Curr. Opin. Immunol. 34, 75–82.

Garton, K.J., Gough, P.J., Raines, E.W., 2006. Emerging roles for ectodomain shedding in the regulation of inflammatory responses. J. Leukoc. Biol. 79, 1105–1116.
Gautheron, J., Vucur, M., Reisinger, F., Cardenas, D.V., Roderburg, C., Koppe, C., Kreggenwinkel, K., Schneider, A.T., Bartneck, M., Neumann, U.P., Canbay, A., Reeves, H.L., Luedde, M., Tacke, F., Trautwein, C., Heikenwalder, M., Luedde, T., 2014. A positive feedback loop between RIP3 and JNK controls non-alcoholic steatohepatitis. EMBO Mol. Med. 6, 1062–1074.
Geng, J., Ito, Y., Shi, L., Amin, P., Chu, J., Ouchida, A.T., Mookhtiar, A.K., Zhao, H., Xu, D., Shan, B., Najafov, A., Gao, G., Akira, S., Yuan, J., 2017. Regulation of RIPK1 activation by TAK1-mediated phosphorylation dictates apoptosis and necroptosis. Nat. Commun. 8, 359.
Gentle, I.E., Wong, W.W., Evans, J.M., Bankovacki, A., Cook, W.D., Khan, N.R., Nachbur, U., Rickard, J., Anderton, H., Moulin, M., Lluis, J.M., Moujalled, D.M., Silke, J., Vaux, D.L., 2011. In TNF-stimulated cells, RIPK1 promotes cell survival by stabilizing TRAF2 and cIAP1, which limits induction of non-canonical NF-kappaB and activation of caspase-8. J. Biol. Chem. 286, 13282–13291.
Gérard, C., Hubeau, C., Carnet, O., Bellefroid, M., Sounni, N.E., Blacher, S., Bendavid, G., Moser, M., Fässler, R., Noel, A., Cataldo, D., Rocks, N., 2018. Microenvironment-derived ADAM28 prevents cancer dissemination. Oncotarget 9, 37185–37199.
Gerbe, F., Sidot, E., Smyth, D.J., Ohmoto, M., Matsumoto, I., Dardalhon, V., Cesses, P., Garnier, L., Pouzolles, M., Brulin, B., Bruschi, M., Harcus, Y., Zimmermann, V.S., Taylor, N., Maizels, R.M., Jay, P., 2016. Intestinal epithelial tuft cells initiate type 2 mucosal immunity to helminth parasites. Nature 529, 226–230.
Gerlach, B., Cordier, S.M., Schmukle, A.C., Emmerich, C.H., Rieser, E., Haas, T.L., Webb, A.I., Rickard, J.A., Anderton, H., Wong, W.W., Nachbur, U., Gangoda, L., Warnken, U., Purcell, A.W., Silke, J., Walczak, H., 2011. Linear ubiquitination prevents inflammation and regulates immune signalling. Nature 471, 591–596.
Gersemann, M., Stange, E.F., Wehkamp, J., 2011. From intestinal stem cells to inflammatory bowel diseases. World J. Gastroenterol. 17, 3198–3203.
Geserick, P., Hupe, M., Moulin, M., Wong, W.W., Feoktistova, M., Kellert, B., Gollnick, H., Silke, J., Leverkus, M., 2009. Cellular IAPs inhibit a cryptic CD95-induced cell death by limiting RIP1 kinase recruitment. J. Cell Biol. 187, 1037–1054.
Gong, Y.N., Guy, C., Olauson, H., Becker, J.U., Yang, M., Fitzgerald, P., Linkermann, A., Green, D.R., 2017. Escrt-III acts downstream of MLKL to regulate necroptotic cell death and its consequences. Cell 169, 286–300.
Gonzales, P.E., Solomon, A., Miller, A.B., Leesnitzer, M.A., Sagi, I., Milla, M.E., 2004. Inhibition of the tumor necrosis factor-alpha-converting enzyme by its pro domain. J. Biol. Chem. 279, 31638–31645.
González-Juarbe, N., Gilley, R.P., Hinojosa, C.A., Bradley, K.M., Kamei, A., Gao, G., Dube, P.H., Bergman, M.A., Orihuela, C.J., 2015. Pore-forming toxins induce macrophage necroptosis during acute bacterial pneumonia. PLoS Pathog. 11, e1005337.
Gribble, F.M., Reimann, F., 2016. Enteroendocrine cells: chemosensors in the intestinal epithelium. Annu. Rev. Physiol. 78, 277–299.
Grieve, A.G., Xu, H., Künzel, U., Bambrough, P., Sieber, B., Freeman, M., 2017. Phosphorylation of iRhom2 at the plasma membrane controls mammalian TACE-dependent inflammatory and growth factor signalling. Elife 6, e23968.
Günther, C., Martini, E., Wittkopf, N., Amann, K., Weigmann, B., Neumann, H., Waldner, M.J., Hedrick, S.M., Tenzer, S., Neurath, M.F., Becker, C., 2011. Caspase-8 regulates TNF-alpha-induced epithelial necroptosis and terminal ileitis. Nature 477, 335–339.

Günther, C., Buchen, B., Neurath, M.F., Becker, C., 2014. Regulation and pathophysiological role of epithelial turnover in the gut. Semin. Cell Dev. Biol. 35, 40–50.

Guo, H., Omoto, S., Harris, P.A., Finger, J.N., Bertin, J., Gough, P.J., Kaiser, W.J., Mocarski, E.S., 2015. Herpes simplex virus suppresses necroptosis in human cells. Cell Host Microbe 17, 243–251.

Haas, T.L., Emmerich, C.H., Gerlach, B., Schmukle, A.C., Cordier, S.M., Rieser, E., Feltham, R., Vince, J., Warnken, U., Wenger, T., Koschny, R., Komander, D., Silke, J., Walczak, H., 2009. Recruitment of the linear ubiquitin chain assembly complex stabilizes the TNF-R1 signaling complex and is required for TNF-mediated gene induction. Mol. Cell 36, 831–844.

Hanus, J., Zhang, H., Wang, Z., Liu, Q., Zhou, Q., Wang, S., 2013. Induction of necrotic cell death by oxidative stress in retinal pigment epithelial cells. Cell Death Dis. 4 e965.

Harris, F.I., Bell, G., Brunn, H., 1933. Chronic cicatrizing enteritis—regional ileitis (Crohn). Surg. Gynecol. Obstet. 57, 637–645.

Hartmann, D., de Strooper, B., Serneels, L., Craessaerts, K., Herreman, A., Annaert, W., Umans, L., Lübke, T., Lena Illert, A., von Figura, K., Saftig, P., 2002. The disintegrin/metalloprotease ADAM 10 is essential for Notch signalling but not for alpha-secretase activity in fibroblasts. Hum. Mol. Genet. 11, 2615–2624.

He, S., Wang, L., Miao, L., Wang, T., Du, F., Zhao, L., Wang, X., 2009. Receptor interacting protein kinase-3 determines cellular necrotic response to TNF-alpha. Cell 137, 1100–1111.

He, S., Liang, Y., Shao, F., Wang, X., 2011. Toll-like receptors activate programmed necrosis in macrophages through a receptor-interacting kinase-3-mediated pathway. Proc. Natl. Acad. Sci. U. S. A. 108, 20054–20059.

He, L., Peng, K., Liu, Y., Xiong, J., Zhu, F.F., 2013. Low expression of mixed lineage kinase domain-like protein is associated with poor prognosis in ovarian cancer patients. OncoTargets Ther. 6, 1539–1543.

He, G.W., Günther, C., Thonn, V., Yu, Y.Q., Martini, E., Buchen, B., Neurath, M.F., Stürzl, M., Becker, C., 2017. Regression of apoptosis-resistant colorectal tumors by induction of necroptosis in mice. J. Exp. Med. 214, 1655–1662.

Helander, H.F., Fändriks, L., 2014. Surface area of the digestive tract—revisited. Scand. J. Gastroenterol. 49, 681–689.

Hemler, M.E., 2005. Tetraspanin functions and associated microdomains. Nat. Rev. Mol. Cell Biol. 6, 801–811.

Herat, L., Rudnicka, C., Okada, Y., Mochizuki, S., Schlaich, M., Matthews, V., 2017. The metalloproteinase ADAM28 promotes metabolic dysfunction in mice. Int. J. Mol. Sci. 18, 884.

Hildebrand, J.M., Tanzer, M.C., Lucet, I.S., Young, S.N., Spall, S.K., Sharma, P., Pierotti, C., Garnier, J.M., Dobson, R.C., Webb, A.I., Tripaydonis, A., Babon, J.J., Mulcair, M.D., Scanlon, M.J., Alexander, W.S., Wilks, A.F., Czabotar, P.E., Lessene, G., Murphy, J.M., Silke, J., 2014. Activation of the pseudokinase MLKL unleashes the four-helix bundle domain to induce membrane localization and necroptotic cell death. Proc. Natl. Acad. Sci. U. S. A. 111, 15072–15077.

Hoffmann, F.S., Kuhn, P.H., Laurent, S.A., Hauck, S.M., Berer, K., Wendlinger, S.A., Krumbholz, M., Khademi, M., Olsson, T., Dreyling, M., Pfister, H.W., Alexander, T., Hiepe, F., Kümpfel, T., Crawford, H.C., Wekerle, H., Hohlfeld, R., Lichtenthaler, S.F., Meinl, E., 2015. The immunoregulator soluble TACI is released by ADAM10 and reflects B cell activation in autoimmunity. J. Immunol. 194, 542–552.

Holler, N., Zaru, R., Micheau, O., Thomé, M., Attinger, A., Valitutti, S., Bodmer, J.L., Schneider, P., Seed, B., Tschopp, J., 2000. Fas triggers an alternative, caspase-8-independent cell death pathway using the kinase RIP as effector molecule. Nat. Immunol. 1, 489–495.

Honda, H., Takamura, M., Yamagiwa, S., Genda, T., Horigome, R., Kimura, N., Setsu, T., Tominaga, K., Kamimura, H., Matsuda, Y., Wakai, T., Aoyagi, Y., Terai, S., 2017. Overexpression of a disintegrin and metalloproteinase 21 is associated with motility, metastasis, and poor prognosis in hepatocellular carcinoma. Sci. Rep. 7, 15485.

Horiuchi, K., Zhou, H.M., Kelly, K., Manova, K., Blobel, C.P., 2005. Evaluation of the contributions of ADAMs 9, 12, 15, 17, and 19 to heart development and ectodomain shedding of neuregulins beta1 and beta2. Dev. Biol. 283, 459–471.

Horiuchi, K., Kimura, T., Miyamoto, T., Takaishi, H., Okada, Y., Toyama, Y., Blobel, C.P., 2007. Cutting edge: TNF-alpha-converting enzyme (TACE/ADAM17) inactivation in mouse myeloid cells prevents lethality from endotoxin shock. J. Immunol. 179, 2686–2689.

Horiuchi, K., Kimura, T., Miyamoto, T., Miyamoto, K., Akiyama, H., Takaishi, H., Morioka, H., Nakamura, T., Okada, Y., Blobel, C.P., Toyama, Y., 2009. Conditional inactivation of TACE by a Sox9 promoter leads to osteoporosis and increased granulopoiesis via dysregulation of IL-17 and G-CSF. J. Immunol. 182, 2093–2101.

Howitt, M.R., Lavoie, S., Michaud, M., Blum, A.M., Tran, S.V., Weinstock, J.V., Gallini, C.A., Redding, K., Margolskee, R.F., Osborne, L.C., Artis, D., Garrett, W.S., 2016. Tuft cells, taste-chemosensory cells, orchestrate parasite type 2 immunity in the gut. Science 351, 1329–1333.

Hsia, H.E., Tüshaus, J., Brummer, T., Zheng, Y., Scilabra, S.D., Lichtenthaler, S.F., 2019. Functions of 'A disintegrin and metalloproteases (ADAMs)' in the mammalian nervous system. Cell. Mol. Life Sci. 76, 3055–3081.

Hsu, H., Xiong, J., Goeddel, D.V., 1995. The TNF receptor 1-associated protein TRADD signals cell death and NF-kappa B activation. Cell 81, 495–504.

Huang, Z., Wu, S.Q., Liang, Y., Zhou, X., Chen, W., Li, L., Wu, J., Zhuang, Q., Chen, C., Li, J., Zhong, C.Q., Xia, W., Zhou, R., Zheng, C., Han, J., 2015. RIP1/RIP3 binding to HSV-1 ICP6 initiates necroptosis to restrict virus propagation in mice. Cell Host Microbe 17, 229–242.

Huang, D., Zheng, X., Wang, Z.A., Chen, X., He, W.T., Zhang, Y., Xu, J.G., Zhao, H., Shi, W., Wang, X., Zhu, Y., Han, J., 2017. The MLKL channel in necroptosis is an octamer formed by tetramers in a dyadic process. Mol. Cell. Biol. 37, e00497-16.

Iba, K., Albrechtsen, R., Gilpin, B., Fröhlich, C., Loechel, F., Zolkiewska, A., Ishiguro, K., Kojima, T., Liu, W., Langford, J.K., Sanderson, R.D., Brakebusch, C., Fässler, R., Wewer, U.M., 2000. The cysteine-rich domain of human ADAM 12 supports cell adhesion through syndecans and triggers signaling events that lead to beta1 integrin-dependent cell spreading. J. Cell Biol. 149, 1143–1156.

Ireland, H., Houghton, C., Howard, L., Winton, D.J., 2005. Cellular inheritance of a Cre-activated reporter gene to determine paneth cell longevity in the murine small intestine. Dev. Dyn. 233, 1332–1336.

Jaco, I., Annibaldi, A., Lalaoui, N., Wilson, R., Tenev, T., Laurien, L., Kim, C., Jamal, K., Wicky John, S., Liccardi, G., Chau, D., Murphy, J.M., Brumatti, G., Feltham, R., Pasparakis, M., Silke, J., Meier, P., 2017. MK2 phosphorylates ripk1 to prevent TNF-induced cell death. Mol. Cell 66, 698–710.

Jacobsen, A.V., Lowes, K.N., Tanzer, M.C., Lucet, I.S., Hildebrand, J.M., Petrie, E.J., van Delft, M.F., Liu, Z., Conos, S.A., Zhang, J.G., Huang, D.C., Silke, J., Lessene, G., Murphy, J.M., 2016. HSP90 activity is required for MLKL oligomerisation and membrane translocation and the induction of necroptotic cell death. Cell Death Dis. 7, e2051.

Janes, P.W., Saha, N., Barton, W.A., Kolev, M.V., Wimmer-Kleikamp, S.H., Nievergall, E., Blobel, C.P., Himanen, J.P., Lackmann, M., Nikolov, D.B., 2005. Adam meets Eph: an ADAM substrate recognition module acts as a molecular switch for ephrin cleavage in trans. Cell 123, 291–304.

Jones, A.V., Lambert, D.W., Speight, P.M., Whawell, S.A., 2013. ADAM 10 is over expressed in oral squamous cell carcinoma and contributes to invasive behaviour through a functional association with alphavbeta6 integrin. FEBS Lett. 587, 3529–3534.

Jones, J.C., Rustagi, S., Dempsey, P.J., 2016. ADAM proteases and gastrointestinal function. Annu. Rev. Physiol. 78, 243–276.

Jostins, L., Ripke, S., Weersma, R.K., Duerr, R.H., McGovern, D.P., Hui, K.Y., Lee, J.C., Schumm, L.P., Sharma, Y., Anderson, C.A., Essers, J., Mitrovic, M., Ning, K., Cleynen, I., Theatre, E., Spain, S.L., Raychaudhuri, S., Goyette, P., Wei, Z., Abraham, C., Achkar, J.P., Ahmad, T., Amininejad, L., Ananthakrishnan, A.N., Andersen, V., Andrews, J.M., Baidoo, L., Balschun, T., Bampton, P.A., Bitton, A., Boucher, G., Brand, S., Buning, C., Cohain, A., Cichon, S., D'Amato, M., De Jong, D., Devaney, K.L., Dubinsky, M., Edwards, C., Ellinghaus, D., Ferguson, L.R., Franchimont, D., Fransen, K., Gearry, R., Georges, M., Gieger, C., Glas, J., Haritunians, T., Hart, A., Hawkey, C., Hedl, M., Hu, X., Karlsen, T.H., Kupcinskas, L., Kugathasan, S., Latiano, A., Laukens, D., Lawrance, I.C., Lees, C.W., Louis, E., Mahy, G., Mansfield, J., Morgan, A.R., Mowat, C., Newman, W., Palmieri, O., Ponsioen, C.Y., Potocnik, U., Prescott, N.J., Regueiro, M., Rotter, J.I., Russell, R.K., Sanderson, J.D., Sans, M., Satsangi, J., Schreiber, S., Simms, L.A., Sventoraityte, J., Targan, S.R., Taylor, K.D., Tremelling, M., Verspaget, H.W., De Vos, M., Wijmenga, C., Wilson, D.C., Winkelmann, J., Xavier, R.J., Zeissig, S., Zhang, B., Zhang, C.K., Zhao, H., Silverberg, M.S., Annese, V., Hakonarson, H., Brant, S.R., Radford-Smith, G., Mathew, C.G., Rioux, J.D., Schadt, E.E., Daly, M.J., Franke, A., Parkes, M., Vermeire, S., Barrett, J.C., Cho, J.H., 2012. Host-microbe interactions have shaped the genetic architecture of inflammatory bowel disease. Nature 491, 119–124.

Kaczmarek, A., Vandenabeele, P., Krysko, D.V., 2013. Necroptosis: the release of damage-associated molecular patterns and its physiological relevance. Immunity 38, 209–223.

Kaiser, W.J., Upton, J.W., Long, A.B., Livingston-Rosanoff, D., Daley-Bauer, L.P., Hakem, R., Caspary, T., Mocarski, E.S., 2011. RIP3 mediates the embryonic lethality of caspase-8-deficient mice. Nature 471, 368–372.

Kaiser, W.J., Sridharan, H., Huang, C., Mandal, P., Upton, J.W., Gough, P.J., Sehon, C.A., Marquis, R.W., Bertin, J., Mocarski, E.S., 2013. Toll-like receptor 3-mediated necrosis via TRIF, RIP3, and MLKL. J. Biol. Chem. 288, 31268–31279.

Kaiser, W.J., Daley-Bauer, L.P., Thapa, R.J., Mandal, P., Berger, S.B., Huang, C., Sundararajan, A., Guo, H., Roback, L., Speck, S.H., Bertin, J., Gough, P.J., Balachandran, S., Mocarski, E.S., 2014. RIP1 suppresses innate immune necrotic as well as apoptotic cell death during mammalian parturition. Proc. Natl. Acad. Sci. U. S. A. 111, 7753–7758.

Kalai, M., van Loo, G., Vanden Berghe, T., Meeus, A., Burm, W., Saelens, X., Vandenabeele, P., 2002. Tipping the balance between necrosis and apoptosis in human and murine cells treated with interferon and dsRNA. Cell Death Differ. 9, 981–994.

Karl, I., Jossberger-Werner, M., Schmidt, N., Horn, S., Goebeler, M., Leverkus, M., Wajant, H., Giner, T., 2014. TRAF2 inhibits TRAIL- and CD95L-induced apoptosis and necroptosis. Cell Death Dis. 5 e1444.

Kawahara, A., Ohsawa, Y., Matsumura, H., Uchiyama, Y., Nagata, S., 1998. Caspase-independent cell killing by Fas-associated protein with death domain. J. Cell Biol. 143, 1353–1360.

Killock, D.J., Ivetić, A., 2010. The cytoplasmic domains of TNFalpha-converting enzyme (TACE/ADAM17) and L-selectin are regulated differently by p38 MAPK and PKC to promote ectodomain shedding. Biochem. J. 428, 293–304.

Kim, S.J., Li, J., 2013. Caspase blockade induces RIP3-mediated programmed necrosis in Toll-like receptor-activated microglia. Cell Death Dis. 4, e716.

King, M.D., Whitaker-Lea, W.A., Campbell, J.M., Alleyne Jr., C.H., Dhandapani, K.M., 2014. Necrostatin-1 reduces neurovascular injury after intracerebral hemorrhage. Int. J. Cell Biol. 2014, 495817.

Kinzler, K.W., Vogelstein, B., 1996. Lessons from hereditary colorectal cancer. Cell 87, 159–170.

Koehler, H., Cotsmire, S., Langland, J., Kibler, K.V., Kalman, D., Upton, J.W., Mocarski, E.S., Jacobs, B.L., 2017. Inhibition of DAI-dependent necroptosis by the Z-DNA binding domain of the vaccinia virus innate immune evasion protein, E3. Proc. Natl. Acad. Sci. U. S. A. 114, 11506–11511.

Kohutek, Z.A., diPierro, C.G., Redpath, G.T., Hussaini, I.M., 2009. ADAM-10-mediated N-cadherin cleavage is protein kinase C-alpha dependent and promotes glioblastoma cell migration. J. Neurosci. 29, 4605–4615.

Kuhn, P.H., Colombo, A.V., Schusser, B., Dreymueller, D., Wetzel, S., Schepers, U., Herber, J., Ludwig, A., Kremmer, E., Montag, D., Müller, U., Schweizer, M., Saftig, P., Bräse, S., Lichtenthaler, S.F., 2016. Systematic substrate identification indicates a central role for the metalloprotease ADAM10 in axon targeting and synapse function. Elife 5, e12748.

Künzel, U., Grieve, A.G., Meng, Y., Sieber, B., Cowley, S.A., Freeman, M., 2018. FRMD8 promotes inflammatory and growth factor signalling by stabilising the iRhom/ADAM17 sheddase complex. Elife 7, e35012.

Kuriakose, T., Man, S.M., Malireddi, R.K., Karki, R., Kesavardhana, S., Place, D.E., Neale, G., Vogel, P., Kanneganti, T.D., 2016. ZBP1/DAI is an innate sensor of influenza virus triggering the NLRP3 inflammasome and programmed cell death pathways. Sci. Immunol. 1, aag2045.

Lafont, E., Kantari-Mimoun, C., Draber, P., De Miguel, D., Hartwig, T., Reichert, M., Kupka, S., Shimizu, Y., Taraborrelli, L., Spit, M., Sprick, M.R., Walczak, H., 2017. The linear ubiquitin chain assembly complex regulates TRAIL-induced gene activation and cell death. EMBO J. 36, 1147–1166.

Lafont, E., Draber, P., Rieser, E., Reichert, M., Kupka, S., de Miguel, D., Draberova, H., von Mässenhausen, A., Bhamra, A., Henderson, S., Wojdyla, K., Chalk, A., Surinova, S., Linkermann, A., Walczak, H., 2018. TBK1 and IKKepsilon prevent TNF-induced cell death by RIPK1 phosphorylation. Nat. Cell Biol. 20, 1389–1399.

Lambrecht, B.N., Vanderkerken, M., Hammad, H., 2018. The emerging role of ADAM metalloproteinases in immunity. Nat. Rev. Immunol. 18, 745–758.

Lammich, S., Kojro, E., Postina, R., Gilbert, S., Pfeiffer, R., Jasionowski, M., Haass, C., Fahrenholz, F., 1999. Constitutive and regulated alpha-secretase cleavage of Alzheimer's amyloid precursor protein by a disintegrin metalloprotease. Proc. Natl. Acad. Sci. U. S. A. 96, 3922–3927.

Laster, S.M., Wood, J.G., Gooding, L.R., 1988. Tumor necrosis factor can induce both apoptic and necrotic forms of cell lysis. J. Immunol. 141, 2629–2634.

Le Gall, S.M., Bobé, P., Reiss, K., Horiuchi, K., Niu, X.D., Lundell, D., Gibb, D.R., Conrad, D., Saftig, P., Blobel, C.P., 2009. ADAMs 10 and 17 represent differentially regulated components of a general shedding machinery for membrane proteins such as transforming growth factor alpha, L-selectin, and tumor necrosis factor alpha. Mol. Biol. Cell 20, 1785–1794.

Lee, S.B., Kim, J.J., Han, S.A., Fan, Y., Guo, L.S., Aziz, K., Nowsheen, S., Kim, S.S., Park, S.Y., Luo, Q., Chung, J.O., Choi, S.I., Aziz, A., Yin, P., Tong, S.Y., Fiesel, F.C., Springer, W., Zhang, J.S., Lou, Z., 2019. The AMPK-Parkin axis negatively regulates necroptosis and tumorigenesis by inhibiting the necrosome. Nat. Cell Biol. 21, 940–951.

Leonard, J.D., Lin, F., Milla, M.E., 2005. Chaperone-like properties of the prodomain of TNFalpha-converting enzyme (TACE) and the functional role of its cysteine switch. Biochem. J. 387, 797–805.

Letronne, F., Laumet, G., Ayral, A.M., Chapuis, J., Demiautte, F., Laga, M., Vandenberghe, M.E., Malmanche, N., Leroux, F., Eysert, F., Sottejeau, Y., Chami, L., Flaig, A., Bauer, C., Dourlen, P., Lesaffre, M., Delay, C., Huot, L., Dumont, J., Werkmeister, E., Lafont, F., Mendes, T., Hansmannel, F., Dermaut, B., Deprez, B., Hérard, A.S., Dhenain, M., Souedet, N., Pasquier, F., Tulasne, D., Berr, C., Hauw, J.J., Lemoine, Y., Amouyel, P., Mann, D., Déprez, R., Checler, F., Hot, D., Delzescaux, T., Gevaert, K., Lambert, J.C., 2016. ADAM30 downregulates APP-linked defects through cathepsin D activation in Alzheimer's disease. EBioMedicine 9, 278–292.

Lewis, J., Devin, A., Miller, A., Lin, Y., Rodriguez, Y., Neckers, L., Liu, Z.G., 2000. Disruption of hsp90 function results in degradation of the death domain kinase, receptor-interacting protein (RIP), and blockage of tumor necrosis factor-induced nuclear factor-kappaB activation. J. Biol. Chem. 275, 10519–10526.

Li, H., Kobayashi, M., Blonska, M., You, Y., Lin, X., 2006. Ubiquitination of RIP is required for tumor necrosis factor alpha-induced NF-kappaB activation. J. Biol. Chem. 281, 13636–13643.

Li, J., McQuade, T., Siemer, A.B., Napetschnig, J., Moriwaki, K., Hsiao, Y.S., Damko, E., Moquin, D., Walz, T., McDermott, A., Chan, F.K., Wu, H., 2012. The RIP1/RIP3 necrosome forms a functional amyloid signaling complex required for programmed necrosis. Cell 150, 339–350.

Li, D., Xiao, Z., Wang, G., Song, X., 2015a. Knockdown of ADAM10 inhibits migration and invasion of fibroblast-like synoviocytes in rheumatoid arthritis. Mol. Med. Rep. 12, 5517–5523.

Li, D., Xu, T., Cao, Y., Wang, H., Li, L., Chen, S., Wang, X., Shen, Z., 2015b. A cytosolic heat shock protein 90 and cochaperone CDC37 complex is required for RIP3 activation during necroptosis. Proc. Natl. Acad. Sci. U. S. A. 112, 5017–5022.

Li, X., Guo, J., Ding, A.P., Qi, W.W., Zhang, P.H., Lv, J., Qiu, W.S., Sun, Z.Q., 2017. Association of mixed lineage kinase domain-like protein expression with prognosis in patients with colon cancer. Technol. Cancer Res. Treat. 16, 428–434.

Li, X., Gong, W., Wang, H., Li, T., Attri, K.S., Lewis, R.E., Kalil, A.C., Bhinderwala, F., Powers, R., Yin, G., Herring, L.E., Asara, J.M., Lei, Y.L., Yang, X., Rodriguez, D.A., Yang, M., Green, D.R., Singh, P.K., Wen, H., 2019. O-GlcNAc transferase suppresses inflammation and necroptosis by targeting receptor-interacting serine/threonine-protein kinase 3. Immunity 50, 576–590.

Lim, S.Y., Davidson, S.M., Mocanu, M.M., Yellon, D.M., Smith, C.C., 2007. The cardioprotective effect of necrostatin requires the cyclophilin-D component of the mitochondrial permeability transition pore. Cardiovasc. Drugs Ther. 21, 467–469.

Lin, Y., Devin, A., Rodriguez, Y., Liu, Z.G., 1999. Cleavage of the death domain kinase RIP by Caspase-8 prompts TNF-induced apoptosis. Genes Dev. 13, 2514–2526.

Lin, J., Li, H., Yang, M., Ren, J., Huang, Z., Han, F., Huang, J., Ma, J., Zhang, D., Zhang, Z., Wu, J., Huang, D., Qiao, M., Jin, G., Wu, Q., Huang, Y., Du, J., Han, J., 2013. A role of RIP3-mediated macrophage necrosis in atherosclerosis development. Cell Rep. 3, 200–210.

Lin, J., Kumari, S., Kim, C., Van, T.M., Wachsmuth, L., Polykratis, A., Pasparakis, M., 2016. RIPK1 counteracts ZBP1-mediated necroptosis to inhibit inflammation. Nature 540, 124–128.

Linkermann, A., Bräsen, J.H., De Zen, F., Weinlich, R., Schwendener, R.A., Green, D.R., Kunzendorf, U., Krautwald, S., 2012. Dichotomy between RIP1- and RIP3-mediated necroptosis in tumor necrosis factor-alpha-induced shock. Mol. Med. 18, 577–586.

Linkermann, A., Bräsen, J.H., Darding, M., Jin, M.K., Sanz, A.B., Heller, J.O., De Zen, F., Weinlich, R., Ortiz, A., Walczak, H., Weinberg, J.M., Green, D.R., Kunzendorf, U., Krautwald, S., 2013. Two independent pathways of regulated necrosis mediate ischemia-reperfusion injury. Proc. Natl. Acad. Sci. U. S. A. 110, 12024–12029.

Liu, J.Z., van Sommeren, S., Huang, H., Ng, S.C., Alberts, R., Takahashi, A., Ripke, S., Lee, J.C., Jostins, L., Shah, T., Abedian, S., Cheon, J.H., Cho, J., Dayani, N.E., Franke, L., Fuyuno, Y., Hart, A., Juyal, R.C., Juyal, G., Kim, W.H., Morris, A.P., Poustchi, H., Newman, W.G., Midha, V., Orchard, T.R., Vahedi, H., Sood, A., Sung, J.Y., Malekzadeh, R., Westra, H.J., Yamazaki, K., Yang, S.K., International Multiple Sclerosis Genetics Consortium, International I. B. D. Genetics Consortium, Barrett, J.C., Alizadeh, B.Z., Parkes, M., Bk, T., Daly, M.J., Kubo, M., Anderson, C.A., Weersma, R.K., 2015a. Association analyses identify 38 susceptibility loci for inflammatory bowel disease and highlight shared genetic risk across populations. Nat. Genet. 47, 979–986.

Liu, M., Wu, W., Li, H., Li, S., Huang, L.T., Yang, Y.Q., Sun, Q., Wang, C.X., Yu, Z., Hang, C.H., 2015b. Necroptosis, a novel type of programmed cell death, contributes to early neural cells damage after spinal cord injury in adult mice. J. Spinal Cord Med. 38, 745–753.

Liu, Z.Y., Wu, B., Guo, Y.S., Zhou, Y.H., Fu, Z.G., Xu, B.Q., Li, J.H., Jing, L., Jiang, J.L., Tang, J., Chen, Z.N., 2015c. Necrostatin-1 reduces intestinal inflammation and colitis-associated tumorigenesis in mice. Am. J. Cancer Res. 5, 3174–3185.

Liu, S., Liu, H., Johnston, A., Hanna-Addams, S., Reynoso, E., Xiang, Y., Wang, Z., 2017. MLKL forms disulfide bond-dependent amyloid-like polymers to induce necroptosis. Proc. Natl. Acad. Sci. U. S. A. 114, E7450–E7459.

Liu, W., Chen, B., Wang, Y., Meng, C., Huang, H., Huang, X.R., Qin, J., Mulay, S.R., Anders, H.J., Qiu, A., Yang, B., Freeman, G.J., Lu, H.J., Lin, H.Y., Zheng, Z.H., Lan, H.Y., Huang, Y., Xia, Y., 2018. RGMb protects against acute kidney injury by inhibiting tubular cell necroptosis via an MLKL-dependent mechanism. Proc. Natl. Acad. Sci. U. S. A. 115, E1475–E1484.

Liu, Z.Y., Zheng, M., Li, Y.M., Fan, X.Y., Wang, J.C., Li, Z.C., Yang, H.J., Yu, J.M., Cui, J., Jiang, J.L., Tang, J., Chen, Z.N., 2019. RIP3 promotes colitis-associated colorectal cancer by controlling tumor cell proliferation and CXCL1-induced immune suppression. Theranostics 9, 3659–3673.

Lorenzen, I., Lokau, J., Korpys, Y., Oldefest, M., Flynn, C.M., Künzel, U., Garbers, C., Freeman, M., Grötzinger, J., Düsterhoft, S., 2016. Control of ADAM17 activity by regulation of its cellular localisation. Sci. Rep. 6, 35067.

Lu, J.V., Weist, B.M., van Raam, B.J., Marro, B.S., Nguyen, L.V., Srinivas, P., Bell, B.D., Luhrs, K.A., Lane, T.E., Salvesen, G.S., Walsh, C.M., 2011. Complementary roles of Fas-associated death domain (FADD) and receptor interacting protein kinase-3 (RIPK3) in T-cell homeostasis and antiviral immunity. Proc. Natl. Acad. Sci. U. S. A. 108, 15312–15317.

Luedde, M., Lutz, M., Carter, N., Sosna, J., Jacoby, C., Vucur, M., Gautheron, J., Roderburg, C., Borg, N., Reisinger, F., Hippe, H.J., Linkermann, A., Wolf, M.J., Rose-John, S., Lüllmann-Rauch, R., Adam, D., Flögel, U., Heikenwalder, M., Luedde, T., Frey, N., 2014. RIP3, a kinase promoting necroptotic cell death, mediates adverse remodelling after myocardial infarction. Cardiovasc. Res. 103, 206–216.

Maelfait, J., Liverpool, L., Bridgeman, A., Ragan, K.B., Upton, J.W., Rehwinkel, J., 2017. Sensing of viral and endogenous RNA by ZBP1/DAI induces necroptosis. EMBO J. 36, 2529–2543.

Maloy, K.J., Powrie, F., 2011. Intestinal homeostasis and its breakdown in inflammatory bowel disease. Nature 474, 298–306.

Marcello, E., Gardoni, F., Di Luca, M., Pérez-Otaño, I., 2010. An arginine stretch limits ADAM10 exit from the endoplasmic reticulum. J. Biol. Chem. 285, 10376–10384.

Maretzky, T., McIlwain, D.R., Issuree, P.D., Li, X., Malapeira, J., Amin, S., Lang, P.A., Mak, T.W., Blobel, C.P., 2013. iRhom2 controls the substrate selectivity of stimulated

ADAM17-dependent ectodomain shedding. Proc. Natl. Acad. Sci. U. S. A. 110, 11433–11438.

Marino, M.W., Dunn, A., Grail, D., Inglese, M., Noguchi, Y., Richards, E., Jungbluth, A., Wada, H., Moore, M., Williamson, B., Basu, S., Old, L.J., 1997. Characterization of tumor necrosis factor-deficient mice. Proc. Natl. Acad. Sci. U. S. A. 94, 8093–8098.

Martins, M., Mansinho, A., Cruz-Duarte, R., Martins, S.L., Costa, L., 2018. Anti-EGFR therapy to treat metastatic colorectal cancer: not for all. Adv. Exp. Med. Biol. 1110, 113–131.

Marzia, M., Guaiquil, V., Horne, W.C., Blobel, C.P., Baron, R., Chiusaroli, R., 2011. Lack of ADAM15 in mice is associated with increased osteoblast function and bone mass. Biol. Chem. 392, 877–885.

Masaki, M., Kurisaki, T., Shirakawa, K., Sehara-Fujisawa, A., 2005. Role of meltrin {alpha} (ADAM12) in obesity induced by high-fat diet. Endocrinology 146, 1752–1763.

Matsuzawa-Ishimoto, Y., Shono, Y., Gomez, L.E., Hubbard-Lucey, V.M., Cammer, M., Neil, J., Dewan, M.Z., Lieberman, S.R., Lazrak, A., Marinis, J.M., Beal, A., Harris, P.A., Bertin, J., Liu, C., Ding, Y., van den Brink, M.R.M., Cadwell, K., 2017. Autophagy protein ATG16L1 prevents necroptosis in the intestinal epithelium. J. Exp. Med. 214, 3687–3705.

Mazzocca, A., Coppari, R., De Franco, R., Cho, J.Y., Libermann, T.A., Pinzani, M., Toker, A., 2005. A secreted form of ADAM9 promotes carcinoma invasion through tumor-stromal interactions. Cancer Res. 65, 4728–4738.

McComb, S., Cheung, H.H., Korneluk, R.G., Wang, S., Krishnan, L., Sad, S., 2012. cIAP1 and cIAP2 limit macrophage necroptosis by inhibiting Rip1 and Rip3 activation. Cell Death Differ. 19, 1791–1801.

McComb, S., Cessford, E., Alturki, N.A., Joseph, J., Shutinoski, B., Startek, J.B., Gamero, A.M., Mossman, K.L., Sad, S., 2014a. Type-I interferon signaling through ISGF3 complex is required for sustained Rip3 activation and necroptosis in macrophages. Proc. Natl. Acad. Sci. U. S. A. 111, E3206–E3213.

McComb, S., Shutinoski, B., Thurston, S., Cessford, E., Kumar, K., Sad, S., 2014b. Cathepsins limit macrophage necroptosis through cleavage of Rip1 kinase. J. Immunol. 192, 5671–5678.

McCormick, A.L., Skaletskaya, A., Barry, P.A., Mocarski, E.S., Goldmacher, V.S., 2003. Differential function and expression of the viral inhibitor of caspase 8-induced apoptosis (vICA) and the viral mitochondria-localized inhibitor of apoptosis (vMIA) cell death suppressors conserved in primate and rodent cytomegaloviruses. Virology 316, 221–233.

McIlwain, D.R., Lang, P.A., Maretzky, T., Hamada, K., Ohishi, K., Maney, S.K., Berger, T., Murthy, A., Duncan, G., Xu, H.C., Lang, K.S., Häussinger, D., Wakeham, A., Itie-Youten, A., Khokha, R., Ohashi, P.S., Blobel, C.P., Mak, T.W., 2012. iRhom2 regulation of TACE controls TNF-mediated protection against Listeria and responses to LPS. Science 335, 229–232.

McNamara, D.E., Dovey, C.M., Hale, A.T., Quarato, G., Grace, C.R., Guibao, C.D., Diep, J., Nourse, A., Cai, C.R., Wu, H., Kalathur, R.C., Green, D.R., York, J.D., Carette, J.E., Moldoveanu, T., 2019. Direct activation of human MLKL by a select repertoire of inositol phosphate metabolites. Cell Chem. Biol. 26, 863–877.

Meng, L., Jin, W., Wang, X., 2015. RIP3-mediated necrotic cell death accelerates systematic inflammation and mortality. Proc. Natl. Acad. Sci. U. S. A. 112, 11007–11012.

Menon, M.B., Gropengiesser, J., Fischer, J., Novikova, L., Deuretzbacher, A., Lafera, J., Schimmeck, H., Czymmeck, N., Ronkina, N., Kotlyarov, A., Aepfelbacher, M., Gaestel, M., Ruckdeschel, K., 2017. p38MAPK/MK2-dependent phosphorylation controls cytotoxic RIPK1 signalling in inflammation and infection. Nat. Cell Biol. 19, 1248–1259.

Merchant, N.B., Voskresensky, I., Rogers, C.M., Lafleur, B., Dempsey, P.J., Graves-Deal, R., Revetta, F., Foutch, A.C., Rothenberg, M.L., Washington, M.K., Coffey, R.J., 2008. TACE/ADAM-17: a component of the epidermal growth factor receptor axis and a promising therapeutic target in colorectal cancer. Clin. Cancer Res. 14, 1182–1191.

Micheau, O., Tschopp, J., 2003. Induction of TNF receptor I-mediated apoptosis via two sequential signaling complexes. Cell 114, 181–190.

Mitsui, Y., Mochizuki, S., Kodama, T., Shimoda, M., Ohtsuka, T., Shiomi, T., Chijiiwa, M., Ikeda, T., Kitajima, M., Okada, Y., 2006. ADAM28 is overexpressed in human breast carcinomas: implications for carcinoma cell proliferation through cleavage of insulin-like growth factor binding protein-3. Cancer Res. 66, 9913–9920.

Moghadaszadeh, B., Albrechtsen, R., Guo, L.T., Zaik, M., Kawaguchi, N., Borup, R.H., Kronqvist, P., Schroder, H.D., Davies, K.E., Voit, T., Nielsen, F.C., Engvall, E., Wewer, U.M., 2003. Compensation for dystrophin-deficiency: ADAM12 overexpression in skeletal muscle results in increased alpha 7 integrin, utrophin and associated glycoproteins. Hum. Mol. Genet. 12, 2467–2479.

Mompeán, M., Li, W., Li, J., Laage, S., Siemer, A.B., Bozkurt, G., Wu, H., McDermott, A.E., 2018. The structure of the necrosome RIPK1-RIPK3 core, a human hetero-amyloid signaling complex. Cell 173, 1244–1253.

Moquin, D.M., McQuade, T., Chan, F.K., 2013. CYLD deubiquitinates RIP1 in the TNFalpha-induced necrosome to facilitate kinase activation and programmed necrosis. PLoS One 8, e76841.

Morgan, J.E., Prola, A., Mariot, V., Pini, V., Meng, J., Hourde, C., Dumonceaux, J., Conti, F., Relaix, F., Authier, F.J., Tiret, L., Muntoni, F., Bencze, M., 2018. Necroptosis mediates myofibre death in dystrophin-deficient mice. Nat. Commun. 9, 3655.

Morioka, S., Broglie, P., Omori, E., Ikeda, Y., Takaesu, G., Matsumoto, K., Ninomiya-Tsuji, J., 2014. TAK1 kinase switches cell fate from apoptosis to necrosis following TNF stimulation. J. Cell Biol. 204, 607–623.

Moriwaki, K., Chan, F.K., 2016. Regulation of RIPK3- and RHIM-dependent necroptosis by the proteasome. J. Biol. Chem. 291, 5948–5959.

Moriwaki, K., Chan, F.K., 2017. The inflammatory signal adaptor RIPK3: functions beyond necroptosis. Int. Rev. Cell Mol. Biol. 328, 253–275.

Moriwaki, K., Balaji, S., McQuade, T., Malhotra, N., Kang, J., Chan, F.K., 2014. The necroptosis adaptor RIPK3 promotes injury-induced cytokine expression and tissue repair. Immunity 41, 567–578.

Moser, A.R., Luongo, C., Gould, K.A., McNeley, M.K., Shoemaker, A.R., Dove, W.F., 1995. ApcMin: a mouse model for intestinal and mammary tumorigenesis. Eur. J. Cancer 31A, 1061–1064.

Moss, M.L., Jin, S.L., Milla, M.E., Bickett, D.M., Burkhart, W., Carter, H.L., Chen, W.J., Clay, W.C., Didsbury, J.R., Hassler, D., Hoffman, C.R., Kost, T.A., Lambert, M.H., Leesnitzer, M.A., McCauley, P., McGeehan, G., Mitchell, J., Moyer, M., Pahel, G., Rocque, W., Overton, L.K., Schoenen, F., Seaton, T., Su, J.L., Warner, J., Willard, D., Becherer, J.D., 1997. Cloning of a disintegrin metalloproteinase that processes precursor tumour-necrosis factor-alpha. Nature 385, 733–736.

Moss, M.L., Bomar, M., Liu, Q., Sage, H., Dempsey, P., Lenhart, P.M., Gillispie, P.A., Stoeck, A., Wildeboer, D., Bartsch, J.W., Palmisano, R., Zhou, P., 2007. The ADAM10 prodomain is a specific inhibitor of ADAM10 proteolytic activity and inhibits cellular shedding events. J. Biol. Chem. 282, 35712–35721.

Moujalled, D.M., Cook, W.D., Okamoto, T., Murphy, J., Lawlor, K.E., Vince, J.E., Vaux, D.L., 2013. TNF can activate RIPK3 and cause programmed necrosis in the absence of RIPK1. Cell Death Dis. 4, e465.

Müller, M., Wetzel, S., Köhn-Gaone, J., Chalupsky, K., Lüllmann-Rauch, R., Barikbin, R., Bergmann, J., Wöhner, B., Zbodakova, O., Leuschner, I., Martin, G., Tiegs, G.,

Rose-John, S., Sedlacek, R., Tirnitz-Parker, J.E., Saftig, P., Schmidt-Arras, D., 2016. A disintegrin and metalloprotease 10 (ADAM10) is a central regulator of murine liver tissue homeostasis. Oncotarget 7, 17431–17441.

Mullooly, M., McGowan, P.M., Kennedy, S.A., Madden, S.F., Crown, J., O'Donovan, N., Duffy, M.J., 2015. ADAM10: a new player in breast cancer progression? Br. J. Cancer 113, 945–951.

Mullooly, M., McGowan, P.M., Crown, J., Duffy, M.J., 2016. The ADAMs family of proteases as targets for the treatment of cancer. Cancer Biol. Ther. 17, 870–880.

Murakami, Y., Matsumoto, H., Roh, M., Suzuki, J., Hisatomi, T., Ikeda, Y., Miller, J.W., Vavvas, D.G., 2012. Receptor interacting protein kinase mediates necrotic cone but not rod cell death in a mouse model of inherited degeneration. Proc. Natl. Acad. Sci. U. S. A. 109, 14598–14603.

Murakami, Y., Matsumoto, H., Roh, M., Giani, A., Kataoka, K., Morizane, Y., Kayama, M., Thanos, A., Nakatake, S., Notomi, S., Hisatomi, T., Ikeda, Y., Ishibashi, T., Connor, K.M., Miller, J.W., Vavvas, D.G., 2014. Programmed necrosis, not apoptosis, is a key mediator of cell loss and DAMP-mediated inflammation in dsRNA-induced retinal degeneration. Cell Death Differ. 21, 270–277.

Murphy, G., 2008. The ADAMs: signalling scissors in the tumour microenvironment. Nat. Rev. Cancer 8, 929–941.

Murthy, A., Defamie, V., Smookler, D.S., Di Grappa, M.A., Horiuchi, K., Federici, M., Sibilia, M., Blobel, C.P., Khokha, R., 2010. Ectodomain shedding of EGFR ligands and TNFR1 dictates hepatocyte apoptosis during fulminant hepatitis in mice. J. Clin. Invest. 120, 2731–2744.

Murumkar, P.R., DasGupta, S., Chandani, S.R., Giridhar, R., Yadav, M.R., 2010. Novel TACE inhibitors in drug discovery: a review of patented compounds. Expert Opin. Ther. Pat. 20, 31–57.

Negroni, A., Cucchiara, S., Stronati, L., 2015. Apoptosis, necrosis, and necroptosis in the gut and intestinal homeostasis. Mediators Inflamm. 2015, 250762.

Negroni, A., Colantoni, E., Pierdomenico, M., Palone, F., Costanzo, M., Oliva, S., Tiberti, A., Cucchiara, S., Stronati, L., 2017. RIP3 AND pMLKL promote necroptosis-induced inflammation and alter membrane permeability in intestinal epithelial cells. Dig. Liver Dis. 49, 1201–1210.

Newton, K., Manning, G., 2016. Necroptosis and inflammation. Annu. Rev. Biochem. 85, 743–763.

Newton, K., Dugger, D.L., Maltzman, A., Greve, J.M., Hedehus, M., Martin-McNulty, B., Carano, R.A., Cao, T.C., van Bruggen, N., Bernstein, L., Lee, W.P., Wu, X., DeVoss, J., Zhang, J., Jeet, S., Peng, I., McKenzie, B.S., Roose-Girma, M., Caplazi, P., Diehl, L., Webster, J.D., Vucic, D., 2016a. RIPK3 deficiency or catalytically inactive RIPK1 provides greater benefit than MLKL deficiency in mouse models of inflammation and tissue injury. Cell Death Differ. 23, 1565–1576.

Newton, K., Wickliffe, K.E., Maltzman, A., Dugger, D.L., Strasser, A., Pham, V.C., Lill, J.R., Roose-Girma, M., Warming, S., Solon, M., Ngu, H., Webster, J.D., Dixit, V.M., 2016b. RIPK1 inhibits ZBP1-driven necroptosis during development. Nature 540, 129–133.

Newton, K., Wickliffe, K.E., Dugger, D.L., Maltzman, A., Roose-Girma, M., Dohse, M., Kőműves, L., Webster, J.D., Dixit, V.M., 2019. Cleavage of RIPK1 by caspase-8 is crucial for limiting apoptosis and necroptosis. Nature 574, 428–431.

Nicolaou, A., Zhao, Z., Northoff, B.H., Sass, K., Herbst, A., Kohlmaier, A., Chalaris, A., Wolfrum, C., Weber, C., Steffens, S., Rose-John, S., Teupser, D., Holdt, L.M., 2017. Adam17 deficiency promotes atherosclerosis by enhanced TNFR2 signaling in mice. Arterioscler. Thromb. Vasc. Biol. 37, 247–257.

Noah, T.K., Shroyer, N.F., 2013. Notch in the intestine: regulation of homeostasis and pathogenesis. Annu. Rev. Physiol. 75, 263–288.

Nomura, M., Ueno, A., Saga, K., Fukuzawa, M., Kaneda, Y., 2014. Accumulation of cytosolic calcium induces necroptotic cell death in human neuroblastoma. Cancer Res. 74, 1056–1066.
Oberst, A., Dillon, C.P., Weinlich, R., McCormick, L.L., Fitzgerald, P., Pop, C., Hakem, R., Salvesen, G.S., Green, D.R., 2011. Catalytic activity of the caspase-8-FLIP(L) complex inhibits RIPK3-dependent necrosis. Nature 471, 363–367.
O'Donnell, M.A., Perez-Jimenez, E., Oberst, A., Ng, A., Massoumi, R., Xavier, R., Green, D.R., Ting, A.T., 2011. Caspase 8 inhibits programmed necrosis by processing CYLD. Nat. Cell Biol. 13, 1437–1442.
Oerlemans, M.I., Liu, J., Arslan, F., den Ouden, K., van Middelaar, B.J., Doevendans, P.A., Sluijter, J.P., 2012. Inhibition of RIP1-dependent necrosis prevents adverse cardiac remodeling after myocardial ischemia-reperfusion in vivo. Basic Res. Cardiol. 107, 270.
Ofengeim, D., Ito, Y., Najafov, A., Zhang, Y., Shan, B., DeWitt, J.P., Ye, J., Zhang, X., Chang, A., Vakifahmetoglu-Norberg, H., Geng, J., Py, B., Zhou, W., Amin, P., Berlink Lima, J., Qi, C., Yu, Q., Trapp, B., Yuan, J., 2015. Activation of necroptosis in multiple sclerosis. Cell Rep. 10, 1836–1849.
O'Hara, A.M., Shanahan, F., 2006. The gut flora as a forgotten organ. EMBO Rep. 7, 688–693.
Ohtsuka, T., Shiomi, T., Shimoda, M., Kodama, T., Amour, A., Murphy, G., Ohuchi, E., Kobayashi, K., Okada, Y., 2006. ADAM28 is overexpressed in human non-small cell lung carcinomas and correlates with cell proliferation and lymph node metastasis. Int. J. Cancer 118, 263–273.
Oikonomidi, I., Burbridge, E., Cavadas, M., Sullivan, G., Collis, B., Naegele, H., Clancy, D., Brezinova, J., Hu, T., Bileck, A., Gerner, C., Bolado, A., von Kriegsheim, A., Martin, S.J., Steinberg, F., Strisovsky, K., Adrain, C., 2018. iTAP, a novel iRhom interactor, controls TNF secretion by policing the stability of iRhom/TACE. Elife 7, e35032.
Omoto, S., Guo, H., Talekar, G.R., Roback, L., Kaiser, W.J., Mocarski, E.S., 2015. Suppression of RIP3-dependent necroptosis by human cytomegalovirus. J. Biol. Chem. 290, 11635–11648.
Onizawa, M., Oshima, S., Schulze-Topphoff, U., Oses-Prieto, J.A., Lu, T., Tavares, R., Prodhomme, T., Duong, B., Whang, M.I., Advincula, R., Agelidis, A., Barrera, J., Wu, H., Burlingame, A., Malynn, B.A., Zamvil, S.S., Ma, A., 2015. The ubiquitin-modifying enzyme A20 restricts ubiquitination of the kinase RIPK3 and protects cells from necroptosis. Nat. Immunol. 16, 618–627.
Orme, J.J., Du, Y., Vanarsa, K., Mayeux, J., Li, L., Mutwally, A., Arriens, C., Min, S., Hutcheson, J., Davis, L.S., Chong, B.F., Satterthwaite, A.B., Wu, T., Mohan, C., 2016. Heightened cleavage of Axl receptor tyrosine kinase by ADAM metalloproteases may contribute to disease pathogenesis in SLE. Clin. Immunol. 169, 58–68.
Orozco, S., Yatim, N., Werner, M.R., Tran, H., Gunja, S.Y., Tait, S.W., Albert, M.L., Green, D.R., Oberst, A., 2014. RIPK1 both positively and negatively regulates RIPK3 oligomerization and necroptosis. Cell Death Differ. 21, 1511–1521.
Osborn, S.L., Diehl, G., Han, S.J., Xue, L., Kurd, N., Hsieh, K., Cado, D., Robey, E.A., Winoto, A., 2010. Fas-associated death domain (FADD) is a negative regulator of T-cell receptor-mediated necroptosis. Proc. Natl. Acad. Sci. U. S. A. 107, 13034–13039.
Pan, T., Wu, S., He, X., Luo, H., Zhang, Y., Fan, M., Geng, G., Ruiz, V.C., Zhang, J., Mills, L., Bai, C., Zhang, H., 2014. Necroptosis takes place in human immunodeficiency virus type-1 (HIV-1)-infected CD4+ T lymphocytes. PLoS One 9, e93944.
Park, S.Y., Shim, J.H., Cho, Y.S., 2014. Distinctive roles of receptor-interacting protein kinases 1 and 3 in caspase-independent cell death of L929. Cell Biochem. Funct. 32, 62–69.
Parry, D.A., Toomes, C., Bida, L., Danciger, M., Towns, K.V., McKibbin, M., Jacobson, S.G., Logan, C.V., Ali, M., Bond, J., Chance, R., Swendeman, S.,

Daniele, L.L., Springell, K., Adams, M., Johnson, C.A., Booth, A.P., Jafri, H., Rashid, Y., Banin, E., Strom, T.M., Farber, D.B., Sharon, D., Blobel, C.P., Pugh Jr., E.N., Pierce, E.A., Inglehearn, C.F., 2009. Loss of the metalloprotease ADAM9 leads to cone-rod dystrophy in humans and retinal degeneration in mice. Am. J. Hum. Genet. 84, 683–691.

Pasparakis, M., Vandenabeele, P., 2015. Necroptosis and its role in inflammation. Nature 517, 311–320.

Pasparakis, M., Alexopoulou, L., Episkopou, V., Kollias, G., 1996. Immune and inflammatory responses in TNF alpha-deficient mice: a critical requirement for TNF alpha in the formation of primary B cell follicles, follicular dendritic cell networks and germinal centers, and in the maturation of the humoral immune response. J. Exp. Med. 184, 1397–1411.

Pearson, J.S., Murphy, J.M., 2017. Down the rabbit hole: is necroptosis truly an innate response to infection? Cell. Microbiol. 19, e12750.

Pearson, J.S., Giogha, C., Mühlen, S., Nachbur, U., Pham, C.L., Zhang, Y., Hildebrand, J.M., Oates, C.V., Lung, T.W., Ingle, D., Dagley, L.F., Bankovacki, A., Petrie, E.J., Schroeder, G.N., Crepin, V.F., Frankel, G., Masters, S.L., Vince, J., Murphy, J.M., Sunde, M., Webb, A.I., Silke, J., Hartland, E.L., 2017. EspL is a bacterial cysteine protease effector that cleaves RHIM proteins to block necroptosis and inflammation. Nat. Microbiol. 2, 16258.

Peschon, J.J., Slack, J.L., Reddy, P., Stocking, K.L., Sunnarborg, S.W., Lee, D.C., Russell, W.E., Castner, B.J., Johnson, R.S., Fitzner, J.N., Boyce, R.W., Nelson, N., Kozlosky, C.J., Wolfson, M.F., Rauch, C.T., Cerretti, D.P., Paxton, R.J., March, C.J., Black, R.A., 1998. An essential role for ectodomain shedding in mammalian development. Science 282, 1281–1284.

Petersen, S.L., Chen, T.T., Lawrence, D.A., Marsters, S.A., Gonzalvez, F., Ashkenazi, A., 2015. TRAF2 is a biologically important necroptosis suppressor. Cell Death Differ. 22, 1846–1857.

Petrie, E.J., Sandow, J.J., Jacobsen, A.V., Smith, B.J., Griffin, M.D.W., Lucet, I.S., Dai, W., Young, S.N., Tanzer, M.C., Wardak, A., Liang, L.Y., Cowan, A.D., Hildebrand, J.M., Kersten, W.J.A., Lessene, G., Silke, J., Czabotar, P.E., Webb, A.I., Murphy, J.M., 2018. Conformational switching of the pseudokinase domain promotes human MLKL tetramerization and cell death by necroptosis. Nat. Commun. 9, 2422.

Petrie, E.J., Czabotar, P.E., Murphy, J.M., 2019. The structural basis of necroptotic cell death signaling. Trends Biochem. Sci. 44, 53–63.

Pierdomenico, M., Negroni, A., Stronati, L., Vitali, R., Prete, E., Bertin, J., Gough, P.J., Aloi, M., Cucchiara, S., 2014. Necroptosis is active in children with inflammatory bowel disease and contributes to heighten intestinal inflammation. Am. J. Gastroenterol. 109, 279–287.

Pobezinskaya, Y.L., Kim, Y.S., Choksi, S., Morgan, M.J., Li, T., Liu, C., Liu, Z., 2008. The function of TRADD in signaling through tumor necrosis factor receptor 1 and TRIF-dependent Toll-like receptors. Nat. Immunol. 9, 1047–1054.

Polykratis, A., Hermance, N., Zelic, M., Roderick, J., Kim, C., Van, T.M., Lee, T.H., Chan, F.K., Pasparakis, M., Kelliher, M.A., 2014. Cutting edge: RIPK1 Kinase inactive mice are viable and protected from TNF-induced necroptosis in vivo. J. Immunol. 193, 1539–1543.

Posselt, G., Crabtree, J.E., Wessler, S., 2017. Proteolysis in Helicobacter pylori-Induced Gastric Cancer. Toxins (Basel) 9, 134.

Postina, R., Schroeder, A., Dewachter, I., Bohl, J., Schmitt, U., Kojro, E., Prinzen, C., Endres, K., Hiemke, C., Blessing, M., Flamez, P., Dequenne, A., Godaux, E., van Leuven, F., Fahrenholz, F., 2004. A disintegrin-metalloproteinase prevents amyloid plaque formation and hippocampal defects in an Alzheimer disease mouse model. J. Clin. Invest. 113, 1456–1464.

Pouillon, L., Bossuyt, P., Peyrin-Biroulet, L., 2016. Considerations, challenges and future of anti-TNF therapy in treating inflammatory bowel disease. Expert Opin. Biol. Ther. 16, 1277–1290.

Prox, J., Willenbrock, M., Weber, S., Lehmann, T., Schmidt-Arras, D., Schwanbeck, R., Saftig, P., Schwake, M., 2012. Tetraspanin15 regulates cellular trafficking and activity of the ectodomain sheddase ADAM10. Cell. Mol. Life Sci. 69, 2919–2932.

Qinli, Z., Meiqing, L., Xia, J., Li, X., Weili, G., Xiuliang, J., Junwei, J., Hailan, Y., Ce, Z., Qiao, N., 2013. Necrostatin-1 inhibits the degeneration of neural cells induced by aluminum exposure. Restor. Neurol. Neurosci. 31, 543–555.

Re, D.B., Le Verche, V., Yu, C., Amoroso, M.W., Politi, K.A., Phani, S., Ikiz, B., Hoffmann, L., Koolen, M., Nagata, T., Papadimitriou, D., Nagy, P., Mitsumoto, H., Kariya, S., Wichterle, H., Henderson, C.E., Przedborski, S., 2014. Necroptosis drives motor neuron death in models of both sporadic and familial ALS. Neuron 81, 1001–1008.

Reddy, P., Slack, J.L., Davis, R., Cerretti, D.P., Kozlosky, C.J., Blanton, R.A., Shows, D., Peschon, J.J., Black, R.A., 2000. Functional analysis of the domain structure of tumor necrosis factor-alpha converting enzyme. J. Biol. Chem. 275, 14608–14614.

Reedijk, M., Odorcic, S., Zhang, H., Chetty, R., Tennert, C., Dickson, B.C., Lockwood, G., Gallinger, S., Egan, S.E., 2008. Activation of Notch signaling in human colon adenocarcinoma. Int. J. Oncol. 33, 1223–1229.

Reiss, K., Saftig, P., 2009. The "a disintegrin and metalloprotease" (ADAM) family of sheddases: physiological and cellular functions. Semin. Cell Dev. Biol. 20, 126–137.

Reyat, J.S., Chimen, M., Noy, P.J., Szyroka, J., Rainger, G.E., Tomlinson, M.G., 2017. ADAM10-interacting tetraspanins tspan5 and tspan17 regulate VE-cadherin expression and promote t lymphocyte transmigration. J. Immunol. 199, 666–676.

Reynoso, E., Liu, H., Li, L., Yuan, A.L., Chen, S., Wang, Z., 2017. Thioredoxin-1 actively maintains the pseudokinase MLKL in a reduced state to suppress disulfide bond-dependent MLKL polymer formation and necroptosis. J. Biol. Chem. 292, 17514–17524.

Rickard, J.A., Anderton, H., Etemadi, N., Nachbur, U., Darding, M., Peltzer, N., Lalaoui, N., Lawlor, K.E., Vanyai, H., Hall, C., Bankovacki, A., Gangoda, L., Wong, W.W., Corbin, J., Huang, C., Mocarski, E.S., Murphy, J.M., Alexander, W.S., Voss, A.K., Vaux, D.L., Kaiser, W.J., Walczak, H., Silke, J., 2014a. TNFR1-dependent cell death drives inflammation in Sharpin-deficient mice. Elife 3, e03464.

Rickard, J.A., O'Donnell, J.A., Evans, J.M., Lalaoui, N., Poh, A.R., Rogers, T., Vince, J.E., Lawlor, K.E., Ninnis, R.L., Anderton, H., Hall, C., Spall, S.K., Phesse, T.J., Abud, H.E., Cengia, L.H., Corbin, J., Mifsud, S., Di Rago, L., Metcalf, D., Ernst, M., Dewson, G., Roberts, A.W., Alexander, W.S., Murphy, J.M., Ekert, P.G., Masters, S.L., Vaux, D.L., Croker, B.A., Gerlic, M., Silke, J., 2014b. RIPK1 regulates RIPK3-MLKL-driven systemic inflammation and emergency hematopoiesis. Cell 157, 1175–1188.

Robinson, N., McComb, S., Mulligan, R., Dudani, R., Krishnan, L., Sad, S., 2012. Type I interferon induces necroptosis in macrophages during infection with Salmonella enterica serovar Typhimurium. Nat. Immunol. 13, 954–962.

Roca, F.J., Ramakrishnan, L., 2013. TNF dually mediates resistance and susceptibility to mycobacteria via mitochondrial reactive oxygen species. Cell 153, 521–534.

Rodriguez, D.A., Weinlich, R., Brown, S., Guy, C., Fitzgerald, P., Dillon, C.P., Oberst, A., Quarato, G., Low, J., Cripps, J.G., Chen, T., Green, D.R., 2016. Characterization of RIPK3-mediated phosphorylation of the activation loop of MLKL during necroptosis. Cell Death Differ. 23, 76–88.

Rose-John, S., Heinrich, P.C., 1994. Soluble receptors for cytokines and growth factors: generation and biological function. Biochem. J. 300 (Pt. 2), 281–290.

Rose-John, S., Winthrop, K., Calabrese, L., 2017. The role of IL-6 in host defence against infections: immunobiology and clinical implications. Nat. Rev. Rheumatol. 13, 399–409.

Rosenbaum, D.M., Degterev, A., David, J., Rosenbaum, P.S., Roth, S., Grotta, J.C., Cuny, G.D., Yuan, J., Savitz, S.I., 2010. Necroptosis, a novel form of caspase-independent cell death, contributes to neuronal damage in a retinal ischemia-reperfusion injury model. J. Neurosci. Res. 88, 1569–1576.

Rothe, M., Pan, M.G., Henzel, W.J., Ayres, T.M., Goeddel, D.V., 1995. The TNFR2-TRAF signaling complex contains two novel proteins related to baculoviral inhibitor of apoptosis proteins. Cell 83, 1243–1252.

Roychowdhury, S., McMullen, M.R., Pisano, S.G., Liu, X., Nagy, L.E., 2013. Absence of receptor interacting protein kinase 3 prevents ethanol-induced liver injury. Hepatology 57, 1773–1783.

Rudnicka, C., Mochizuki, S., Okada, Y., McLaughlin, C., Leedman, P.J., Stuart, L., Epis, M., Hoyne, G., Boulos, S., Johnson, L., Schlaich, M., Matthews, V., 2016. Overexpression and knock-down studies highlight that a disintegrin and metalloproteinase 28 controls proliferation and migration in human prostate cancer. Medicine (Baltimore) 95, e5085.

Sagane, K., Yamazaki, K., Mizui, Y., Tanaka, I., 1999. Cloning and chromosomal mapping of mouse ADAM11, ADAM22 and ADAM23. Gene 236, 79–86.

Sagane, K., Hayakawa, K., Kai, J., Hirohashi, T., Takahashi, E., Miyamoto, N., Ino, M., Oki, T., Yamazaki, K., Nagasu, T., 2005. Ataxia and peripheral nerve hypomyelination in ADAM22-deficient mice. BMC Neurosci. 6, 33.

Sahin, U., Weskamp, G., Kelly, K., Zhou, H.M., Higashiyama, S., Peschon, J., Hartmann, D., Saftig, P., Blobel, C.P., 2004. Distinct roles for ADAM10 and ADAM17 in ectodomain shedding of six EGFR ligands. J. Cell Biol. 164, 769–779.

Sanderson, M.P., Erickson, S.N., Gough, P.J., Garton, K.J., Wille, P.T., Raines, E.W., Dunbar, A.J., Dempsey, P.J., 2005. ADAM10 mediates ectodomain shedding of the betacellulin precursor activated by p-aminophenylmercuric acetate and extracellular calcium influx. J. Biol. Chem. 280, 1826–1837.

Saraceno, C., Marcello, E., Di Marino, D., Borroni, B., Claeysen, S., Perroy, J., Padovani, A., Tramontano, A., Gardoni, F., Di Luca, M., 2014. SAP97-mediated ADAM10 trafficking from Golgi outposts depends on PKC phosphorylation. Cell Death Dis. 5 e1547.

Sarhan, M., Land, W.G., Tonnus, W., Hugo, C.P., Linkermann, A., 2018. Origin and consequences of necroinflammation. Physiol. Rev. 98, 727–780.

Sato, K., Li, S., Gordon, W.C., He, J., Liou, G.I., Hill, J.M., Travis, G.H., Bazan, N.G., Jin, M., 2013. Receptor interacting protein kinase-mediated necrosis contributes to cone and rod photoreceptor degeneration in the retina lacking interphotoreceptor retinoid-binding protein. J. Neurosci. 33, 17458–17468.

Scharfenberg, F., Helbig, A., Sammel, M., Benzel, J., Schlomann, U., Peters, F., Wichert, R., Bettendorff, M., Schmidt-Arras, D., Rose-John, S., Moali, C., Lichtenthaler, S.F., Pietrzik, C.U., Bartsch, J.W., Tholey, A., Becker-Pauly, C., 2020. Degradome of soluble ADAM10 and ADAM17 metalloproteases. Cell. Mol. Life Sci. 77, 331–350.

Scheller, J., Chalaris, A., Garbers, C., Rose-John, S., 2011a. ADAM17: a molecular switch to control inflammation and tissue regeneration. Trends Immunol. 32, 380–387.

Scheller, J., Chalaris, A., Schmidt-Arras, D., Rose-John, S., 2011b. The pro- and anti-inflammatory properties of the cytokine interleukin-6. Biochim. Biophys. Acta 1813, 878–888.

Schildkopf, P., Frey, B., Mantel, F., Ott, O.J., Weiss, E.M., Sieber, R., Janko, C., Sauer, R., Fietkau, R., Gaipl, U.S., 2010. Application of hyperthermia in addition to ionizing irradiation fosters necrotic cell death and HMGB1 release of colorectal tumor cells. Biochem. Biophys. Res. Commun. 391, 1014–1020.

Schlomann, U., Wildeboer, D., Webster, A., Antropova, O., Zeuschner, D., Knight, C.G., Docherty, A.J., Lambert, M., Skelton, L., Jockusch, H., Bartsch, J.W., 2002. The metalloprotease disintegrin ADAM8. Processing by autocatalysis is required for proteolytic activity and cell adhesion. J. Biol. Chem. 277, 48210–48219.

Schmidt, S., Schumacher, N., Schwarz, J., Tangermann, S., Kenner, L., Schlederer, M., Sibilia, M., Linder, M., Altendorf-Hofmann, A., Knösel, T., Gruber, E.S., Oberhuber, G., Bolik, J., Rehman, A., Sinha, A., Lokau, J., Arnold, P., Cabron, A.S., Zunke, F., Becker-Pauly, C., Preaudet, A., Nguyen, P., Huynh, J., Afshar-Sterle, S., Chand, A.L., Westermann, J., Dempsey, P.J., Garbers, C., Schmidt-Arras, D., Rosenstiel, P., Putoczki, T., Ernst, M., Rose-John, S., 2018. ADAM17 is required for EGF-R-induced intestinal tumors via IL-6 trans-signaling. J. Exp. Med. 215, 1205–1225.

Schock, S.N., Chandra, N.V., Sun, Y., Irie, T., Kitagawa, Y., Gotoh, B., Coscoy, L., Winoto, A., 2017. Induction of necroptotic cell death by viral activation of the RIG-I or STING pathway. Cell Death Differ. 24, 615–625.

Schumacher, N., Rose-John, S., 2019. ADAM17 activity and IL-6 trans-signaling in inflammation and cancer. Cancers (Basel) 11, 1736.

Schütte, A., Ermund, A., Becker-Pauly, C., Johansson, M.E., Rodriguez-Pineiro, A.M., Bäckhed, F., Müller, S., Lottaz, D., Bond, J.S., Hansson, G.C., 2014. Microbial-induced meprin beta cleavage in MUC2 mucin and a functional CFTR channel are required to release anchored small intestinal mucus. Proc. Natl. Acad. Sci. U. S. A. 111, 12396–12401.

Schwarz, J., Broder, C., Helmstetter, A., Schmidt, S., Yan, I., Müller, M., Schmidt-Arras, D., Becker-Pauly, C., Koch-Nolte, F., Mittrücker, H.W., Rabe, B., Rose-John, S., Chalaris, A., 2013. Short-term TNFalpha shedding is independent of cytoplasmic phosphorylation or furin cleavage of ADAM17. Biochim. Biophys. Acta 1833, 3355–3367.

Schwarz, J., Schmidt, S., Will, O., Koudelka, T., Köhler, K., Boss, M., Rabe, B., Tholey, A., Scheller, J., Schmidt-Arras, D., Schwake, M., Rose-John, S., Chalaris, A., 2014. Polo-like kinase 2, a novel ADAM17 signaling component, regulates tumor necrosis factor alpha ectodomain shedding. J. Biol. Chem. 289, 3080–3093.

Seo, J., Lee, E.W., Sung, H., Seong, D., Dondelinger, Y., Shin, J., Jeong, M., Lee, H.K., Kim, J.H., Han, S.Y., Lee, C., Seong, J.K., Vandenabeele, P., Song, J., 2016. CHIP controls necroptosis through ubiquitylation- and lysosome-dependent degradation of RIPK3. Nat. Cell Biol. 18, 291–302.

Shaw, D., Gohil, K., Basson, M.D., 2012. Intestinal mucosal atrophy and adaptation. World J. Gastroenterol. 18, 6357–6375.

Shimoda, M., Principe, S., Jackson, H.W., Luga, V., Fang, H., Molyneux, S.D., Shao, Y.W., Aiken, A., Waterhouse, P.D., Karamboulas, C., Hess, F.M., Ohtsuka, T., Okada, Y., Ailles, L., Ludwig, A., Wrana, J.L., Kislinger, T., Khokha, R., 2014. Loss of the Timp gene family is sufficient for the acquisition of the CAF-like cell state. Nat. Cell Biol. 16, 889–901.

Shirakabe, K., Wakatsuki, S., Kurisaki, T., Fujisawa-Sehara, A., 2001. Roles of Meltrin beta/ADAM19 in the processing of neuregulin. J. Biol. Chem. 276, 9352–9358.

Shu, H.B., Takeuchi, M., Goeddel, D.V., 1996. The tumor necrosis factor receptor 2 signal transducers TRAF2 and c-IAP1 are components of the tumor necrosis factor receptor 1 signaling complex. Proc. Natl. Acad. Sci. U. S. A. 93, 13973–13978.

Sibilia, M., Kroismayr, R., Lichtenberger, B.M., Natarajan, A., Hecking, M., Holcmann, M., 2007. The epidermal growth factor receptor: from development to tumorigenesis. Differentiation 75, 770–787.

Siegel, R.L., Miller, K.D., Fedewa, S.A., Ahnen, D.J., Meester, R.G.S., Barzi, A., Jemal, A., 2017. Colorectal cancer statistics, 2017. CA-Cancer J. Clin. 67, 177–193.

Skovronsky, D.M., Fath, S., Lee, V.M., Milla, M.E., 2001. Neuronal localization of the TNFalpha converting enzyme (TACE) in brain tissue and its correlation to amyloid plaques. J. Neurobiol. 49, 40–46.

Smith, C.C., Davidson, S.M., Lim, S.Y., Simpkin, J.C., Hothersall, J.S., Yellon, D.M., 2007. Necrostatin: a potentially novel cardioprotective agent? Cardiovasc. Drugs Ther. 21, 227–233.

Snyder, S.H., 2006. Neuroscience. Adam finds an exciting mate. Science 313, 1744–1745.

Sommer, A., Kordowski, F., Büch, J., Maretzky, T., Evers, A., Andrä, J., Düsterhoft, S., Michalek, M., Lorenzen, I., Somasundaram, P., Tholey, A., Sönnichsen, F.D., Kunzelmann, K., Heinbockel, L., Nehls, C., Gutsmann, T., Grötzinger, J., Bhakdi, S., Reiss, K., 2016. Phosphatidylserine exposure is required for ADAM17 sheddase function. Nat. Commun. 7, 11523.

Soond, S.M., Everson, B., Riches, D.W., Murphy, G., 2005. ERK-mediated phosphorylation of Thr735 in TNFalpha-converting enzyme and its potential role in TACE protein trafficking. J. Cell Sci. 118, 2371–2380.

Sosna, J., Voigt, S., Mathieu, S., Kabelitz, D., Trad, A., Janssen, O., Meyer-Schwesinger, C., Schütze, S., Adam, D., 2013. The proteases HtrA2/Omi and UCH-L1 regulate TNF-induced necroptosis. Cell Commun. Signal 11, 76.

Sosna, J., Philipp, S., Fuchslocher Chico, J., Saggau, C., Fritsch, J., Föll, A., Plenge, J., Arenz, C., Pinkert, T., Kalthoff, H., Trauzold, A., Schmitz, I., Schütze, S., Adam, D., 2016. Differences and similarities in TRAIL- and TNF-mediated necroptotic signaling in cancer cells. Mol. Cell. Biol. 36, 2626–2644.

Specian, R.D., Oliver, M.G., 1991. Functional biology of intestinal goblet cells. Am. J. Physiol. 260, C183–C193.

Sridharan, H., Ragan, K.B., Guo, H., Gilley, R.P., Landsteiner, V.J., Kaiser, W.J., Upton, J.W., 2017. Murine cytomegalovirus IE3-dependent transcription is required for DAI/ZBP1-mediated necroptosis. EMBO Rep. 18, 1429–1441.

Srivatsa, S., Paul, M.C., Cardone, C., Holcmann, M., Amberg, N., Pathria, P., Diamanti, M.A., Linder, M., Timelthaler, G., Dienes, H.P., Kenner, L., Wrba, F., Prager, G.W., Rose-John, S., Eferl, R., Liguori, G., Botti, G., Martinelli, E., Greten, F.R., Ciardiello, F., Sibilia, M., 2017. EGFR in tumor-associated myeloid cells promotes development of colorectal cancer in mice and associates with outcomes of patients. Gastroenterology 153, 178–190.

Stöcker, W., Grams, F., Baumann, U., Reinemer, P., Gomis-Rüth, F.X., McKay, D.B., Bode, W., 1995. The metzincins—topological and sequential relations between the astacins, adamalysins, serralysins, and matrixins (collagenases) define a superfamily of zinc-peptidases. Protein Sci. 4, 823–840.

Stoeck, A., Keller, S., Riedle, S., Sanderson, M.P., Runz, S., Le Naour, F., Gutwein, P., Ludwig, A., Rubinstein, E., Altevogt, P., 2006. A role for exosomes in the constitutive and stimulus-induced ectodomain cleavage of L1 and CD44. Biochem. J. 393, 609–618.

Strelow, A., Bernardo, K., Adam-Klages, S., Linke, T., Sandhoff, K., Krönke, M., Adam, D., 2000. Overexpression of acid ceramidase protects from tumor necrosis factor-induced cell death. J. Exp. Med. 192, 601–612.

Strilic, B., Yang, L., Albarrán-Juárez, J., Wachsmuth, L., Han, K., Müller, U.C., Pasparakis, M., Offermanns, S., 2016. Tumour-cell-induced endothelial cell necroptosis via death receptor 6 promotes metastasis. Nature 536, 215–218.

Sulis, M.L., Saftig, P., Ferrando, A.A., 2011. Redundancy and specificity of the metalloprotease system mediating oncogenic NOTCH1 activation in T-ALL. Leukemia 25, 1564–1569.

Sun, X.Q., Yin, J.P., Starovasnik, M.A., Fairbrother, W.J., Dixit, V.M., 2002. Identification of a novel homotypic interaction motif required for the phosphorylation of receptor-interacting protein (RIP) by RIP3. J. Biol. Chem. 277, 9505–9511.

Sun, L., Wang, H., Wang, Z., He, S., Chen, S., Liao, D., Wang, L., Yan, J., Liu, W., Lei, X., Wang, X., 2012. Mixed lineage kinase domain-like protein mediates necrosis signaling downstream of RIP3 kinase. Cell 148, 213–227.

Takahashi, N., Vereecke, L., Bertrand, M.J., Duprez, L., Berger, S.B., Divert, T., Gonçalves, A., Sze, M., Gilbert, B., Kourula, S., Goossens, V., Lefebvre, S., Günther, C., Becker, C., Bertin, J., Gough, P.J., Declercq, W., van Loo, G., Vandenabeele, P., 2014. RIPK1 ensures intestinal homeostasis by protecting the epithelium against apoptosis. Nature 513, 95–99.

Takeda, S., 2009. Three-dimensional domain architecture of the ADAM family proteinases. Semin. Cell Dev. Biol. 20, 146–152.

Takeda, S., 2016. ADAM and ADAMTS family proteins and snake venom metalloproteinases: a structural overview. Toxins (Basel) 8, 155.

Tamura, R., Nemoto, T., Maruta, T., Onizuka, S., Yanagita, T., Wada, A., Murakami, M., Tsuneyoshi, I., 2014. Up-regulation of NaV1.7 sodium channels expression by tumor necrosis factor-alpha in cultured bovine adrenal chromaffin cells and rat dorsal root ganglion neurons. Anesth. Analg. 118, 318–324.

Tanabe, C., Hotoda, N., Sasagawa, N., Sehara-Fujisawa, A., Maruyama, K., Ishiura, S., 2007. ADAM19 is tightly associated with constitutive Alzheimer's disease APP alpha-secretase in A172 cells. Biochem. Biophys. Res. Commun. 352, 111–117.

Tenev, T., Bianchi, K., Darding, M., Broemer, M., Langlais, C., Wallberg, F., Zachariou, A., Lopez, J., MacFarlane, M., Cain, K., Meier, P., 2011. The ripoptosome, a signaling platform that assembles in response to genotoxic stress and loss of IAPs. Mol. Cell 43, 432–448.

Thapa, R.J., Nogusa, S., Chen, P., Maki, J.L., Lerro, A., Andrake, M., Rall, G.F., Degterev, A., Balachandran, S., 2013. Interferon-induced RIP1/RIP3-mediated necrosis requires PKR and is licensed by FADD and caspases. Proc. Natl. Acad. Sci. U. S. A. 110, E3109–E3118.

Thapa, R.J., Ingram, J.P., Ragan, K.B., Nogusa, S., Boyd, D.F., Benitez, A.A., Sridharan, H., Kosoff, R., Shubina, M., Landsteiner, V.J., Andrake, M., Vogel, P., Sigal, L.J., tenOever, B.R., Thomas, P.G., Upton, J.W., Balachandran, S., 2016. DAI senses influenza A virus genomic RNA and activates RIPK3-dependent cell death. Cell Host Microbe 20, 674–681.

Thon, L., Möhlig, H., Mathieu, S., Lange, A., Bulanova, E., Winoto-Morbach, S., Schütze, S., Bulfone-Paus, S., Adam, D., 2005. Ceramide mediates caspase-independent programmed cell death. FASEB J. 19, 1945–1956.

Thon, L., Mathieu, S., Kabelitz, D., Adam, D., 2006. The murine TRAIL receptor signals caspase-independent cell death through ceramide. Exp. Cell Res. 312, 3808–3821.

Torres, J., Mehandru, S., Colombel, J.F., Peyrin-Biroulet, L., 2017. Crohn's disease. Lancet 389, 1741–1755.

Tousseyn, T., Thathiah, A., Jorissen, E., Raemaekers, T., Konietzko, U., Reiss, K., Maes, E., Snellinx, A., Serneels, L., Nyabi, O., Annaert, W., Saftig, P., Hartmann, D., De Strooper, B., 2009. ADAM10, the rate-limiting protease of regulated intramembrane proteolysis of Notch and other proteins, is processed by ADAMS-9, ADAMS-15, and the gamma-secretase. J. Biol. Chem. 284, 11738–11747.

Trichonas, G., Murakami, Y., Thanos, A., Morizane, Y., Kayama, M., Debouck, C.M., Hisatomi, T., Miller, J.W., Vavvas, D.G., 2010. Receptor interacting protein kinases mediate retinal detachment-induced photoreceptor necrosis and compensate for inhibition of apoptosis. Proc. Natl. Acad. Sci. U. S. A. 107, 21695–21700.

Tripathi, P., Awasthi, S., Gao, P., 2014. ADAM metallopeptidase domain 33 (ADAM33): a promising target for asthma. Mediators Inflamm. 2014, 572025.

Tsai, Y.H., Feng, Y., Wade, A.W., Teitelbaum, D.H., Dempsey, P.J., 2012. Intestinal cell responses in different injury/regeneration models are differentially dependent on ADAM17. Gastroenterology 142, S61.

Tsai, Y.H., Feng, Y., Rakshit, R., Fearon, E.R., Dempsey, P., 2013. Cell-autonomous ADAM10 signaling is required for adenoma initiation after APC mutation. Gastroenterology 144, S51–S52.

Tsai, Y.H., VanDussen, K.L., Sawey, E.T., Wade, A.W., Kasper, C., Rakshit, S., Bhatt, R.G., Stoeck, A., Maillard, I., Crawford, H.C., Samuelson, L.C., Dempsey, P.J., 2014. ADAM10 regulates Notch function in intestinal stem cells of mice. Gastroenterology 147, 822–834.

Tsukerman, P., Eisenstein, E.M., Chavkin, M., Schmiedel, D., Wong, E., Werner, M., Yaacov, B., Averbuch, D., Molho-Pessach, V., Stepensky, P., Kaynan, N., Bar-On, Y., Seidel, E., Yamin, R., Sagi, I., Elpeleg, O., Mandelboim, O., 2015. Cytokine secretion and NK cell activity in human ADAM17 deficiency. Oncotarget 6, 44151–44160.

Tucher, J., Linke, D., Koudelka, T., Cassidy, L., Tredup, C., Wichert, R., Pietrzik, C., Becker-Pauly, C., Tholey, A., 2014. LC-MS based cleavage site profiling of the proteases ADAM10 and ADAM17 using proteome-derived peptide libraries. J. Proteome Res. 13, 2205–2214.

Ungaro, R., Mehandru, S., Allen, P.B., Peyrin-Biroulet, L., Colombel, J.F., 2017. Ulcerative colitis. Lancet 389, 1756–1770.

Upton, J.W., Kaiser, W.J., Mocarski, E.S., 2008. Cytomegalovirus M45 cell death suppression requires receptor-interacting protein (RIP) homotypic interaction motif (RHIM)-dependent interaction with RIP1. J. Biol. Chem. 283, 16966–16970.

Upton, J.W., Kaiser, W.J., Mocarski, E.S., 2010. Virus inhibition of RIP3-dependent necrosis. Cell Host Microbe 7, 302–313.

Upton, J.W., Kaiser, W.J., Mocarski, E.S., 2012. DAI/ZBP1/DLM-1 complexes with RIP3 to mediate virus-induced programmed necrosis that is targeted by murine cytomegalovirus vIRA. Cell Host Microbe 11, 290–297.

van der Flier, L.G., Clevers, H., 2009. Stem cells, self-renewal, and differentiation in the intestinal epithelium. Annu. Rev. Physiol. 71, 241–260.

van der Vorst, E.P., Jeurissen, M., Wolfs, I.M., Keijbeck, A., Theodorou, K., Wijnands, E., Schurgers, L., Weber, S., Gijbels, M.J., Hamers, A.A., Dreymueller, D., Rose-John, S., de Winther, M.P., Ludwig, A., Saftig, P., Biessen, E.A., Donners, M.M., 2015. Myeloid A disintegrin and metalloproteinase domain 10 deficiency modulates atherosclerotic plaque composition by shifting the balance from inflammation toward fibrosis. Am. J. Pathol. 185, 1145–1155.

Van Eerdewegh, P., Little, R.D., Dupuis, J., Del Mastro, R.G., Falls, K., Simon, J., Torrey, D., Pandit, S., McKenny, J., Braunschweiger, K., Walsh, A., Liu, Z., Hayward, B., Folz, C., Manning, S.P., Bawa, A., Saracino, L., Thackston, M., Benchekroun, Y., Capparell, N., Wang, M., Adair, R., Feng, Y., Dubois, J., FitzGerald, M.G., Huang, H., Gibson, R., Allen, K.M., Pedan, A., Danzig, M.R., Umland, S.P., Egan, R.W., Cuss, F.M., Rorke, S., Clough, J.B., Holloway, J.W., Holgate, S.T., Keith, T.P., 2002. Association of the ADAM33 gene with asthma and bronchial hyperresponsiveness. Nature 418, 426–430.

Van Emburgh, B.O., Sartore-Bianchi, A., Di Nicolantonio, F., Siena, S., Bardelli, A., 2014. Acquired resistance to EGFR-targeted therapies in colorectal cancer. Mol. Oncol. 8, 1084–1094.

van Es, J.H., van Gijn, M.E., Riccio, O., van den Born, M., Vooijs, M., Begthel, H., Cozijnsen, M., Robine, S., Winton, D.J., Radtke, F., Clevers, H., 2005. Notch/gamma-secretase inhibition turns proliferative cells in intestinal crypts and adenomas into goblet cells. Nature 435, 959–963.

Van Hoecke, L., Van Lint, S., Roose, K., Van Parys, A., Vandenabeele, P., Grooten, J., Tavernier, J., De Koker, S., Saelens, X., 2018. Treatment with mRNA coding for the necroptosis mediator MLKL induces antitumor immunity directed against neoepitopes. Nat. Commun. 9, 3417.

van Raam, B.J., Ehrnhoefer, D.E., Hayden, M.R., Salvesen, G.S., 2013. Intrinsic cleavage of receptor-interacting protein kinase-1 by caspase-6. Cell Death Differ. 20, 86–96.

Vanden Berghe, T., Kalai, M., van Loo, G., Declercq, W., Vandenabeele, P., 2003. Disruption of HSP90 function reverts tumor necrosis factor-induced necrosis to apoptosis. J. Biol. Chem. 278, 5622–5629.

Vandenabeele, P., Grootjans, S., Callewaert, N., Takahashi, N., 2013. Necrostatin-1 blocks both RIPK1 and IDO: consequences for the study of cell death in experimental disease models. Cell Death Differ. 20, 185–187.

Varfolomeev, E., Blankenship, J.W., Wayson, S.M., Fedorova, A.V., Kayagaki, N., Garg, P., Zobel, K., Dynek, J.N., Elliott, L.O., Wallweber, H.J., Flygare, J.A., Fairbrother, W.J., Deshayes, K., Dixit, V.M., Vucic, D., 2007. IAP antagonists induce autoubiquitination of c-IAPs, NF-kappaB activation, and TNFalpha-dependent apoptosis. Cell 131, 669–681.

Varfolomeev, E., Goncharov, T., Fedorova, A.V., Dynek, J.N., Zobel, K., Deshayes, K., Fairbrother, W.J., Vucic, D., 2008. c-IAP1 and c-IAP2 are critical mediators of tumor necrosis factor alpha (TNFalpha)-induced NF-kappaB activation. J. Biol. Chem. 283, 24295–24299.

Vedamurthy, A., Ananthakrishnan, A.N., 2019. Influence of environmental factors in the development and outcomes of inflammatory bowel disease. Gastroenterol. Hepatol. (N. Y.) 15, 72–82.

Vercammen, D., Vandenabeele, P., Beyaert, R., Declercq, W., Fiers, W., 1997. Tumour necrosis factor-induced necrosis versus anti-Fas-induced apoptosis in L929 cells. Cytokine 9, 801–808.

Vercammen, D., Beyaert, R., Denecker, G., Goossens, V., van Loo, G., Declercq, W., Grooten, J., Fiers, W., Vandenabeele, P., 1998a. Inhibition of caspases increases the sensitivity of L929 cells to necrosis mediated by tumor necrosis factor. J. Exp. Med. 187, 1477–1485.

Vercammen, D., Brouckaert, G., Denecker, G., VandeCraen, M., Declercq, W., Fiers, W., Vandenabeele, P., 1998b. Dual signaling of the Fas receptor: initiation of both apoptotic and necrotic cell death pathways. J. Exp. Med. 188, 919–930.

Vitner, E.B., Salomon, R., Farfel-Becker, T., Meshcheriakova, A., Ali, M., Klein, A.D., Platt, F.M., Cox, T.M., Futerman, A.H., 2014. RIPK3 as a potential therapeutic target for Gaucher's disease. Nat. Med. 20, 204–208.

von Moltke, J., Ji, M., Liang, H.E., Locksley, R.M., 2016. Tuft-cell-derived IL-25 regulates an intestinal ILC2-epithelial response circuit. Nature 529, 221–225.

Wakatsuki, S., Kurisaki, T., Sehara-Fujisawa, A., 2004. Lipid rafts identified as locations of ectodomain shedding mediated by Meltrin beta/ADAM19. J. Neurochem. 89, 119–123.

Wakatsuki, S., Yumoto, N., Komatsu, K., Araki, T., Sehara-Fujisawa, A., 2009. Roles of meltrin-beta/ADAM19 in progression of Schwann cell differentiation and myelination during sciatic nerve regeneration. J. Biol. Chem. 284, 2957–2966.

Waldner, M.J., Neurath, M.F., 2009. Colitis-associated cancer: the role of T cells in tumor development. Semin. Immunopathol. 31, 249–256.

Wang, L., Du, F., Wang, X., 2008. TNF-alpha induces two distinct caspase-8 activation pathways. Cell 133, 693–703.

Wang, H., Sun, L., Su, L., Rizo, J., Liu, L., Wang, L.F., Wang, F.S., Wang, X., 2014a. Mixed lineage kinase domain-like protein MLKL causes necrotic membrane disruption upon phosphorylation by RIP3. Mol. Cell 54, 133–146.

Wang, Y., Wang, H., Tao, Y., Zhang, S., Wang, J., Feng, X., 2014b. Necroptosis inhibitor necrostatin-1 promotes cell protection and physiological function in traumatic spinal cord injury. Neuroscience 266, 91–101.
Wang, S., Ni, H.M., Dorko, K., Kumer, S.C., Schmitt, T.M., Nawabi, A., Komatsu, M., Huang, H., Ding, W.X., 2016. Increased hepatic receptor interacting protein kinase 3 expression due to impaired proteasomal functions contributes to alcohol-induced steatosis and liver injury. Oncotarget 7, 17681–17698.
Wang, H., Meng, H., Li, X., Zhu, K., Dong, K., Mookhtiar, A.K., Wei, H., Li, Y., Sun, S.C., Yuan, J., 2017. PELI1 functions as a dual modulator of necroptosis and apoptosis by regulating ubiquitination of RIPK1 and mRNA levels of c-FLIP. Proc. Natl. Acad. Sci. U. S. A. 114, 11944–11949.
Watson, A.J., Duckworth, C.A., Guan, Y., Montrose, M.H., 2009. Mechanisms of epithelial cell shedding in the Mammalian intestine and maintenance of barrier function. Ann. N. Y. Acad. Sci. 1165, 135–142.
Weber, S., Saftig, P., 2012. Ectodomain shedding and ADAMs in development. Development 139, 3693–3709.
Weinlich, R., Oberst, A., Dillon, C.P., Janke, L.J., Milasta, S., Lukens, J.R., Rodriguez, D.A., Gurung, P., Savage, C., Kanneganti, T.D., Green, D.R., 2013. Protective roles for caspase-8 and cFLIP in adult homeostasis. Cell Rep. 5, 340–348.
Welz, P.S., Wullaert, A., Vlantis, K., Kondylis, V., Fernández-Majada, V., Ermolaeva, M., Kirsch, P., Sterner-Kock, A., van Loo, G., Pasparakis, M., 2011. FADD prevents RIP3-mediated epithelial cell necrosis and chronic intestinal inflammation. Nature 477, 330–334.
Wen, S., Ling, Y., Yang, W., Shen, J., Li, C., Deng, W., Liu, W., Liu, K., 2017. Necroptosis is a key mediator of enterocytes loss in intestinal ischaemia/reperfusion injury. J. Cell. Mol. Med. 21, 432–443.
Weng, D., Marty-Roix, R., Ganesan, S., Proulx, M.K., Vladimer, G.I., Kaiser, W.J., Mocarski, E.S., Pouliot, K., Chan, F.K., Kelliher, M.A., Harris, P.A., Bertin, J., Gough, P.J., Shayakhmetov, D.M., Goguen, J.D., Fitzgerald, K.A., Silverman, N., Lien, E., 2014. Caspase-8 and RIP kinases regulate bacteria-induced innate immune responses and cell death. Proc. Natl. Acad. Sci. U. S. A. 111, 7391–7396.
White, J.M., 2003. ADAMs: modulators of cell-cell and cell-matrix interactions. Curr. Opin. Cell Biol. 15, 598–606.
Wichert, R., Ermund, A., Schmidt, S., Schweinlin, M., Ksiazek, M., Arnold, P., Knittler, K., Wilkens, F., Potempa, B., Rabe, B., Stirnberg, M., Lucius, R., Bartsch, J.W., Nikolaus, S., Falk-Paulsen, M., Rosenstiel, P., Metzger, M., Rose-John, S., Potempa, J., Hansson, G.C., Dempsey, P.J., Becker-Pauly, C., 2017. Mucus detachment by host metalloprotease meprin beta requires shedding of its inactive pro-form, which is abrogated by the pathogenic protease RgpB. Cell Rep. 21, 2090–2103.
Wilks, S., 1859. Morbid appearances in the intestines of Miss Bankes. Lond. Med. Gaz. 2, 264–265.
Wilson, C.A., Browning, J.L., 2002. Death of HT29 adenocarcinoma cells induced by TNF family receptor activation is caspase-independent and displays features of both apoptosis and necrosis. Cell Death Differ. 9, 1321–1333.
Xie, Y., Zhu, S., Zhong, M., Yang, M., Sun, X., Liu, J., Kroemer, G., Lotze, M., Zeh 3rd, H.J., Kang, R., Tang, D., 2017. Inhibition of aurora kinase A induces necroptosis in pancreatic carcinoma. Gastroenterology 153, 1429–1443.
Xu, P., Derynck, R., 2010. Direct activation of TACE-mediated ectodomain shedding by p38 MAP kinase regulates EGF receptor-dependent cell proliferation. Mol. Cell 37, 551–566.
Xu, X., Chua, K.W., Chua, C.C., Liu, C.F., Hamdy, R.C., Chua, B.H., 2010. Synergistic protective effects of humanin and necrostatin-1 on hypoxia and ischemia/reperfusion injury. Brain Res. 1355, 189–194.

Xu, P., Liu, J., Sakaki-Yumoto, M., Derynck, R., 2012. TACE activation by MAPK-mediated regulation of cell surface dimerization and TIMP3 association. Sci. Signal. 5 ra34.

Xu, Y.L., Tang, H.L., Zhu, S.Y., Peng, H.R., Qi, Z.T., Wang, W., 2017. RIP3 deficiency exacerbates inflammation in dextran sodium sulfate-induced ulcerative colitis mice model. Cell Biochem. Funct. 35, 156–163.

Yagami-Hiromasa, T., Sato, T., Kurisaki, T., Kamijo, K., Nabeshima, Y., Fujisawa-Sehara, A., 1995. A metalloprotease-disintegrin participating in myoblast fusion. Nature 377, 652–656.

Yamada, Y., Mori, H., 2007. Multistep carcinogenesis of the colon in Apc(Min/+) mouse. Cancer Sci. 98, 6–10.

Yan, F., Liu, L., Dempsey, P.J., Tsai, Y.H., Raines, E.W., Wilson, C.L., Cao, H., Cao, Z., Liu, L., Polk, D.B., 2013. A Lactobacillus rhamnosus GG-derived soluble protein, p40, stimulates ligand release from intestinal epithelial cells to transactivate epidermal growth factor receptor. J. Biol. Chem. 288, 30742–30751.

Yan, I., Schwarz, J., Lücke, K., Schumacher, N., Schumacher, V., Schmidt, S., Rabe, B., Saftig, P., Donners, M., Rose-John, S., Mittrücker, H.W., Chalaris, A., 2016. ADAM17 controls IL-6 signaling by cleavage of the murine IL-6Ralpha from the cell surface of leukocytes during inflammatory responses. J. Leukoc. Biol. 99, 749–760.

Yang, P., Baker, K.A., Hagg, T., 2005. A disintegrin and metalloprotease 21 (ADAM21) is associated with neurogenesis and axonal growth in developing and adult rodent CNS. J. Comp. Neurol. 490, 163–179.

Yang, M., Li, Y., Chilukuri, K., Brady, O.A., Boulos, M.I., Kappes, J.C., Galileo, D.S., 2011. L1 stimulation of human glioma cell motility correlates with FAK activation. J. Neurooncol 105, 27–44.

Yang, S., Wang, B., Tang, L.S., Siednienko, J., Callanan, J.J., Moynagh, P.N., 2013. Pellino3 targets RIP1 and regulates the pro-apoptotic effects of TNF-alpha. Nat. Commun. 4, 2583.

Yatim, N., Jusforgues-Saklani, H., Orozco, S., Schulz, O., Barreira da Silva, R., Reis e Sousa, C., Green, D.R., Oberst, A., Albert, M.L., 2015. RIPK1 and NF-kappaB signaling in dying cells determines cross-priming of CD8(+) T cells. Science 350, 328–334.

Yoon, S., Kovalenko, A., Bogdanov, K., Wallach, D., 2017. MLKL, the protein that mediates necroptosis, also regulates endosomal trafficking and extracellular vesicle generation. Immunity 47, 51–65.

Yoshida, S., Setoguchi, M., Higuchi, Y., Akizuki, S., Yamamoto, S., 1990. Molecular cloning of cDNA encoding MS2 antigen, a novel cell surface antigen strongly expressed in murine monocytic lineage. Int. Immunol. 2, 585–591.

You, Z., Savitz, S.I., Yang, J., Degterev, A., Yuan, J., Cuny, G.D., Moskowitz, M.A., Whalen, M.J., 2008. Necrostatin-1 reduces histopathology and improves functional outcome after controlled cortical impact in mice. J. Cereb. Blood Flow Metab. 28, 1564–1573.

Yuan, J., Kroemer, G., 2010. Alternative cell death mechanisms in development and beyond. Genes Dev. 24, 2592–2602.

Zagouras, P., Stifani, S., Blaumueller, C.M., Carcangiu, M.L., Artavanis-Tsakonas, S., 1995. Alterations in Notch signaling in neoplastic lesions of the human cervix. Proc. Natl. Acad. Sci. U. S. A. 92, 6414–6418.

Zargarian, S., Shlomovitz, I., Erlich, Z., Hourizadeh, A., Ofir-Birin, Y., Croker, B.A., Regev-Rudzki, N., Edry-Botzer, L., Gerlic, M., 2017. Phosphatidylserine externalization, "necroptotic bodies" release, and phagocytosis during necroptosis. PLoS Biol. 15 e2002711.

Zhang, D.W., Shao, J., Lin, J., Zhang, N., Lu, B.J., Lin, S.C., Dong, M.Q., Han, J., 2009. RIP3, an energy metabolism regulator that switches TNF-induced cell death from apoptosis to necrosis. Science 325, 332–336.

Zhang, Y., Dubé, P.E., Washington, M.K., Yan, F., Polk, D.B., 2012. ErbB2 and ErbB3 regulate recovery from dextran sulfate sodium-induced colitis by promoting mouse colon epithelial cell survival. Lab. Invest. 92, 437–450.

Zhang, T., Zhang, Y., Cui, M., Jin, L., Wang, Y., Lv, F., Liu, Y., Zheng, W., Shang, H., Zhang, J., Zhang, M., Wu, H., Guo, J., Zhang, X., Hu, X., Cao, C.M., Xiao, R.P., 2016a. CaMKII is a RIP3 substrate mediating ischemia- and oxidative stress-induced myocardial necroptosis. Nat. Med. 22, 175–182.

Zhang, X., Fan, C., Zhang, H., Zhao, Q., Liu, Y., Xu, C., Xie, Q., Wu, X., Yu, X., Zhang, J., Zhang, H., 2016b. MLKL and FADD are critical for suppressing progressive lymphoproliferative disease and activating the NLRP3 inflammasome. Cell Rep. 16, 3247–3259.

Zhang, Y., Su, S.S., Zhao, S., Yang, Z., Zhong, C.Q., Chen, X., Cai, Q., Yang, Z.H., Huang, D., Wu, R., Han, J., 2017. RIP1 autophosphorylation is promoted by mitochondrial ROS and is essential for RIP3 recruitment into necrosome. Nat. Commun. 8, 14329.

Zhang, C., He, A., Liu, S., He, Q., Luo, Y., He, Z., Chen, Y., Tao, A., Yan, J., 2019a. Inhibition of HtrA2 alleviated dextran sulfate sodium (DSS)-induced colitis by preventing necroptosis of intestinal epithelial cells. Cell Death Dis. 10, 344.

Zhang, C., Luo, Y., He, Q., Liu, S., He, A., Yan, J., 2019b. A pan-RAF inhibitor LY3009120 inhibits necroptosis by preventing phosphorylation of RIPK1 and alleviates dextran sulfate sodium-induced colitis. Clin. Sci. (Lond.) 133, 919–932.

Zhang, J., Qin, D., Yang, Y.J., Hu, G.Q., Qin, X.X., Du, C.T., Chen, W., 2019c. MLKL deficiency inhibits DSS-induced colitis independent of intestinal microbiota. Mol. Immunol. 107, 132–141.

Zhang, J.M., Wang, C.C., Zhang, G.C., Jiang, Q., Yang, S.M., Fu, H.X., Wang, Q.M., Zhu, X.L., Zhu, H.H., Jiang, H., Wang, Y., Lv, M., Lu, J., Chen, H., Han, W., Chang, Y.J., Kong, Y., Xu, L.P., Liu, K.Y., Huang, X.J., Zhang, X.H., 2019d. ADAM28 promotes tumor growth and dissemination of acute myeloid leukemia through IGFBP-3 degradation and IGF-I-induced cell proliferation. Cancer Lett. 442, 193–201.

Zhao, X.M., Chen, Z., Zhao, J.B., Zhang, P.P., Pu, Y.F., Jiang, S.H., Hou, J.J., Cui, Y.M., Jia, X.L., Zhang, S.Q., 2016. Hsp90 modulates the stability of MLKL and is required for TNF-induced necroptosis. Cell Death Dis. 7, e2089.

Zhao, Q., Yu, X., Li, M., Liu, Y., Han, Y., Zhang, X., Li, X.M., Wu, X., Qin, J., Fang, J., Zhang, H., 2019. MLKL attenuates colon inflammation and colitis-tumorigenesis via suppression of inflammatory responses. Cancer Lett. 459, 100–111.

Zheng, X., Tsuchiya, K., Okamoto, R., Iwasaki, M., Kano, Y., Sakamoto, N., Nakamura, T., Watanabe, M., 2011. Suppression of hath1 gene expression directly regulated by hes1 via notch signaling is associated with goblet cell depletion in ulcerative colitis. Inflamm. Bowel Dis. 17, 2251–2260.

Zhou, H.M., Weskamp, G., Chesneau, V., Sahin, U., Vortkamp, A., Horiuchi, K., Chiusaroli, R., Hahn, R., Wilkes, D., Fisher, P., Baron, R., Manova, K., Basson, C.T., Hempstead, B., Blobel, C.P., 2004. Essential role for ADAM19 in cardiovascular morphogenesis. Mol. Cell. Biol. 24, 96–104.

Zhou, B.B., Peyton, M., He, B., Liu, C., Girard, L., Caudler, E., Lo, Y., Baribaud, F., Mikami, I., Reguart, N., Yang, G., Li, Y., Yao, W., Vaddi, K., Gazdar, A.F., Friedman, S.M., Jablons, D.M., Newton, R.C., Fridman, J.S., Minna, J.D., Scherle, P.A., 2006. Targeting ADAM-mediated ligand cleavage to inhibit HER3 and EGFR pathways in non-small cell lung cancer. Cancer Cell 10, 39–50.

Zhu, S., Zhang, Y., Bai, G., Li, H., 2011. Necrostatin-1 ameliorates symptoms in R6/2 transgenic mouse model of Huntington's disease. Cell Death Dis. 2, e115.

Zhu, X., Tao, L., Tejima-Mandeville, E., Qiu, J., Park, J., Garber, K., Ericsson, M., Lo, E.H., Whalen, M.J., 2012. Plasmalemma permeability and necrotic cell death phenotypes after intracerebral hemorrhage in mice. Stroke 43, 524–531.
Zou, B., Liu, X., Gong, Y., Cai, C., Li, P., Xing, S., Pokhrel, B., Zhang, B., Li, J., 2018. A novel 12-marker panel of cancer-associated fibroblasts involved in progression of hepatocellular carcinoma. Cancer Manag. Res. 10, 5303–5311.
Zunke, F., Rose-John, S., 2017. The shedding protease ADAM17: physiology and pathophysiology. Biochim. Biophys. Acta 1864, 2059–2070.

CHAPTER FOUR

Regulation of cell death in the cardiovascular system

Pooja Patel[a], Jason Karch[a,b,*]
[a]Department of Molecular Physiology and Biophysics, Baylor College of Medicine, Houston, TX, United States
[b]Cardiovascular Research Institute, Baylor College of Medicine, Houston, TX, United States
*Corresponding author: e-mail address: jason.karch@bcm.edu

Contents

Abstract

The adult heart is a post-mitotic terminally differentiated organ; therefore, beyond development, cardiomyocyte cell death is maladaptive. Heart disease is the leading cause of death in the world and aberrant cardiomyocyte cell death is the underlying problem for most cardiovascular-related diseases and fatalities. In this chapter, we will discuss the different cell death mechanisms that engage during normal cardiac development, aging, and disease states. The most abundant loss of cardiomyocytes occurs during a myocardial infarction, when the blood supply to the heart is obstructed, and the affected myocardium succumbs to cell death. Originally, this form of cell death was considered to be unregulated; however, research from the last half a century clearly demonstrates that this form of cell death is multifaceted and employees various degrees of regulation. We will explore all of the cell death pathways that have been implicated in this disease state and the potential interplay between them. Beyond myocardial infarction, we also explore the role and mechanisms of cardiomyocyte cell death in heart failure, myocarditis, and chemotherapeutic-induced cardiotoxicity. Inhibition of cardiomyocyte cell death has extensive therapeutic potential that will increase the longevity and health of the human heart.

International Review of Cell and Molecular Biology, Volume 353
ISSN 1937-6448
https://doi.org/10.1016/bs.ircmb.2019.11.005

1. Introduction

The chambers of the human heart are formed by 8 weeks of gestation. As the fetus matures, cardiomyocytes proliferate, and the heart grows in size. Upon birth, cardiomyocytes maintain their proliferative potential, which declines overtime until adulthood. Additionally, there is little evidence for the existence of an endogenous cardiac stem cell pool that can naturally rejuvenate the heart (Maliken and Molkentin, 2018; Mollova et al., 2013). Therefore, the cardiomyocytes present in early adulthood are the same cardiomyocytes that are present in old age (Bergmann et al., 2015). Perhaps the most apparent evidence for the lack of endogenous cardiomyocyte renewal, either through proliferation or through a stem cell population, is that following a cardiac injury, where a substantial amount of cardiomyocytes are lost, the heart forms a fibrotic scar to serve as a patch where cardiomyocytes use to reside (Xin et al., 2013). Notably, scar tissue does not maintain the ability to effectively contract as well as healthy myocardium, but still serves an essential structural role in the injured heart (Xin et al., 2013). Over time, newly formed cardiomyocytes do not infiltrate the scar and replace it with contractile tissue. Instead, the scarring persists, and the heart compensates for the decrease in functionality by other means that eventually result in heart failure (Xin et al., 2013). Efforts are underway to enhance the proliferative potential of adult cardiomyocytes in order to repair the injured heart as well as to convert the fibrotic scar into functioning cardiomyocytes. Additionally, numerous clinical trials have explored the efficacy of cardiac stem cell injections into the injured heart (Chugh et al., 2012); however, the injected stem cells have an extremely low life span and do not significantly contribute to new myocardium at the site of injury (Tang et al., 2018). Alternatively, preventative research is underway to explore how and why atherosclerotic plaques form as well as how to pre-emptively detect and safely eliminate them to prevent an ischemic event. Central to these strategies is the crux of the problem: the cell death of cardiomyocytes, arguably, the largest cell death-related issue in the world. Heart disease is the leading cause of death in the developed world and the majority of heart disease cases are initiated by the loss of cardiomyocytes (Benjamin et al., 2019). Elucidation of the cell death pathways that engage in the diseased heart is paramount for the identification of novel therapeutic approaches. Here, we will discuss the variety of cell death pathways that occur in the heart in development and in disease states.

2. Cell death in heart development

Programmed cell death is crucial to development; it eliminates vestigial anatomy, manages the number of cells in the developing tissue, and aids in the reorganization of structures (Vaux and Korsmeyer, 1999). Remarkably, from embryonic day 20 to embryonic day 35, the human heart develops from two endocardial tubes into a four-chambered heart (Okamoto and Satow, 1975). During this transformation, large amounts of cell migration, proliferation, and death systematically occur. Cell death was first observed in the developing chicken embryonic heart in the late 1960s by transmission electron microscopy (EM) (Manasek, 1969). Extensive characterization of the areas of cell death deemed "zones of cell death" have been established in the chick embryo. Subsequently, research has since extensively recognized 32 zones of death from day 1 to day 10 in the developing chick heart, 21 zones of death between embryonic day 13–17 in rats, and 16 zones in humans (Van Den Hoff et al., 2000). Each zone of death has a specific spatial and temporal location that plays a critical role in cardiac morphogenesis (Van Den Hoff et al., 2000). For example, this cell death aids in fusion of arterial septum with the interventricular septum (Van Den Hoff et al., 2000). By day 10 in the chick, which is equivalent to day 50 in the human, the zones of cell death are no longer detectable and programmed cell death in the heart is complete (Van Den Hoff et al., 2000).

Although programmed cell death takes place in the developing heart, genetic deletions of critical apoptotic regulators (B-cell lymphoma type 2 (BCL2) family members, death receptors, initiating and effector caspases, etc.) exhibit normal cardiac morphogenesis, suggesting that the apoptotic machinery is dispensable for this process (Van Den Hoff et al., 2000). Alternatively, mice deficient in growth factors and growth factor receptors showed significant developmental cardiac phenotypes such as vessel malformations as well as ventricular and atrial defects resulting in perinatal lethality (Donovan et al., 1996; Ghyselinck et al., 1998; Kurihara et al., 1995; Sanford et al., 1997; Soriano, 1997; Yoshida et al., 1996). These phenotypes suggest that growth factors are a more influential regulator of cell death than the core apoptotic machinery, during heart development. Growth factors are typically associated with the autophagic pathway as growth factor withdrawal is a classic method to induce autophagic flux (Lum et al., 2005). Deletion of autophagy related (*Atg*) 13 in mice resulted in perinatal lethality accompanied with myocardial growth defects (Kaizuka and Mizushima, 2016).

In zebra fish, the knockdown of key autophagy regulating genes (*Atg5*, *Atg7*, and *Beclin 1*) all resulted in defects in cardiac morphogenesis and heart structure, thus suggesting that the autophagic, and not the apoptotic, cell death pathway may play a dominant role in heart development.

Two anomalies that argue against this hypothesis are caspase-8 (*Casp8*)-deficient mice and myeloid cell leukemia 1 (*Mcl1*)-deficient mice. Pro-CASP8 is a critical regulator of extrinsic apoptosis that, once cleaved, becomes activated as a result of death receptor stimulation and leads to the cleavage and activation of the downstream effector caspases (Oberst and Green, 2011). *Casp8*-null mice display underdeveloped ventricles as well as thin and disorganized trabeculae carneae, and these mice display a lethal phenotype at embryonic day 12.5 (Varfolomeev et al., 1998). Therefore, the cardiac phenotype observed *Casp8*-deficient mice implicates a role for the extrinsic apoptotic pathway in cardiac development. However, CASP8 also plays an inhibitory role in the necroptotic cell death pathway through the cleavage of receptor interacting protein kinase 1 (RIPK1), which initiates the necroptotic cell death pathway (Oberst et al., 2011). The *Casp8*-deficiency related cardiac defects and embryonic lethality were rescued when crossed with *Ripk3*-null mice, which is downstream of RIPK1 in the necroptotic pathway (Kaiser et al., 2011). Therefore, the defects observed in the *Casp8*-null embryos were due to aberrant necroptosis. Since the hearts in the *Casp8* and *Ripk3* double-knockout mice develop normally, both extrinsic apoptosis and necroptosis are dispensable forms of cell death during cardiac development (Kaiser et al., 2011).

MCL1 is an anti-apoptotic BCL2 family member that is essential for the inhibition of the apoptotic pathway through the inhibition of BCL2 associated X (BAX) and BCL2 Antagonist/Killer 1 (BAK) activation. BAX and BAK are pro-apoptotic BCL2 family members that result in mitochondrial dysregulation during cell death (apoptosis and necrosis) (Karch et al., 2013; Wei et al., 2001). *Mcl1*-deleted mice result in a peri-implantation embryonic lethality (Rinkenberger et al., 2000). Since the death of these mice occurs before cardiac development, the role that MCL1 may play in the cell death observed during cardiac morphogenesis remains to be determined. However, it is clear that MCL1 plays a vital role in the heart post-cardiac development. Cardiac and skeletal muscle-specific deletion of *Mcl1* at embryonic day 13.5, after the heart has developed and programmed cell death in the heart is no longer detectable, results in rapid cardiomyopathy and fatality roughly 10 days after birth (Thomas et al., 2013; Wang et al., 2013). This fatality was rescued when these mice were crossed with *Bax/Bak1* cardiac-specific null mice, suggesting that MCL-1 is a critical anti-apoptotic regulator in the heart (Wang et al., 2013). In summary, it is clear that MCL1 plays a vital role in cardiac mitochondrial function, however due to the severe phenotype observed with complete deletion of *Mcl1* it remains undetermined if MCL1 is involved with the programed cell death that occurs during morphogenesis of the developing heart.

3. The adult heart is naturally death-resistant

Since cardiomyocytes are long-lived cells with an insignificant amount of turnover, they have to be inherently resistant to regulated cell death. Indeed, healthy adult human hearts express altered levels of cell death regulating factors that render them death-resistant (Fig. 1). Critical regulators of apoptosis, necroptosis, and pyroptosis are generally downregulated while anti-cell death-related proteins are expressed at comparable levels in the heart compared to all other tissues in the body except for the brain and skeletal muscle (Fig. 1A and B).

Apoptosis consists of two main pathways: extrinsic and intrinsic (Riedl and Salvesen, 2007). The extrinsic apoptotic pathway relies on extracellular stimulation through the death receptors and downstream machinery, which eventually leads to the cleavage of pro-CASP8 and, subsequently, to apoptotic cell death (Fig. 2 and Table 1) (Riedl and Salvesen, 2007). In contrast, the intrinsic apoptotic pathway is regulated by the BCL2 family which controls mitochondrial outer membrane permeabilization (MOMP), resulting in the release of cytochrome *c* (CytC) which initiates the formation of the apoptosome and the activation of the caspase cascade (Fig. 2 and Table 1) (Riedl and Salvesen, 2007). There is also evidence that the extrinsic pathway can feed into the intrinsic pathway through the BCL2 family (Li et al., 1998; Roy and Nicholson, 2000). The distinctiveness of both apoptotic pathway stems from their dependency on caspase activation (Tait and Green, 2010). The activation of the executioner caspase 3 (CASP3) and caspase 7 (CASP7) during apoptosis leads to DNA fragmentation, nuclear condensation, and membrane blebbing, all of which are classical hallmarks of an apoptotic cell (Tait and Green, 2010). The heart, brain, and skeletal muscle all express reduced levels of the pro-apoptotic BCL2 family members, caspases, and death receptor machinery (Fig. 1A). In contrast, the anti-apoptotic BCL2

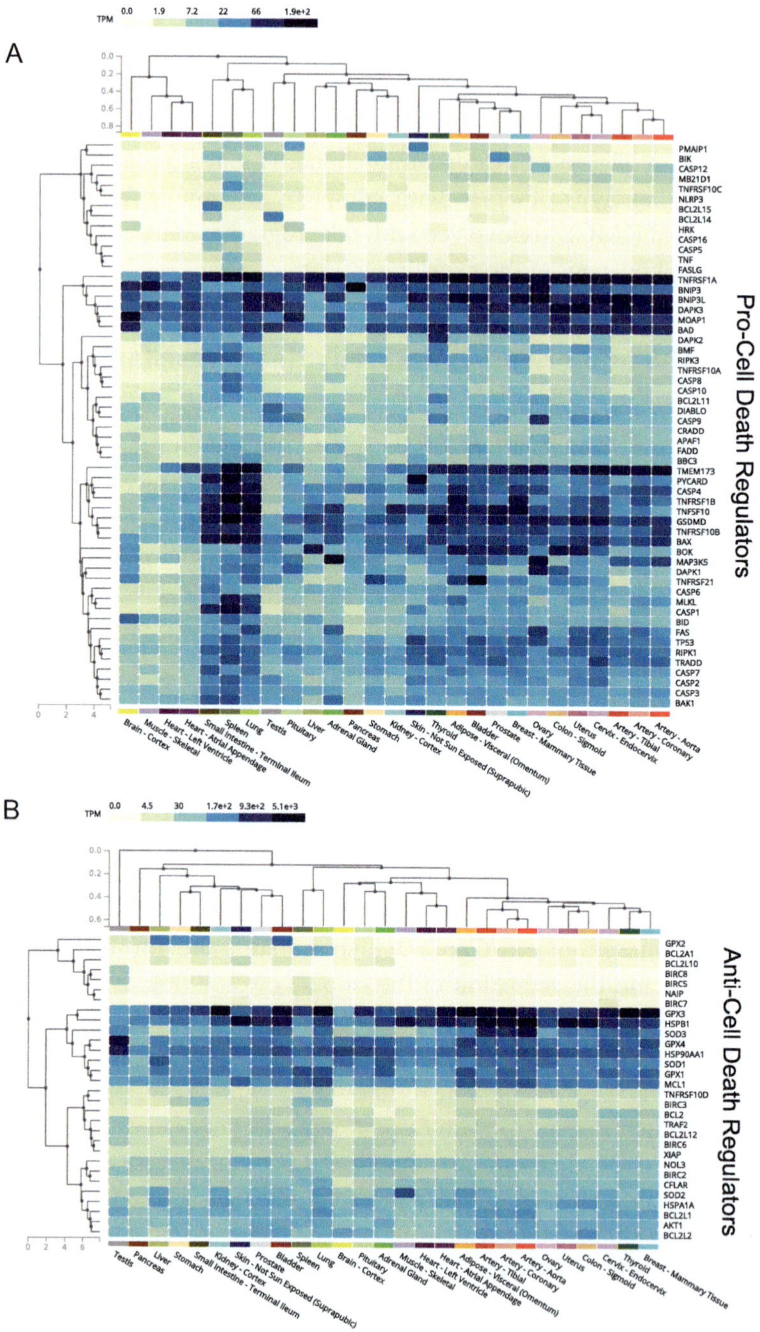

Fig. 1 See legend on opposite page

family members, baculoviral inhibitors of apoptosis repeat containing (BIRC) proteins, anti-apoptotic heat shock proteins, and other apoptosis inhibitors (CASP8 and FADD-like apoptosis regulator (CFLAR) and apoptosis repressor with CARD domain (ARC, a.k.a. NOL3)) are expressed at similar levels to most other tissues, with a few exceptions (Fig. 1B). Overall, the downregulation of the apoptotic pathway regulators suggest that the heart, brain, and skeletal muscle are prewired to resist both extrinsic and intrinsic apoptotic cell death.

Necroptosis and pyroptosis are two other forms of cell death that involve the extrinsic apoptotic regulatory components and caspases that are also involved in the immune response. In contrast to apoptotic cell death, these forms of cell death result in a necrotic morphology, which include loss of mitochondrial membrane potential, mitochondrial and organelle swelling, and cellular rupture (Tang et al., 2019). The combination of death receptor stimulation and caspase inhibition leads to necroptotic cell death through RIPK1 and RIPK3. Phosphorylation and dimerization of RIPK1 and RIPK3 lead to the phosphorylation of mixed lineage kinase domain like pseudokinase (MLKL), which leads to necrotic cell death by causing either increased cytosolic Ca^{2+}, mitochondria dysfunction, or direct plasma membrane permeability, depending on the cell type (Fig. 2 and Table 1) (Cai et al., 2014; Karch et al., 2015; Sun et al., 2012). Similar to the apoptotic regulators, the necroptotic regulators (RIPK1, RIPK3, and MLKL) are all expressed at low levels in the brain, heart, and skeletal muscle compared to the vast majority of other tissues in the body (Fig. 1A).

Pyroptosis is another type of necrotic cell death involved with pathogenic immune responses (Bergsbaken et al., 2009). During pyroptosis, the presentation of pathogen-associated molecular patterns (PAMPs) or damage-associated molecular patterns (DAMPs) leads to the formation

Fig. 1 The heart is naturally cell death-resistant. Heat map of human RNA expression across various tissues of (A) pro-cell death (55 genes) and (B) anti-cell death (30 genes) regulating factors. Data are represented as transcripts per million (TPM) and colored accordingly to the scales. (A) Cluster analysis reveals that the samples isolated from the heart, brain, and skeletal muscle generally express the various pro-cell death genes at lower levels compared to all other tissues. (B) Unlike the pro-cell death regulators, the heart expresses the anti-cell death genes at comparable levels to all other tissues and clusters accordingly. Therefore, the balance between cell death genes in the heart is tipped toward the anti-cell death side rendering the heart death-resistant. Both heat maps were generated using the Genotype-Tissue Expression (GTEx) Project Multi Gene Query.

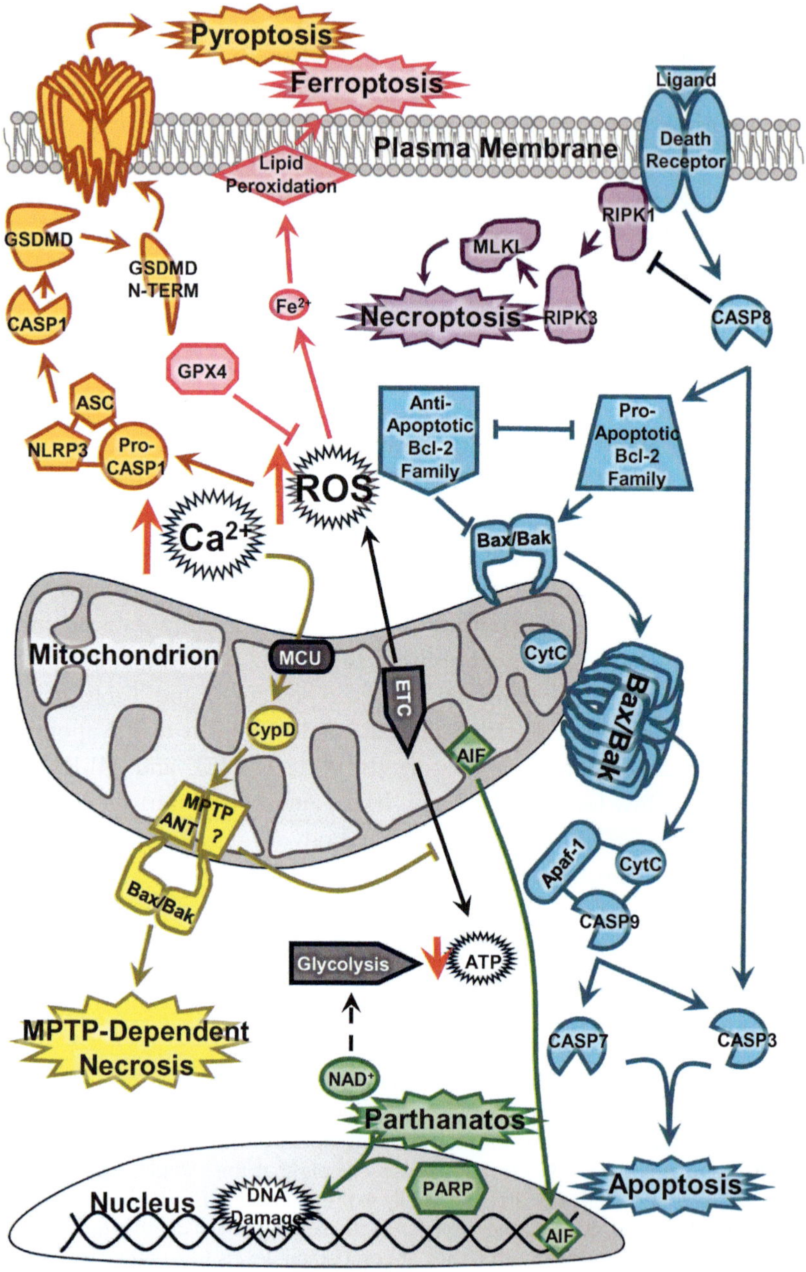

Fig. 2 See legend on opposite page

of the inflammasome which is formed by NOD-like receptor protein 3 (NLRP3), apoptosis-associated speck-like protein (ASC), and pro-caspase1 (pro-CASP1) (Bergsbaken et al., 2009; Fernandes-Alnemri et al., 2007; Jia et al., 2019). Inflammasome formation leads to the activation of CASP1 which is responsible for the cleavage of Gasdermin D (GSDMD), which allows the N-terminal of Gasdermin D to translocate to the plasma membrane, where it forms a homo-oligomeric pore that directly permeabilizes the plasma membrane, resulting in necrotic cell death (Fig. 2 and Table 1) (Ding et al., 2016; Jia et al., 2019). Similar to the previously mentioned forms of cell death, all regulators of pyroptosis are present at exceptionally low levels in the heart, brain, and skeletal muscle compared to most other tissues in the body (Fig. 1A).

Table 1 Cell death pathways and their defining characteristics.

	Morphology	Distinctive characteristic
MPTP-dependent necrosis	Necrotic	MPTP opening
Parthanatos	Necrotic	PARP-1 over-activation
Necroptosis	Necrotic	RIPK activation
Oxytosis/ferroptosis	Necrotic	Lipid peroxidation
Pyroptosis	Necrotic	Gasdermin D pore formation
Autophagic cell death	Autophagic	Overabundant autophagic vacuoles
Extrinsic apoptosis	Apoptotic	Death receptor activation
Intrinsic apoptosis	Apoptotic	MOMP

Fig. 2 The cell death pathways engaged during cardiac I/R injury. Schematic representation of the multiple forms of cell death that are triggered during I/R injury. During ischemia, the affected cardiomyocytes are deprived of oxygen, which results in a decrease in ATP production. Following reperfusion, the affected cardiomyocytes are overloaded with Ca^{2+} and ROS. Both of these stressors can trigger various forms of cell death. High levels of mitochondrial matrix Ca^{2+} and ROS directly lead to the opening of the MPTP and eventually MPTP-dependent necrosis (yellow). The BCL2 family are the master regulators of apoptotic cell death (blue), but they have also been implicated in regulating the MPTP through Bax and Bak expression and activity. ROS accumulation can also lead to inflammasome formation, which may initiate pyroptosis (pink) or excess ROS can cause DNA damage, which leads to the over-activation of PARP and parthanatos (green). Once cardiomyocytes begin to die in the affected area, neighboring cells may be susceptible to extrinsic forms of cell death through death receptor activation, which can lead to extrinsic apoptosis (blue) or necroptosis (purple).

The combination of baseline low expression of apoptotic and necrotic regulators with the abundant expression of endogenous cell death inhibitors creates a cell death-resistant environment in the heart. Even though cardiomyocytes are inherently resistant to cell death, heart disease remains the leading cause of death in the world, with the loss of cardiomyocytes being the foremost facilitating factor for the majority of heart diseases. Cardiomyocytes depend heavily on mitochondrial respiration as a source of ATP, so mitochondrial dysfunction is a key contributor to cardiomyocyte cell death during disease states. Thus, it is important to note that mitochondrial proteins essential for normal mitochondrial function that are abundantly expressed in cardiomyocytes take on alternative roles under stressful conditions that contribute to mitochondrial dysfunction and subsequent cardiomyocyte death.

4. Cell death in the aging heart

As hearts age, changes that occur in the heart decrease cardiac integrity, predisposing the heart to diseases such as arrhythmia, hypertension, atherosclerosis, myocardial ischemia, and, subsequently, heart failure (Strait and Lakatta, 2012). Under normal conditions, an estimated 30% of the cardiomyocytes in the heart drop out overtime from the age 20–75 in humans (Olivetti et al., 1991). This underlying cardiomyocyte drop out may be the driving factor for most of the compensatory mechanisms that correlate with age including increased individual cardiomyocyte volume, increased fibrosis, and ventricular wall thickening (Strait and Lakatta, 2012). As cells age, there is a build-up of damaged proteins and organelles that negatively affect homeostasis if not cleared (Shirakabe et al., 2016).

Autophagy is a clearance/recycling and cell death-related mechanism that is cardioprotective through its ability to delay the aging process. This non-apoptotic cellular degradation pathway is activated in response to stress signals such as oxidative, metabolic, genotoxic, and endoplasmic reticulum (ER) stresses (Sciarretta et al., 2018). In normal heart physiology, autophagy serves a housekeeping function that helps the cell adapt to the aforementioned stresses, which aids in cardiomyocyte survival (Sciarretta et al., 2018). The autophagic pathway sustains cardiac integrity by recycling misfolded proteins and dysfunctional organelles. Throughout aging, autophagic activity decreases, which primes the heart for disease (Taneike et al., 2010).

There are three major types of autophagy: microautophagy, chaperone-mediated autophagy, and macroautophagy. During microautophagy, cellular components are directly shuttled into lysosomes to be broken down (Mizushima and Komatsu, 2011). Chaperone-mediated autophagy occurs when misfolded proteins or protein aggregates containing a KFERQ sequence are recognized by HSP70, which binds to and shuttles the protein to the lysosome (Mizushima and Komatsu, 2011). Macroautophagy is mediated by double-membrane vesicles called autophagosomes, which engulf cellular components and fuse with lysosomes to form autolysosomes, where the constituents of autophagosomes are degraded (Mizushima and Komatsu, 2011). Mitophagy is a form of macroautophagy that specifically targets mitochondria to be degraded by the lysosome (Saito and Sadoshima, 2015). Within cardiomyocytes, mitophagy is an important pathway that clears the dysfunctional mitochondria that either slowly build-up over time or occur during acute cardiac stress (Saito and Sadoshima, 2015).

Key molecular regulators of autophagy include the ATG family, which encode proteins that initiate and mature autophagosomes as well as aid in the fusion of autophagosomes with lysosomes (Mizushima and Komatsu, 2011). The unc-51-like autophagy-activating kinase (Ulk) complex regulates the initiation of autophagosome formation, whose membrane predominantly originates from the ER, Golgi, mitochondria, or cellular plasma membrane (Bernales et al., 2007; Sciarretta et al., 2018). The Ulk complex along with Beclin 1, ATG14L, and vacuolar protein sorting family members activate the production of phosphatidylinositol-3-phosphate, which is recruited onto the autophagosome to promote autophagosome expansion (Burman and Ktistakis, 2010). The phagophore further matures upon the assembly of ATG12, ATG10, and ATG5, which is moderated by ATG7 (Sciarretta et al., 2018). This ATG12-ATG5 complex leads to the growth of phagophore through the enzymatic processing of the microtubule-associated protein light chain 3A (LC3A) to LC3B (Kabeya et al., 2000). LC3B is a ubiquitin-like protein that becomes embedded in the autophagosome membrane, indicating the final maturation state of the autophagosome and is frequently used as a marker to measure autophagic activity (Kabeya et al., 2000).

Inhibition of the removal of damaged organelles or misfolded proteins results in an accelerated-aging phenotypes (Nakai et al., 2007). Mice lacking autophagic machinery such as *Atg4*, *Atg4c*, *Lc3b*, GABA type A receptor-associated protein (*Gabarap*), *Ulk1*, and *Ulk2* genes showed no distinct phenotypes potentially due to functional redundancy (Kuma et al., 2017).

However, *Atg13*-deficient mice, as previously mentioned, are embryonically lethal due to improper cardiomyocyte growth and thinning of the cardiac walls (Kaizuka and Mizushima, 2016). Cardiomyocyte-specific deletion of *Atg5* exacerbates the hypertrophic response to pressure overload in mice. Left ventricular dilation occurs as a response to increased cell death that accumulates from the lack of autophagy occurring within these hearts, suggesting that the autophagic pathway is protective against heart failure (Nakai et al., 2007).

Mitochondrial quality control is another major mechanism through which the aging heart maintains homeostasis (Sciarretta et al., 2018). This quality control mechanism depends heavily on mitochondrial fission and fusion (Saito and Sadoshima, 2015). Portions of the mitochondrial network can undergo mitochondrial fission to segregate dysfunctional mitochondria from healthy mitochondria, where they can be digested through mitophagy (Liesa and Shirihai, 2013). Normally, myocardial mitochondria are maintained in a highly-fused state to maximize oxidative phosphorylation (Liesa and Shirihai, 2013). Due to reduced autophagic flux that occurs during the aging process, mitochondrial turnover decreases, leading to an increase in dysfunctional mitochondria (Linton et al., 2015).

Mitofusin 1 and 2 (MFN1/2) are involved in mitochondrial fusion by maintaining cardiomyocyte integrity and regulating mitochondrial fusion of the outer mitochondrial membrane (Chen et al., 2011). Mice lacking both *Mfn1* and *Mfn2* in cardiomyocytes develop cardiomyopathy as a result of a build-up of malfunctioning mitochondria (Chen et al., 2011). Deletion of *Mfn2* in the heart results in increased cardiomyocyte size as well as decreased left ventricular function, whereas cardiac deletion of *Mfn1* displays little to no phenotype under physiological conditions (Papanicolaou et al., 2011, 2012). Optic atrophy 1 (OPA1) regulates mitochondrial inner membrane fusion (Cipolat et al., 2004). Cardiomyocytes deficient in OPA1 develop mechanical stress resulting in cardiac hypertrophy as well as abnormal mitochondrial morphology (Piquereau et al., 2012). OPA1 is processed by proteases, YME1-like 1 ATPase (YME1L1) and OMA1, a zinc metallopeptidase (Wai et al., 2015). Mice whose cardiomyocytes lack YME1L1 have decreased life-spans, reduced heart function, and abnormal mitochondrial morphology, due to activation of OMA1 (Wai et al., 2015). Indeed, deletion of *Oma1* on the *Yme1l1*-null background restored normal cardiac function as well as mitochondrial morphology (Wai et al., 2015).

Mitochondrial fission often occurs when damaged portions of mitochondria need to be removed via mitophagy (Youle and Van Der Bliek, 2012).

The heart responds to pressure overload by activating mitophagy; increased mitophagy can cause decreased energy production, which can lead to cell death that morphologically resembles autophagic cell death (Saito and Sadoshima, 2015). A major protein that regulates mitochondrial fission is dynamin-regulated protein-1 (DRP1), a GTPase. DRP1 regulates mitochondrial fission by binding to mitochondrial fission factor (MFF), mitochondrial fission 1 protein (FIS1), and mitochondrial dynamics proteins (MiD) 49 and 51 on the outer mitochondrial membrane (Saito and Sadoshima, 2015). FIS1 recruits DRP1 to the outer mitochondrial membrane. Mitophagy is inhibited in cells lacking FIS1 due to a lack of recruitment of DRP1 to the mitochondria (Shen et al., 2014). *Drp1*-deficient cardiomyocytes have impaired cardiac function associated with an accelerated-aging phenotype and lethality (Ikeda et al., 2015).

Mitophagy is a cargo-specific form of autophagy where mitochondria are selectively targeted for lysosomal degradation (Rodriguez-Enriquez et al., 2006). Mitochondrial damage that activates mitophagy includes membrane depolarization as well as activation of the mitochondrial unfolded protein response (Liesa and Shirihai, 2013). Mitophagy occurs through a Parkin-dependent or Parkin-independent mechanisms. Parkin-dependent mitophagy involves phosphatase and tension homolog (PTEN)-induced kinase 1 (PINK1), a serine-threonine kinase and E3-ubiquitin ligase (Geisler et al., 2010; Matsuda et al., 2010; Narendra et al., 2008, 2010; Vives-Bauza et al., 2010). PINK1 accumulates on the outer mitochondrial membrane in response to mitochondrial damage, enabling the recruitment of Parkin to the outer mitochondrial membrane of the damaged mitochondria (Youle and Narendra, 2011). Upon Parkin recruitment to the outer mitochondrial membrane, PINK1 activates Parkin though phosphorylation (Kondapalli et al., 2012). Activation of Parkin on the outer mitochondrial membrane allows for ubiquitination factors to form interactions between uibiquitinated cellular components and mitophagy receptors and adapters (Lee et al., 2010). Dysfunctional mitochondria bind to LC3B on the autophagosome, through receptor proteins, where they become engulfed by the autophagosome (Youle and Narendra, 2011). *Parkin* knockout animals have impaired cardiac recovery after sepsis and myocardial infarction (Piquereau et al., 2013). These animals also displayed unorganized mitochondrial networks as well as morphologically abnormal mitochondria (Kubli et al., 2013). Cardiomyocyte-specific overexpression of Parkin in mice facilitates mitochondrial resistance to age-dependent decreases in O_2 consumption (Hoshino et al., 2013). Enhancement of the autophagic and mitophagy pathways by subjecting mice to an exercise regimen or caloric restriction are two ways to increase autophagic flux and prolong life span (Niemann et al., 2010; Shinmura et al., 2011; Taffet et al., 1997). Exercise increases the amount of healthy mitochondria in failing hearts by inducing mitophagy through the activation of genes implicated in the mitochondrial quality control pathway (Guan et al., 2019). Caloric restriction also induces autophagy in the heart indicated by increased LC3B/LC3A ratios in both young and old mice (Shirakabe et al., 2016; Wohlgemuth et al., 2007). Both exercise and caloric restriction increase autophagy in the heart, which revitalizes long-lived cardiomyocytes by removing the dysfunctional proteins and organelles that build-up slowly over time (Campos et al., 2017; Wohlgemuth et al., 2007).

Aging is a major risk factor associated with cardiovascular disease. With regards to cardiomyocyte cell death associated with aging, the autophagic pathway plays a protective role by regulating protein and organelle quality control (Linton et al., 2015). The inhibition of autophagic pathway in the heart in aging mice exacerbates the aging process and leads to increased cardiomyocyte cell death, eventually resulting in cardiac dysfunction (Linton et al., 2015). Conversely, if autophagic flux is enhanced in aging animals, overall cardiac function improves (Shirakabe et al., 2016). The cell death associated with aging in the heart requires further molecular and phenotypic characterization, but mitochondrial dysfunction is a major determinant for cardiomyocyte aging as inhibition of mitochondrial dynamism in the heart has detrimental effects over time.

5. Cell death and heart disease

Heart disease is, in essence, the inability for the heart to effectively pump blood through the body. There are many forms of cardiovascular disease (CVD), which can range from cardiac arrhythmias and valve disorders to congenital heart disease, heart failure, hypertrophic cardiomyopathy, and coronary artery disease (CAD) (Strait and Lakatta, 2012). Most types of heart disease do not directly involve the death of cardiomyocytes; however, nearly 50% of CVD-related mortalities are due to CAD (Benjamin et al., 2019). The heart is responsible for pumping blood throughout the body as well as itself; ironically, CAD is the inability of the heart to supply oxygenated blood to portions of the myocardium (Ashley and Niebauer, 2004).

During CAD, an atherosclerotic plaque consisting of connective tissue, immune cells, calcium, cholesterols, and other lipids builds up within the arterial wall (Rafieian-Kopaei et al., 2014). As the plaque grows, it further restricts the blood flow to the heart. CAD usually remains asymptomatic until a rupture of the plaque occurs, upon which the blood within the artery begins to clot and eventually blocks blood flow to the myocardium resulting in a myocardial infarction (MI) (Rafieian-Kopaei et al., 2014). In the United States alone, around 735,000 people have a heart attack each year (Benjamin et al., 2019).

When a myocardial infarction occurs, a race against time ensues to remove the blockage and reperfuse the ischemic area of the heart (Kalogeris et al., 2012). The faster reperfusion occurs, the more cardiomyocytes can be saved; if reperfusion occurs within 15 min after the onset of an ischemic event, nearly 100% of the myocardium can be saved (Vander Heide and Steenbergen, 2013). However, if left blocked for an extended period of time, reperfusion is futile, as all the ischemic myocardium would have perished (Vander Heide and Steenbergen, 2013). Although overall beneficial, myocardial reperfusion has its own complications, which lead to an additional round of cardiomyocyte cell death (Roberta, 2011). Together, the combined damage that occurs during myocardial ischemia followed by reperfusion is referred to as ischemia reperfusion (I/R) injury (Kalogeris et al., 2012). Morphologically, cardiomyocytes within the ischemic area that undergo cell death display total cellular swelling as well as nuclear and mitochondrial swelling, hallmarks of necrotic cell death (Roberta, 2011). Originally, the cell death that occurs during an MI was considered to be unregulated necrotic cell death. Research from over the last 40 years, however, has shown that this type of necrotic cell death is actually regulated since it follows a series of molecular triggers. Nevertheless, this form of cell death should not be considered "programmed necrotic cell death," as "programmed cell death" should be reserved for cells that naturally expire for the improvement of the organism in more physiological contexts, and adult cardiomyocytes do not undergo "programmed cell death" under any circumstances (Galluzzi et al., 2018; Kroemer et al., 2009).

5.1 I/R injury induces regulated necrotic cell death

In 1972, Shen and Jennings were the first to correlate calcium accumulation with tissue damage following I/R injury in canine hearts using radiolabeled $CaCl_2$ (Jennings and Shen, 1972). In this study, they compared the differences in Ca^{2+} uptake between short ischemic (10 min) and long ischemic (40 min) periods and found that nearly the entire myocardium was preserved in the short I/R injury with no Ca^{2+} accumulation. In contrast, upon reperfusion following 40 min of ischemia, there is an influx of extracellular Ca^{2+} taken up into the damaged cardiomyocytes and held within the mitochondrial matrix (Jennings and Shen, 1972). By 1980, the correlation of increased Ca^{2+} accumulation during I/R injury transformed into the causation of the injury itself as rat hearts reperfused under low Ca^{2+} conditions had significantly more functional myocardium than control hearts. In addition, rabbits pretreated with Ca^{2+} channel blocker (Verapamil) and exposed to 90 min of ischemia followed by reperfusion had a significant increase in cardiac function, which correlated with a decrease in mitochondrial Ca^{2+} accumulation and preserved mitochondrial function (Bourdillon and Poole-Wilson, 1982).

The establishment of the role of Ca^{2+} during I/R injury coincided with the role of reactive oxygen species (ROS). In 1973, Hearse et al. first hypothesized that molecular oxygen could be the contributing factor to the negative effects of reperfusion in the myocardium of rats subjected to I/R injury (Hearse et al., 1973). Ten years later, this hypothesis was solidified when the administration of free radical metabolizing enzyme superoxide dismutase (SOD) and catalase prior to reperfusion significantly reduced the reperfusion injury and maintained cardiac function (Gardner et al., 1983). Additionally, inhibition of xanthine oxidase (XO), a superoxide-generating enzyme, also displayed protective effects against I/R injury (Das et al., 1987; Granger and Kvietys, 2015). Both discoveries of the inhibition of Ca^{2+} uptake and ROS generation leading to the preservation of cardiomyocyte viability during I/R injury served as the foundation of the future studies that provided undisputable evidence that the necrotic cell death of cardiomyocytes during I/R injury is a regulated process that can be therapeutically targeted.

5.1.1 Ca^{2+} and ROS overload during I/R injury

Calcium handling in cardiomyocytes is a highly-regulated ATP-dependent process fundamental to excitation-contraction coupling that occurs during normal myocardial contractions (Budoff and Gul, 2008). During ischemia, there is a decrease in O_2 availability, which, in turn, leads to a decrease in ATP production; this decrease in ATP production results in dysregulated Ca^{2+} homeostasis (Kalogeris et al., 2012). A culmination of events leads to increased cytosolic Ca^{2+}, which involves the opening of the L-type calcium channel on the plasma membrane, a decrease in sarco/endoplasmic reticulum Ca^{2+}-ATPase (SERCA) activity on the sarcoplasmic reticulum (SR), SR calcium release through the ryanodine receptor (RYR) or the inositol 1,4,5-triphosphate receptors (IP3Rs), and reverse activity in the sodium-calcium exchanger (NCX) on the plasma membrane (Miyamoto et al., 2000; Sallinen et al., 2007; Talukder et al., 2009; Tani and Neely, 1989). All of these major Ca^{2+} handling proteins have been implicated in I/R injury, and pharmacological inhibition or genetic manipulation of these key components protects cardiomyocytes from cell death during I/R injury (Fauconnier et al., 2013; Miyamoto et al., 2000; Niwano et al., 2008; Ohtsuka et al., 2004). To compensate for the reduction of oxidative phosphorylation during ischemia, cardiomyocytes increase anaerobic respiration in an attempt to prolong homeostasis, resulting in acidosis due to the build-up of lactic acid. This compensation, in turn, stimulates the sodium hydrogen exchanger (NHE) to extrude H^+ and to uptake Na^+, which eventually leads to even more Ca^{2+} uptake as a result of NCX (Brown et al., 2016). Reperfusion exacerbates these events, leading to an extreme level of intracellular Ca^{2+}. Indeed, inhibition of NHE is cardioprotective during I/R injury (Piper et al., 1996). Furthermore, if reperfusion occurs at low pH, there is a significant reduction in cell death, due to decreased Ca^{2+} overload (Inserte et al., 2007). The overall increase in cytosolic Ca^{2+}, in combination with the Ca^{2+} dysregulation specifically occurring at the level of the SR, which is in close proximity to the mitochondria through mitochondrial associated membranes (MAMs), leads to mitochondrial dysfunction through Ca^{2+}-dependent MPTP opening, which has been implicated in multiple forms of necrosis as well as apoptosis (Lemasters et al., 1998; Zhou et al., 2018b). Ca^{2+} accumulation also leads to increased calpain (Ca^{2+}-dependent non-lysosomal cysteine proteases) activity, which also have been implicated in a wide variety of cell death mechanisms (Neuhof and Neuhof, 2014).

In cardiomyocytes, mitochondria are the most critical contributors to ROS production through the electron transport chain (ETC), primarily through complex I and complex III (Murphy, 2009). Under normal physiological conditions, the superoxide converts to hydrogen peroxide (H_2O_2) spontaneously or through the mitochondrial SOD, SOD2 (Murphy, 2009; Weisiger and Fridovich, 1973). The H_2O_2 is then extruded from the mitochondria where it is then converted into O_2 and water by numerous mechanisms including catalases, glutathione peroxidases, thioredoxins, and the peroxiredoxins (Imai and Nakagawa, 2003; Rhee et al., 2001, 2005). ROS is also produced in the cytosol by XO, but, under healthy conditions, XO expression is low and has been shown to increase during heart failure (Berry and Hare, 2004; De Jong et al., 2000). The other source of superoxides is through NADPH oxidase (NOX) on the plasma membrane

(Paravicini and Touyz, 2008). Cardiomyocytes mainly express NOX2 and NOX4 isoforms, and deletion of either *Nox2* or *Nox4* significantly reduces infarct size following I/R injury (Braunersreuther et al., 2013). Upon reperfusion, there is a large burst of ROS production, which is believed to take place in the mitochondria due to the accumulation of succinate during the ischemia period (Cadenas, 2018). The damaging effects of ROS include DNA/RNA damage, lipid peroxidation, and protein oxidation and dysfunction (Schieber and Chandel, 2014). Oxidative damage has been implicated in almost every form of regulated cell death.

During cardiac I/R injury, there are two triggers of cell death: Ca^{2+} and ROS overload. Inhibition of either trigger reduces cardiomyocyte cell death following I/R injury, which indicates that these death triggers are intertwined (Braunersreuther et al., 2013; Das et al., 1987; Fauconnier et al., 2013; Gardner et al., 1983; Miyamoto et al., 2000; Niwano et al., 2008; Ohtsuka et al., 2004; Sallinen et al., 2007). Both Ca^{2+} and ROS have the ability to evoke multiple forms of cell death, which should be considered as a continuum instead of individual pathways (Fig. 2 and Table 1) (Konstantinidis et al., 2012). For simplicity, below we will discuss the evidence for and against the individual cell death pathways that may play a role in the heart during I/R injury.

5.1.2 MPTP-dependent necrosis

The I/R environment is highly conducive to the opening of the MPTP, as MPTP opening is triggered by high levels of mitochondrial matrix calcium and/or ROS (Halestrap and Pasdois, 2009). By 1993, Griffiths and Halestrap connected these correlations by observing a significant reduction in I/R injury in mice treated with cyclosporine A (CsA), an inhibitor of CypD, which is an inducer of MPTP opening (Griffiths and Halestrap, 1993). The MPTP is a non-specific voltage-dependent pore that resides within the inner membrane of the mitochondria that, once opened, is permeable to solutes up to 1.5 kDa (Halestrap and Pasdois, 2009; Halestrap et al., 1998; Petronilli et al., 1994). High levels of matrix Ca^{2+} engages CypD, a *cis-trans* peptidyl-prolyl isomerase, to trigger MPTP opening (Halestrap et al., 1997). Prolonged MPTP opening leads to a collapse in mitochondrial membrane potential, which in turn, diminishes ATP production (Halestrap and Pasdois, 2009) (Fig. 2). Variations of CypD inhibitors are effective in desensitizing MPTP opening and protect against I/R injury (Table 2) (Argaud et al., 2005; Crompton et al., 1988; Halestrap et al., 1997; Hausenloy et al., 2003). Furthermore, genomic deletion of *Ppif*, the gene

Table 2 Cell death regulators and their effect on I/R injury.

Target	Effect	Species	Mode of action	Citation
Apoptosis				
zVAD-FMK	Protective	m, r	Pan-caspase inhibition	Yaoita et al. (1998) and Huang et al. (2000)
BCL2 OE	Protective	m	Inhibition of Bax/Bak	Tanaka et al. (2004)
CASP3 OE	Maladaptive	m	Added to cell death induced by I/R by directly inducing caspase activation	Condorelli et al. (2001)
YVAD-cmk	Protective	r	Pan-caspase inhibition	Kovacs et al. (2001)
TNF-a KO	Protective	m	Suppression of death receptor activation	Maekawa et al. (2002)
ARC OE	Protective	r	Inhibition of caspase activity and/or Bax	Xu et al. (2019)
MMPSI	Protective	rab	Inhibition of caspases 3 and 7	Chapman et al. (2002)
BCLXL OE	Protective	r	Inhibition of Bax/Bak	Huang et al. (2003)
BAX KO	Protective	m	Inhibition of CytC release	Hochhauser et al. (2007)
BNIP3 OE	Maladaptive	m	Bax/Bak activation	Diwan et al. (2007)
BNIP3 KO	Protective	m	Inhibition of Bax/Bak	Diwan et al. (2007)
BH4-peptide	Protective	m	Inhibition of Bax/Bak	Boisguerin et al. (2011)
FADD DN	No effect	m	Did not reduce infarct size through the extrinsic apoptotic pathway	Kristen et al. (2013)
Cap-3/Casp-7 DKO	No effect	m	Deletion of these two major effector caspases did not reduce infarct size	Inserte et al. (2016)
CARD9 KO	Protective	m	Attenuated immune response post I/R injury	Qin et al. (2018)
Bax/Bak DKO	Protective	m	Inhibition of CytC release or mitochondrial dysfunction following MPTP	Whelan et al. (2012) and Karch et al. (2013)

Continued

Table 2 Cell death regulators and their effect on I/R injury.—Cont'd

Target	Effect	Species	Mode of action	Citation
MPTP-dependent necrosis				
CsA	Protective	m, r, rab	CypD inhibitor	Halestrap et al. (1997), Crompton et al. (1988), Argaud et al. (2005)
Sanglifehrin-A	Protective	r	CypD inhibitor	Hausenloy et al. (2003)
NIM811	Protective	rab	CypD inhibitor	Argaud et al. (2005)
CypD KO	Protective	m	Desensitized MPTP	Halestrap and Pasdois (2009)
Bax/Bak DKO and CypD KO	Protective	m	There was no additive effect with the removal of Bax/Bak and CypD	Whelan et al. (2012)
CypD KO plus necrostatin	Protective	m	Nec-1 did not provide an additive protection on the CypD KO background	Lim et al. (2007)
Parthanatos				
PARP KO	Protective	m	Inhibition of NAD+ depletion and nuclear AIF translocation	Pieper et al. (2000)
PJ34	Protective	p	PARP inhibitor	Faro et al. (2002)
INO-1001	Protective	r, p	PARP inhibitor	Faro et al. (2002), Khan et al. (2003)
Imidazoquinolinone	Protective	r	PARP inhibitor	Eltze et al. (2008)
Imidazopyridine	Protective	r	PARP inhibitor	Eltze et al. (2008)
Isoquinolindione	Protective	r	PARP inhibitor	Eltze et al. (2008)

Necroptosis				
Necrostatin-1	Protective	m, p	RIPK1 inhibitor	Koudstaal et al. (2015), Smith et al. (2007), Oerlemans et al. (2012)
RIPK3 KO	Protective	m	Blocked ER stress-dependent calcium overload	Zhu et al. (2018)
RIPK3 KO	Protective	m	Prevents MPTP opening through CAMKII	Zhang et al. (2016)
Nec-1 plus zVAD	Protective	gp	Additive protection with combined RIPK and CASP inhibition	Koshinuma et al. (2014)
Oxytosis/ferroptosis				
GPX1 KO	Maladaptive	m	Enhanced ROS	Lim et al. (2009)
GPX1 OE	Protective	m	Reduced ROS	Shiomi et al. (2004)
Ferrostatin-1	Protective	m	Lipid peroxidation inhibitor	Fang et al. (2019)
DXZ	Protective	m	Iron chelator	Williams et al. (1991)
GPX4 OE	Protective	m	Reduced ROS	Dabkowski et al. (2008)
Pyroptosis				
NLRP3 KO	Protective	m	Inhibits inflammasome formation	Sandanger et al. (2013)
16673-34-0 NLRP3i	Protective	m	NLRP3 inhibitor	Marchetti et al. (2015)
VX-765-casp-1 inhibitor	Protective	r	CASP1 inhibitor	Audia et al. (2018), Do Carmo et al. (2018)
OLT1177	Protective	m	Inflammasome inhibitor	Toldo et al. (2019)
CASP1 KO	Protective	m	Inhibits GSDMD cleavage	Kawaguchi et al. (2011)
ASC KO	Protective	m	Inhibits inflammasome formation	Kawaguchi et al. (2011)

Continued

Table 2 Cell death regulators and their effect on I/R injury.—Cont'd

Target	Effect	Species	Mode of action	Citation
Autophagy related				
mdivi-1	Protective	m	DRP1 inhibitor	Ong et al. (2010)
Rapamycin	Protective	m	mTOR inhibitor	Filippone et al. (2017)
PINK1 KO	Maladaptive	m	Inhibits mitophagic flux	Siddall et al. (2013)
P110-Fis1/Drp1 inhibitor	Protective	r	Inhibition of mitochondrial fission	Disatnik et al. (2013)
DRP1 DN	Protective	r	Inhibition of mitochondrial fission	Zepeda et al. (2014)
Mnf1 and Mfn2 DKO	Protective	m	Inhibition of mitochondrial fusion	Papanicolaou et al. (2012)

Abbreviations: OE, over expression; KO, knockout; DN, dominant negative; m, mouse; r, rat; rab, rabbit; gp, guinea pig; p, pig.

that encodes for CypD, in mice provides an equal amount of protection as CsA-treated mice against I/R injury (Baines et al., 2005; Nakagawa et al., 2005). Importantly, treatment with CsA did not provide additional protection in *Ppif*-null mice, indicating that the mechanism of protection of CsA is inhibition of CypD and subsequent mitochondrial dysfunction (Baines et al., 2005; Nakagawa et al., 2005). These studies provided the first genetic evidence for the existence of an MPTP-dependent regulated necrotic cell death pathway. Unfortunately, the target of CypD in regards to the MPTP pore-forming component remains undetermined.

Beyond CypD, the composition of the MPTP remains controversial. Originally, the literature suggested that the inner membrane pore-forming component of the MPTP was the adenine nucleotide translocator (ANT) due to its association with CypD (Woodfield et al., 1998). ANT is the most abundant protein within the inner mitochondrial membrane and is responsible for the transport of adenine nucleotides (ADP/ATP) across the mitochondrial inner membrane (Kim et al., 2010). In humans, there are four isoforms of the ANT family, each expressed in a tissue-specific manner (Kim et al., 2010). In the heart and skeletal muscle, ANT1 is the most abundant isoform, and ANT2 and ANT3 are generally expressed across all other tissues except the testis, which predominantly express ANT4 (Kim et al., 2010). In addition to binding to CypD, ANT inhibitors, bongkrekic acid (BKA) and atractyloside (ATR), desensitize and sensitize MPTP opening respectively, and adenine nucleotides themselves can also effect MPTP sensitivity (Halestrap and Brenner, 2003). Finally, the ANT, through reconstitution in liposomes, has the ability to generate large channel conductance activity similar to the MPTP (Brustovetsky and Klingenberg, 1996; Brustovetsky et al., 2002; Ruck et al., 1998). Despite the positive support that the ANTs are the pore-forming component of the MPTP, liver mitochondria null for ANT1 and ANT2 still undergo permeability transition similar to CypD-null mitochondria (Kokoszka et al., 2004). Thus, the ANT family was deemed as a regulatory facilitator similar to CypD and not a requisite pore component of the MPTP (Kokoszka et al., 2004). Recently, we reexamined the role of the entire ANT family in regards to MPTP opening. Mitochondria isolated from mouse liver lacking all three isoforms of ANT (mice lack *Ant3*) are extremely resistant to MPTP opening (Karch et al., 2019). Currently, no other inner mitochondrial membrane protein has an impact of this magnitude on MPTP activity. However, at extremely high levels of Ca^{2+}, the MPTP in ANT-null mitochondria still opens (Karch et al., 2019). Surprisingly, treatment of CsA on ANT-null mitochondria completely inhibits MPTP opening, suggesting that CypD or the ANTs alone are partial regulators of the MPTP but together are essential regulators (Karch et al., 2019). Furthermore, mouse mitochondria lacking ANT1, ANT2, ANT4, and CypD do not undergo Ca^{2+}-induced MPTP opening (Karch et al., 2019). These findings suggest that the pore-forming component of the MPTP may be comprised of multiple sources, one being the ANT family members and the other involving something that is absolutely dependent on CypD activity to engage (Karch et al., 2019).

Another putative inner membrane pore-forming component of the MPTP was the mitochondrial phosphate carrier (PiC) (Leung et al., 2008). Similar to the ANTs, PiC is part of the larger mitochondrial solute carrier family (SLC25) of proteins that has also been shown to bind with CypD (Leung et al., 2008). In addition, phosphate itself is essential for Ca^{2+}-induced MPTP opening (Halestrap and Pasdois, 2009). Genetic removal of *Slc25a3* (gene that encodes PiC) from mice was extremely maladaptive to normal cardiac function over time; however, acute removal of *Slc25a3* from adult mouse hearts provided protection against I/R injury (Gutiérrez-Aguilar et al., 2014; Kwong et al., 2014). Mitochondria isolated from PiC-null tissue or cells were still able to undergo permeability transition at fairly normal rates, thus, eliminating PiC as being the essential pore-forming component of the MPTP (Gutiérrez-Aguilar et al., 2014; Halestrap and Pasdois, 2009; Kwong et al., 2014).

More recently, the mitochondrial F_1F_o ATP synthase has been hypothesized to be the pore-forming component of the MPTP by multiple mechanisms (Alavian et al., 2014; Bonora et al., 2013; Giorgio et al., 2013). One hypothesis claims that the c-subunit ring of the F_1F_o ATP synthase loosens to form the MPTP, and another suggests that the MPTP forms between the dimers of the F_1F_o ATP synthase (Alavian et al., 2014; Bonora et al., 2013; Carroll et al., 2019; Karch and Molkentin, 2018). However, both of these hypotheses have been recently invalidated (Carroll et al., 2019; He et al., 2017a,b; Karch and Molkentin, 2018). Besides the unlikely nature of the c-subunit ring to form a pore due to its high lipid content and extreme hydrophobicity, genetic removal of all three isoforms of the c-subunit had little effect on MPTP opening and kinetics (He et al., 2017a; Zhou et al., 2017). Additionally, mitochondria lacking essential F_1F_o ATP synthase subunits of the peripheral stalk maintained normal MPTP opening (He et al., 2017a). Indeed, disruption of the F_1F_o ATP synthase peripheral stalk completely inhibits its ability to form dimers, therefore, arguing against the hypothesis that dimers of the F_1F_o ATP synthase form the MPTP.

The outer mitochondrial membrane component of the MPTP was speculated to be the voltage-dependent anion channel (VDAC) family due to its voltage dependency and its association with ANT and CypD (Crompton et al., 1998). The VDACs are the most abundant outer mitochondrial membrane protein and are responsible for the transport of solutes across the membrane (Weeber et al., 2002). Even though the VDACs seemed likely candidates for the outer membrane pore-forming component of the MPTP, removal of VDAC1 and VDAC3 resulted in no change in MPTP activity. Furthermore, mice lacking either VDAC1 or VDAC3 were not protected from I/R injury (Baines et al., 2007). However, VDAC2 was not considered in either the MPTP mitochondrial assays or in the I/R injury model, and the potential compensation between VDAC1 and VDAC3 in the single deletion models during I/R injury was also not scrutinized (Baines et al., 2007). This gap in knowledge was addressed by the removal of all three isoforms of the VDACs in mouse embryonic fibroblasts (MEFs), which did not change the kinetics of cell death induced by ROS via treatment with H_2O_2 or apoptotic inducer staurosporine (Baines et al., 2007).

In addition to its role in apoptosis, the BCL2 family plays an essential role in MPTP-dependent necrosis (Karch et al., 2013). The primary role of the majority of the BCL2 family proteins is to either inhibit or induce the activity of BAX and BAK. BAX is a BCL2 family member that resides in the cytosol whose activation is mediated by BH3-only pro-apoptotic BCL2 family members (Kalkavan and Green, 2018). Once activated, BAX translocates to the mitochondrial outer membrane through the exposure of its transmembrane domain (Westphal et al., 2011). BAK, like BAX, is a pro-apoptotic BCL2 family member that is constitutively located on the outer mitochondrial membrane, whose activation is mediated by BH3-only pro-apoptotic BCL2 family members (Westphal et al., 2011). Once BAK is activated, it oligomerizes with BAX or itself to form the MOMP, which leads to CytC release and eventual apoptotic cell death (Kalkavan and Green, 2018). However, *Bax/Bak1* double-knockout MEFs are completely resistant to nearly all forms of regulated cell death (apoptosis, MPTP-dependent necrosis, necroptosis, lysosomal-dependent cell death) (Karch et al., 2013, 2015, 2017; Wei et al., 2001). BAX and BAK are functionally redundant with regard to cell death as replacement of either BAX or BAK into the double-null background is able to restore all cell death (Karch and Molkentin, 2015; Wei et al., 2001). BAX was originally thought to play a role in MPTP opening through its association with the VDAC channels; however, BAX-null mitochondria are able to undergo normal MPTP opening, therefore, BAX removal alone is not sufficient to inhibit MPTP opening (Narita et al., 1998; Shimizu et al., 1999). However, the removal of both *Bax* and *Bak1* from MEFs, hepatocytes, or cardiomyocytes all lead to strong MPTP inhibition (Karch et al., 2013). In this same study, we found that BAX and BAK1 in their monomeric states were sufficient to restore MPTP opening and necrotic cell death on the *Bax/Bak1* double-null background (Karch et al., 2013). Notably, removal of *Bax* and *Bak1* on the *Ppif*-null background did not provide additive protection against I/R injury, suggesting the loss of *Bax* and *Bak1* function through the MPTP during I/R injury (Whelan et al., 2012). Additionally, anti-apoptotic BCL2 family members, BCL2 and BCLXL, overexpression in the heart also protects mice against I/R injury (Huang et al., 2000, 2003).

The MPTP clearly plays an essential role in the injury that occurs following reperfusion of the ischemic heart. Key regulators of the MPTP have been identified, but the essential inner membrane pore-forming component remains a mystery. Discovery of this component is paramount for the development of novel methods to therapeutically target and inhibit pore opening. Aside from the MPTP, other cell death pathways have also been hypothesized to play a role in I/R injury.

5.1.3 Necroptosis

Necroptosis, another form of regulated necrotic cell death, is morphologically indistinguishable from all other forms of regulated necrotic cell death (Galluzzi et al., 2018). The necroptotic pathway is thought to be a default mechanism for the extrinsic apoptotic pathway: if the extrinsic apoptotic pathway is blocked through caspase inhibition, RIPK1 becomes phosphorylated and remains intact due to CASP8 inhibition, allowing it to bind to and phosphorylate RIPK3 (Vandenabeele et al., 2010). Downstream from the RIPKs is MLKL, a pseudokinase, that when activated by RIPK3, increases cytosolic Ca^{2+} concentrations by either forming channels in the plasma membrane or associating with transient receptor potential cation channel subfamily c member 7 (TRPM7) (Cai et al., 2014; Sun et al., 2012). In addition, MLKL can translocate to mitochondria where it is involved with inducing mitochondrial dysfunction through the opening of the MPTP (Fig. 2 and Table 1) (Karch et al., 2015).

In the heart, there is evidence that the necroptotic pathway is activated during I/R injury (Zhe-Wei et al., 2018). For instance, the expression levels of MLKL increase during acute ischemic injury, and inhibition of RIPK1 through necrostatin-1 (Nec-1) is protective against I/R injury in mice and pigs (Koudstaal et al., 2015; Oerlemans et al., 2012; Smith et al., 2007).

Furthermore, genetic deletion of *Ripk3* is also protective against I/R injury in the heart; however, in this study, the authors concluded that RIPK3 binds to and activates camp kinase II to mediate cellular necrosis through the MPTP instead of MLKL (Linkermann et al., 2013; Zhang et al., 2016). Interestingly, Nec-1 treatment did not have an additive effect when given to *Ppif*-null mice, suggesting that RIPK-mediated cell death may ultimately be regulated by MPTP opening during I/R injury (Lim et al., 2007). Additionally, RIPK3 promotes ER stress, which leads to increased levels of Ca^{2+} and ROS and eventual MPTP opening during I/R injury (Zhu et al., 2018). Notably, combined inhibition of the RIPKs and caspases through dual treatment of Nec-1 and zVAD-FMK is additively protective against I/R injury, suggesting there are multiple forms of cell death being engaged during I/R injury (Table 2) (Koshinuma et al., 2014). Regardless of the downstream mechanisms, the RIPKs clearly play a role in the heart during I/R injury, however, further investigation is necessary to determine the mechanism of how the RIPKs become phosphorylated and activated during this disease state. Perhaps, since cardiomyocytes endogenously express low levels of caspases and higher levels of caspase inhibitors, death receptor activation is more prone to induce necroptosis than extrinsic apoptosis in the heart.

5.1.4 Ferroptosis/oxytosis

Ferroptosis or oxytosis, is a cell death mechanism involving depletion of the antioxidant activity of glutathione, which results in high levels of lipid peroxidation and necrotic cell death (Yang and Stockwell, 2016). Under normal conditions, glutathione hydroperoxidase 4 (GPX4) reduces lipid peroxides into non-reactive lipid alcohols (Yang and Stockwell, 2016). Glutathione is required for GPX4 activity, and cysteine is the rate-limiting step for glutathione biosynthesis (Brigelius-Flohé and Maiorino, 2013). Ferroptosis/oxytosis is morphologically indistinguishable from other forms of necrotic cell death and has been implicated in cardiac I/R injury (Galluzzi et al., 2018). As previously stated, an abundance of ROS production that occurs during I/R injury results in lipid peroxidation (Stamenkovic et al., 2019). Inhibitors of lipid peroxidation, such as Coenzyme Q10 and ferrostatin-1, significantly reduce cardiac infarct size (Birnbaum et al., 1996; Fang et al., 2019). Additionally, iron chelation by pretreatment with deferoxamine has also been shown to be protective against I/R injury in the heart as ferrous iron promotes lipid peroxidation through the Fenton reaction (Fang et al., 2019; Williams et al., 1991). However, deferoxamine treatment after ischemia and prior to reperfusion was maladaptive (Williams et al., 1991).

Finally, transgenic overexpression of GPX1 or a mitochondrial-targeted version of GPX4 attenuates cardiac dysfunction following I/R injury (Dabkowski et al., 2008; Shiomi et al., 2004). GPX1 knockout mice displayed a maladaptive phenotype during I/R injury (Table 2) (Lim et al., 2009). Future research is required to determine how lipid peroxidation influences MPTP-dependent necrosis or whether MPTP opening is involved with ferroptosis, which is important to determine the sequence of cell death events and possible nodal points that occur during I/R injury in the heart.

5.1.5 Pyroptosis

Pyroptosis is a type of regulated cell death that is primarily associated with the immune response as this cell death pathway is triggered by the presence of PAMPs or DAMPs (Jia et al., 2019). Morphologically, pyroptosis is characterized by features observed in apoptosis, such as nuclear condensation and DNA fragmentation, as well as features observed in necrosis, such as the formation of a pore on the plasma membrane followed by cellular swelling and subsequent rupture (Jia et al., 2019). There are two distinct pathways through which pyroptosis can occur: a canonical pathway involving the activation of CASP1 and a non-canonical pathway involving the activation of caspases 4/5/11 (CASP4/5/11) (Jia et al., 2019; Xu et al., 2018).

The canonical pyroptotic pathway involves the detection of PAMPs and DAMPs by the inflammasome (Jia et al., 2019). The inflammasome consists of (NOD)-like receptors (NLR) family members such as NLRP3, NLRP1, NLRC4, NLRP9, and NLRP6, as well as PYHIN protein families and Pyrin proteins (Jia et al., 2019). The activation of NLR proteins through the presence of PAMPs and DAMPs signals ASC an inflammasome adaptor protein (Gross et al., 2009). ASC then activates CASP1 through the cleavage of pro-CASP1. CASP1 activity leads to the cleavage of GSDMD, which allows the N-terminal of GSDMD to translocate to the plasma membrane where, through homo-oligomerization, it forms a 10–15 nm pore (Chen et al., 2016; Jia et al., 2019). This pore quickly leads to cellular swelling and subsequent cellular membrane rupturing (Chen et al., 2016) (Fig. 2). Non-canonical pyroptosis is primarily induced by the presence of lipopolysaccharides (LPS) in the cytosol (Jia et al., 2019; Shi et al., 2014). The lipid A moiety of LPS is recognized by the CARD domain of caspase-4/5/11 (Cheng et al., 2017; Shi et al., 2014). The activation of these caspases can also lead to the cleavage of GSDMD and subsequent pore formation (Jia et al., 2019).

During I/R injury, pyroptosis can potentially be triggered by neighboring necrotic cells, ROS production, and release of mitochondrial DNA that results from mitochondrial dysfunction and rupture. However, conflicting evidence exists concerning whether pyroptosis actually contributes to cardiomyocyte cell death during I/R injury. According to Kawaguchi et al., *Asc-deleted* mice and *casp-1*-null mice show decreased infarct size in the heart in response to myocardial I/R injury when compared to control mice, suggesting that pyroptosis occurs in cardiomyocytes (Kawaguchi et al., 2011). In addition, the pharmacological inhibition of CASP1 has shown to reduce infarct size in rat myocardial I/R models (Table 2) (Audia et al., 2018; Do Carmo et al., 2018). Conversely, Sandanger et al. showed that infarct size increases in *Nlrp3*-deficient mice and in *Asc*-null mice subjected to I/R injury compared to controls, suggesting that pyroptosis may actually play a protective role during I/R injury (Sandanger et al., 2013). Conversely, NLRP3 inhibitors are cardioprotective in myocardial I/R injury murine models (Marchetti et al., 2014, 2015). Additionally, treatment with OLT1177, an inflammasome inhibitor, in mouse I/R injury models displayed reduced infarct size (Table 2) (Toldo et al., 2019). Since the expression of key regulators of the pyroptotic pathway are very low in cardiomyocytes (Fig. 1), pyroptosis may play a more important role in non-cardiomyocyte populations in the heart during I/R injury. In order to determine the role of pyroptosis in cardiomyocytes during I/R injury, cardiomyocyte-specific deletion of critical pyroptosis regulators must be investigated.

5.1.6 Parthanatos

ROS-induced DNA damage occurs during cardiac I/R injury, and can also result from MPTP opening, leading to mitochondrial dysfunction through high levels of Ca^{2+} and additional production of ROS. DNA damage combined with mitochondrial dysfunction is the perfect storm for the induction of parthanatos, another form of necrotic cell death that is dependent on the over-activation of poly(ADP-ribose) polymerase-1 (PARP-1). PARP-1, an enzyme involved in DNA repair that requires NAD^+ as a substrate to function (Fatokun et al., 2014; Galluzzi et al., 2018). Over-activation of PARP-1 reduces the NAD^+ pool, eventually inhibiting the glycolytic pathway thus further depleting the ATP pool during I/R injury (Fig. 2 and Table 1) (Fatokun et al., 2014). Moreover, PARP-1 activity leads to the accumulation of poly(ADP-ribose) (PAR) (Andrabi et al., 2006). Once enough PAR polymers have accumulated in the nucleus, they enter the cytoplasm, where they can infiltrate depolarized mitochondria (Andrabi et al., 2006). Within the mitochondria, PAR has been shown to associate with apoptosis inducing factor 1 (AIF1) by facilitating its nuclear translocation (Wang et al., 2011). Once within the nucleus, AIF1 promotes nuclear condensation and degradation (Wang et al., 2011). In terms of cardiac ischemia, when the cardiomyocytes are surviving on anaerobic respiration, reduction in the ability to perform glycolysis is maladaptive. Inhibitors of PARP1 and genetic deletion of *Parp1* are protective against cardiac I/R injury in mice (Table 2) (Eltze et al., 2008; Faro et al., 2002; Grupp et al., 1999; Khan et al., 2003; Pacher and Szabó, 2007; Pieper et al., 2000). Additionally, PARP-1 inhibition attenuates AIF nuclear translocation during I/R injury (Xiao et al., 2004). In rat hearts, mitochondrial release of AIF and not CytC was detected following reperfusion injury, suggesting that parthanatos plays a more dominant role than downstream apoptotic regulators (Kim et al., 2003). Due to the nature of parthanatos, it seems to be a secondary cell death mechanism that is engaged during cardiac I/R injury which is preceded by ROS-dependent DNA damage and MPTP-dependent mitochondrial dysfunction.

5.1.7 Autophagy

Baseline autophagy in cardiomyocytes maintains protein and organelle quality control in order to maintain energetic homeostasis (Abdellatif et al., 2018). Originally, autophagic cell death was morphologically defined as cell death in the presence of large amounts of autophagic vacuoles (Lockshin and Beaulaton, 1974). However, autophagic cell death is currently molecularly characterized as cell death that is dependent on autophagic machinery, mainly the ATG family (Berry and Baehrecke, 2008). The occurrence of this type of cell death is rare, and instead usually feeds into other forms of cell death through lysosomal permeability (Galluzzi et al., 2018). Overall, the increase in autophagic flux that occurs during I/R injury is more of a protective rather than maladaptive mechanism.

An increase in autophagy activation occurs during I/R injury, however, autophagy plays different roles during ischemia and reperfusion. During ischemia, autophagy is a mechanism by which the cell tries to maintain energy homeostasis, whereas during reperfusion, autophagy is a mechanism by which the cardiomyocytes clear dysfunctional protein and organelles as a result of from of ROS production and Ca^{2+} overload (Nishida et al., 2009). Some conflict exists concerning whether discrepancies in autophagy are beneficial or detrimental to the heart during I/R injury. Decreased ATP concentrations during ischemia activates adenosine monophosphate-activated protein kinase (AMPK), a sensory protein that responds to energy levels within the cell (Ma et al., 2015). Once activated, AMPK can induce autophagy either directly by phosphorylating ULK1 or indirectly by phosphorylating mechanistic target of rapamycin kinase complex 1 (mTORC1) and thus preventing ULK1 inhibition through mTORC1 (Alers et al., 2012). Inhibition of murine mTOR through treatment of rapamycin is cardioprotective during myocardial I/R injury through the downstream inhibition of pro-apoptotic *Bax* expression (Filippone et al., 2017).

During reperfusion, AMPK is no longer activated, thus it is no longer a key regulator of autophagy in this scenario (Ma et al., 2015). Instead, following reperfusion, ROS is the major driver of autophagy by stimulating the upregulation of *Beclin-1* (Chen-Scarabelli et al., 2014). Upregulation of *Beclin-1* may also have implications in apoptotic or necrotic cell death by its sequestration of BCL2 through its BH3 domain to promote the oligomerization of the BAX and BAK to allow for CytC release or to sensitize MPTP opening (Marquez and Xu, 2012; Pattingre et al., 2005). ROS could also initiate autophagy through the oxidation of ATG4, which would activate LC3 modification (Scherz-Shouval et al., 2007). Although autophagy still occurs during reperfusion, it may be impaired as indicated by decreased autophagosome clearance from the cell, which is detrimental to the cardiomyocyte (Ma et al., 2012). Restoration of autophagosome clearance by overexpression of lysosome-associated membrane protein-2 (LAMP-2) and knockdown of Beclin-1 is protective in vitro against hypoxia/reoxygenation injury (Ma et al., 2012). Following reperfusion, mitophagy also occurs to aid in the clearance of damaged and dysfunctional mitochondria. PINK1-dependent mitophagy plays a protective role in myocardial I/R injury as loss of PINK1 is maladaptive (Siddall et al., 2013). Prevention of mitochondrial fission through DRP1 inhibition is cardioprotective against I/R injury in mice (Table 2) (Disatnik et al., 2013; Ong et al., 2010; Zepeda et al., 2014). In some circumstances, increased autophagy during reperfusion could lead to autophagy-induced cell death. Inhibition of Beclin-1 by urocortin attenuates autophagic cell death in neonatal rat cardiomyocytes exposed to simulated I/R injury (Valentim et al., 2006). Overall, autophagy primarily plays a protective role during cardiac I/R injury by aiding in the maintenance of energy homeostasis during ischemia and by clearing dysfunctional proteins or organelles during reperfusion.

5.1.8 Apoptosis

Apoptotic cell death is morphologically characterized by cellular shrinking, nuclear condensation and fragmentation, and membrane blebbing (Kroemer et al., 2005). The cardiomyocyte cell death that takes place during I/R injury assessed by EM morphologically resembles necrosis, which argues heavily against the apoptotic pathway regulating cardiomyocyte cell death in I/R injury (Kung et al., 2011). During the early characterization of the apoptotic pathway, the term "apoptosis" was synonymous with "cell death," and this misnomer has led to the idea that the apoptotic pathway is the major cell death pathway that regulates I/R injury. More terminology that is appropriate came into use after the discovery that necrotic cell death was also regulated (Galluzzi et al., 2018). Although the resulting morphological features do not resemble apoptotic cell death during I/R injury, many regulators of apoptotic cell death have been implicated in contributing to this disease. Again, cell death is a continuum since there is evidence of crosstalk and overlap between the cell death pathways, as upstream pro-apoptotic regulators may feed into other cell death pathways. Here we will discuss the multiple mechanisms through which the apoptotic pathway feeds into the necrotic pathway to induce cell death during I/R injury.

The BCL2 family proteins are the major facilitators of the apoptotic pathway that are responsible for the release of CytC from the mitochondria, which initiates apoptosome formation and the subsequent caspase cleavage cascade (Fig. 2 and Table 1) (Tang et al., 2019). The BH3-only subfamily members also connect the extrinsic and intrinsic apoptotic pathways downstream of death receptor stimulation to initiate activation BAX and BAK mitochondrial permeabilization (Westphal et al., 2011). Overexpressing anti-apoptotic BCL2 family members, specifically in cardiomyocytes, is protective against I/R injury (Table 2) (Tanaka et al., 2004). Additionally, double-deletion of *Bax* and *Bak1* in cardiomyocytes is also protective against I/R injury (Karch et al., 2013; Whelan et al., 2012). Given this evidence alone, it would be simple to conclude that the BCL2 family regulates apoptosis during I/R injury; however, the end result is still necrotic cell death. Apoptotic peptidase activating factor 1 (APAF1) plays an essential role downstream of CytC release, and, in cardiomyocytes, *Apaf1* expression is low and held in check by endogenous X-linked inhibitor of apoptosis (XIAP), so there is a lack of evidence for apoptosome formation in cardiomyocytes (Potts et al., 2005). This finding disconnects the BCL2 family from the apoptotic execution machinery; therefore, an alternative explanation is required to connect the BCL2 family to necrosis. Originally, BAX and BAK were thought to regulate mitochondrial dysfunction and were hypothesized to bind with the ANT family to regulate the MPTP (Marzo et al., 1998). Both prolonged MPTP opening and BAX/BAK-MOMP can lead to CytC release (Lemasters et al., 1998); there is evidence demonstrating that MOMP-dependent CytC release does not lead to mitochondria inner membrane depolarization and mitochondrial dysfunction (Eskes et al., 1998). Additionally, *Ppif*-null cells are still able to engage BAX/BAK-mediated CytC release and apoptotic cell death (Baines et al., 2005). These findings established the difference between the BCL2 family and mitochondrial dysfunction. Until recently, the BCL2 family were only thought to regulate mitochondrial outer membrane permeabilization and not mitochondrial dysfunction. Work from the past 10 years has reignited the role of the BCL2 family in mitochondrial homeostasis demonstrating that mitochondria deficient in BAX and BAK are extremely resistant to Ca^{2+}-mediated MPTP opening (Karch et al., 2013). Furthermore, BH-3 mimetics have been shown to sensitize mitochondria to MPTP opening (Henz et al., 2019; Karch et al., 2013). Finally, MCL-1, which is now known to localize to the inner mitochondrial membrane and the mitochondrial matrix, clearly plays a vital role in mitochondrial homeostasis in cardiomyocytes as loss of *Mcl-1* in cardiomyocytes leads to rapid cardiomyopathy and death (Thomas et al., 2013; Wang et al., 2013). Recently, it has been shown that BAX over-activation can lead to mitochondrial DNA release (McArthur et al., 2018; Riley et al., 2018). The relevance of this phenomenon occurring in an in vivo setting is still lacking, but the over-activation of BAX is clear evidence that the BCL2 family actively participates in mitochondrial vitality and dysfunction beyond the release of CytC.

Downstream from the mitochondria and the apoptosome formation are the caspases, a family of cysteine proteases that become cleaved and activated during both extrinsic and intrinsic apoptosis, where they are directly responsible for apoptotic execution of a cell (Elmore, 2007). Many studies have shown that treatment with a pan-caspase inhibitor reduces cell death in cardiomyocytes following I/R injury (Table 2) (Huang et al., 2000; Kovacs et al., 2001; Yaoita et al., 1998). However, the genetic double-deletion of *Casp3* and *Casp7* in cardiomyocytes does not reduce infarct size following I/R injury in the mouse heart (Inserte et al., 2016). Together these results suggest, that complete caspase inhibition may not simply be inhibiting apoptosis in the heart, but potentially may be affecting other forms of cell death that involve the caspases during I/R injury.

Perhaps, pan-caspase inhibition could be preventing other types of necrotic cell death such as pyroptosis through the inhibition of CASP1/4/5/11, which may be engaged during I/R injury (Frantz et al., 2003; Merkle et al., 2007). Alternatively, high levels of pan-caspase inhibitors may have off target effects (Waterhouse et al., 1998). At high levels, caspase inhibitors also inhibit calpains, another family of cysteine proteases. Calpains are Ca^{2+}-activated cysteine proteases that also have been implicated in I/R injury (Iwamoto et al., 1999; Mocanu et al., 2000). Indeed, Calpain inhibitors also reduce infarct size post I/R injury (Iwamoto et al., 1999). Additionally, genetic removal of *Calpain1* is protective, and overexpression of Calpain1 is maladaptive during I/R injury (Neuhof and Neuhof, 2014). Furthermore, calpain activity can regulate caspase activity and has been observed to be upstream to caspase cleavage during apoptosis (Harwood et al., 2005). Although cardiomyocytes mainly display a necrotic morphology during I/R injury, apoptotic regulators clearly still play an essential role in regulating cardiomyocyte cell death. Greater understanding of how the apoptotic regulators feed into other cell death pathways would be beneficial for designing therapeutic approaches to inhibit cardiomyocyte cell death during I/R injury.

5.1.9 Ischemic preconditioning

Ischemic preconditioning refers to the phenomenon when the introduction of small acute ischemic bouts prior to a major ischemic event activates signaling pathways that are cardioprotective. Indeed, animals exposed to severe ischemia after ischemic preconditioning have a reduction in myocardial infarct size (Murry et al., 1986). Ischemic preconditioning was first discovered in 1986 by Murry et al., where they performed four 5-min coronary occlusions before subjecting the animals to either a 40-min or 3-h ischemic period (Murry et al., 1986). The dogs subjected to a 40-min ischemic period had significantly reduced infarct size compared to controls, while preconditioning had no effect on infarct size following a 3-h ischemic period. Since this initial observation, identification of the molecular mechanisms that underlie ischemic preconditioning have been under investigation.

One of the mechanisms involved in ischemic preconditioning is utilizing the influx of adenosine into the cell during ischemia to activate the A1 adenosine receptor (A_1AR) (Sheth et al., 2014). The A_1AR receptor activates the phosphatidylinisitol 3-phosphate (PI3K) signaling pathway, leading to subsequent activation of AKT Serine/Threonine Kinase (Akt/PKB) pathway and resulting in the activation of glycogen synthase kinase 3 beta (GSK3β) (Tong et al., 2002). GSK3β is considered protective during ischemia because it activates mTOR and autophagy pathways, promoting cell survival (Zhai et al., 2011).

Another mechanism of cardioprotection from ischemic preconditioning is the generation of nitric oxide (NO) by nitric oxide synthase (NOS) (Wang et al., 2002). Downstream effects of the NO production lead to the activation of signaling pathways implicated in the modification of Ca^{2+}-handling proteins that lead to depletion of Ca^{2+} in cardiomyocytes, which prevents Ca^{2+} overload-induced cell death during myocardial I/R injury (Sun et al., 2006). In addition, NO is a precursor to ROS scavenger peroxynitrite and has also been shown to decrease apoptotic activity through dose-dependent inhibition of caspases (Yu et al., 2018). Finally, exogenous NO leads to a decrease of p53 gene expression, which consequently leads to a decrease in cell death following I/R injury (Phillips et al., 2009).

Conversely, NOS has also been shown to be maladaptive during I/R injury, which hampers the protective effects of NO observed in preconditioning (Yu et al., 2018). There are three isoforms of NOS; endothelial NOS (eNOS) and neuronal NOS (nNOS), which constitutively produce and release NO, and inducible NOS (iNOS), which is active during a pro-inflammatory stimulation (Wang et al., 2002). During myocardial ischemia, an upregulation of nNOS occurs, which inhibits the production of superoxide species (Damy et al., 2003). However, following I/R injury, the iNOS/NO pathway becomes activated, which has been shown to lead to the downregulation of anti-apoptotic BCL2 family members through cGMP signaling (Feng et al., 2001; Yu et al., 2018). Alternatively, increased production of peroxynitrite, from nitric oxide free radical scavenging, correlates with the increase of BCL2 expression in sleep-deprived mice who were subjected to myocardial I/R injury, suggesting a cardioprotective function of NO (Jeddi et al., 2018; Yu et al., 2018).

Ischemic preconditioning also leads to a reduction in the Ca^{2+} overload that normally occurs following I/R injury. During preconditioning, signaling pathways lead to the opening of the sarcolemma K_{ATP} channels (Sadat, 2009). Opening of the K_{ATP} channels increases the speed of rapid repolarization in the cardiac action potentials, preventing the influx of Ca^{2+} into the cell upon reperfusion (Peart and Headrick, 2003; Peart et al., 2003; Sadat, 2009). Consequently, ischemic preconditioning attenuates MPTP opening and prolongs mitochondrial function following I/R injury (Javadov et al., 2003).

5.2 Non-ischemic heart failure

Heart failure is the inability of the heart to maintain its workload. Most heart failure cases are a result of ischemic injury, which clearly involves the cell death pathways. Conversely, the cell death pathways seemingly play less of a leading role in non-ischemic heart failure, particularly pressure overload-induced heart failure. Initially, the heart responds to pressure overload by cardiac hypertrophy; however, long-lasting exposure to this mechanical stress leads to heart failure (Kehat and Molkentin, 2010). The transition from compensated to decompensated hypertrophy and heart failure is characterized by a decrease in cardiac function as well as cardiac wall thinning due to the loss of cardiomyocytes (Kehat and Molkentin, 2010). Although this process is not as abrupt as I/R injury, the subtle loss of cardiomyocytes, has profound effects over time that cannot be naturally reversed due to the lack of endogenous cardiomyocyte regenerative capacity (Maliken and Molkentin, 2018; Van Berlo and Molkentin, 2014).

The mechanisms of cell death that are engaged during decompensated hypertrophy are less characterized than I/R injury. The most common method to induce pressure overload in a rodent model is transverse aortic constriction (TAC) surgery. Unlike I/R injury, there is no evidence for the MPTP opening playing a role in TAC-induced heart failure. Mice lacking the mitochondrial calcium uniporter (MCU), the mechanism for calcium entry into the matrix of the mitochondria, specifically in cardiomyocytes, respond to TAC similarly to control mice (Kwong et al., 2015). Furthermore, *Ppif*-null mice that are highly protective against I/R injury, are actually predisposed to TAC-induced heart failure compared to controls (Elrod et al., 2010). These results suggest that mitochondrial calcium and MPTP opening are not contributing factors to cardiomyocyte drop out post-TAC. Conversely, mice deficient for pro-apoptotic BH-3 only BCL2 proteins, BCL2 interacting protein 3 like (BNIP3L) or P53-upregulated modulator of apoptosis (PUMA), are protected against TAC-induced heart failure (Diwan et al., 2008; Mandl et al., 2011). Furthermore, cardiomyocyte-specific overexpression of CASP3 is sufficient to induce decreased cardiac function similar to heart failure (Narula et al., 1999). Autophagy also contributes to cardiac remodeling as *Atg5*-deficient hearts exacerbate pressure overload induce hypertrophy (Nakai et al., 2007). However, *Beclin-1* heterozygote mice display decreased cardiac remodeling following pressure overload, and BECLIN-1 overexpression leads to diminishing cardiac function following TAC surgery (Zhu et al., 2007). Perhaps the difference between *Atg5*-null mice and the *Beclin-1* heterozygote mice is the ability of BECLIN-1 to crosstalk with the BCL2 family through its BH-3 domain. Due to the low rate of cardiomyocyte drop out and the high rate of fibrosis and infiltrating cells that occur during pressure overload-induced heart failure, identification of the specific cell death mechanisms that lead to the loss of cardiomyocytes during the decomposition phase will be challenging.

5.3 Cancer therapeutics and the heart

Cardiotoxicity is the damage to cardiac tissue, which usually leads to cardiomyopathies and eventually heart failure. A major side effect of anticancer therapeutics has been cardiotoxicity where there is a reduction of the left ventricular ejection fraction, a hallmark of cardiomyopathy and a symptom of heart failure (Albini et al., 2010). Treatment with chemotherapeutic agent doxorubicin (DOX), an anthracycline, correlates with dilated cardiomyopathy and eventual heart failure (Chatterjee et al., 2010; Swain et al., 2003; Takemura and Fujiwara, 2007). In cancer cells, anthracyclines intercalate into DNA and inhibit the function of DNA topoisomerase II (TOP2), which initiates regulated cell death (Shi et al., 2011). In cardiomyocytes, anthracyclines-induced cardiotoxicity results from ROS produced during the reduction of the quinine moiety of anthracycline to semiquinone (Shi et al., 2011). Oxidative stress triggers cardiomyocyte death through lipid peroxidation and/or mitochondrial dysfunction, which are hallmarks of ferroptosis/oxytosis and MPTP-dependent cell death, respectively (Balanehru and Nagarajan, 1992; Fang et al., 2019; Shi et al., 2011). Treatment with antioxidants or iron chelators are effective at reducing DOX-induced cardiotoxicity (Shi et al., 2011). Recently, ferrostatin-1, an inhibitor of ferroptosis through inhibition of lipid peroxidation, was shown to protect against DOX-induced cardiotoxicity (Fang et al., 2019). Additionally, increased ROS production contributes to Ca^{2+} handling dysregulation, which can exacerbate MPTP opening (Gharanei et al., 2013; Kalivendi et al., 2005; Shi et al., 2011). Furthermore, the apoptotic pathway may play a role in DOX-induced cardiomyocyte cell death. Anthracycline can induce intrinsic apoptosis through the downregulation of anti-apoptotic factors and upregulation of BAX expression through activation of the p53 tumor suppressor protein (L'ecuyer et al., 2006; Shi et al., 2011; Shizukuda et al., 2005). DOX can also induce extrinsic apoptotic cardiomyocyte cell death through the activation of pathways and expression of death receptor ligands through increased ROS production (Nakamura et al., 2000; Nitobe et al., 2003;

Niu et al., 2009). Other cancer therapies, including angiogenesis inhibitors, growth factor inhibitors, and radiation therapy, are also maladaptive for the heart, but a clear connection to cell death pathways needs to be further elucidated in these therapeutic models.

5.4 Myocarditis

Myocarditis is the inflammation of the myocardium in response to infections caused by viruses, bacteria, fungi, or protozoa (Caforio et al., 2013). However, viral infections are the most prevalent cause of myocarditis (Fung et al., 2016). Severe myocarditis can lead to cardiomyocyte cell death, potentially resulting in a heart attack, arrhythmia, or heart failure (Fung et al., 2016). The most commonly studied models of myocarditis in mice involve infection with either Coxsackievirus B3 (CVB3) or reovirus (Klingel et al., 2004). CVB3 infects the cell by binding to the coxsackie adenoviral receptor (CXADR) on cells before viral internalization (Fung et al., 2016; Shi et al., 2009). Upon entry into the cell, the virus replicates, evoking an over-reactive autoimmune inflammatory response that persists even after the removal of the viral particles from the heart (Liu and Mason, 2001). Mice genetically lacking *Cxadr* do not develop myocarditis following CVB3 viral infection (Shi et al., 2009). The innate immune system recognizes CVB3 particles as PAMPs, as implicated by the production of CVB3-specific IgG 7 days following injection in mice, which potentially induces pyroptosis (Leipner et al., 2004). Due to the nature of myocarditis, extrinsic apoptosis, necroptosis, and pyroptosis have all been implicated in contributing to this disease.

CVB3 infection correlates with increased expression of BAX and active CASP3 (Chau et al., 2007). In addition, CVB3 viruses contain proteases 2A and 3C within their genome that cleave and activate both pro-CASP8 and pro-CASP9, which can lead to apoptosis (Chau et al., 2007). Upstream from direct caspase activation, the Fas-ligand pathway is also induced during myocarditis, resulting in increased myocardium damage through death receptor activation (Seko et al., 2002). However, activation of the death receptors can also lead to the necroptotic cell death in cardiomyocytes due to their innate caspase inhibition (Zhe-Wei et al., 2018; Zhou et al., 2018a).

Recently, inhibition of necroptosis by treatment with Nec-1 was shown to protect against acute myocarditis induced by CVB3 viral infection (Zhou et al., 2018a). In the same manuscript, Zhou et al. reported elevated levels of RIPK1 and RIPK3 in hearts of CVB3 infected mice, implicating the activation of necroptosis (Zhou et al., 2018a). However, the mode of protection of Nec-1 may not be due to cardiomyocyte cell death inhibition. CVB3 virus mainly infects intestinal epithelial cells, where it was shown that viral infection of CVB3 is dependent on RIP3K activity (Harris et al., 2015). Therefore, treatment of Nec-1 may be protective against viral infectivity and thus the subsequent myocarditis.

Increased pro-inflammatory cytokine release, such as IL-1β, has been observed during myocarditis as has increased inflammasome activation (Wang et al., 2014). Mice infected with CVB3 have increased NLRP3 inflammasome as well as CASP1 activity in their heart tissue (Wang et al., 2014). Furthermore, prevention of inflammasome formation by inhibition of the CASP1 activity leads to improved myocardial function (Wang et al., 2014). Incidentally, vitamin D protects cardiomyocytes from cell death induced by acute myocarditis, and experimental animals exposed to acute myocarditis treated with vitamin D display downregulation in pro-pyroptotic factors ASC and CASP1 (Liu and Mason, 2001; Liu et al., 2017). In order to solidify the role of pyroptosis in myocarditis, genetic removal of key pyroptotic regulators specifically in cardiomyocytes in mice subjected to experimental acute myocarditis would be necessary.

6. Conclusions

Cell death is a key contributing factor for the majority of cardiovascular diseases and associated mortalities. Due to the lack of endogenous cardiomyocyte regeneration and proliferation in the adult heart, cell death in this population has profound long-lasting effects. Naturally, cardiomyocytes are programmed to be death-resistant by expressing relatively low amounts of the majority of pro-cell death regulating genes. Under extreme stress, such as MI, cardiomyocytes experience toxic levels of Ca^{2+} and ROS, which initiates their cellular demise. Multiple cell death pathways have been implicated during this disease state, which heavily involve mitochondrial dysfunction. This common thread suggests the evidence for crosstalk between various forms of cell deaths, especially the multiple necrotic cell death pathways. Further investigation into the common nodal points between these cell death pathways is required in order to fully exploit the therapeutic potential of cardiomyocyte cell death inhibition during I/R injury. Similar therapeutic mechanisms may also apply for the complete inhibition of cardiomyocyte cell death during other forms of cardiovascular disease as well as the cardiotoxicity from cancer therapeutics. Identifying the balance between fortifying the heart, not the cancer, against cell death in patients receiving chemotherapies will be a difficult task without detailed mechanistic characterization of the cell death pathways occurring in both cell types during treatment. Heart disease is the leading cause of death in the modern world, and the biggest risk factor is age. Cardiomyocyte rejuvenation through manipulation of the autophagic pathway is promising for slowing down age-related cardiomyopathies. Further morphological and molecular characterization of age-dependent cardiomyocyte cell death is critical for the identification of novel therapeutic approaches to limit the aging process in the heart. Identification of novel therapeutic approaches that preserve cardiomyocytes in the face of pathological stressors would be extremely beneficial to the preservation of life.

References

Abdellatif, M., Sedej, S., Carmona-Gutierrez, D., Madeo, F., Kroemer, G., 2018. Autophagy in cardiovascular aging. Circ. Res. 123, 803–824.

Alavian, K.N., Beutner, G., Lazrove, E., Sacchetti, S., Park, H.-A., Licznerski, P., Li, H., Nabili, P., Hockensmith, K., Graham, M., 2014. An uncoupling channel within the c-subunit ring of the F1FO ATP synthase is the mitochondrial permeability transition pore. Proc. Natl. Acad. Sci. U. S. A. 111, 10580–10585.

Albini, A., Pennesi, G., Donatelli, F., Cammarota, R., De Flora, S., Noonan, D.M., 2010. Cardiotoxicity of anticancer drugs: the need for cardio-oncology and cardio-oncological prevention. J. Natl. Cancer Inst. 102, 14–25.

Alers, S., Löffler, A.S., Wesselborg, S., Stork, B., 2012. Role of AMPK-mTOR-Ulk1/2 in the regulation of autophagy: cross talk, shortcuts, and feedbacks. Mol. Cell. Biol. 32, 2–11.

Andrabi, S.A., Kim, N.S., Yu, S.-W., Wang, H., Koh, D.W., Sasaki, M., Klaus, J.A., Otsuka, T., Zhang, Z., Koehler, R.C., 2006. Poly (ADP-ribose) (PAR) polymer is a death signal. Proc. Natl. Acad. Sci. U. S. A. 103, 18308–18313.

Argaud, L., Gateau-Roesch, O., Muntean, D., Chalabreysse, L., Loufouat, J., Robert, D., Ovize, M., 2005. Specific inhibition of the mitochondrial permeability transition prevents lethal reperfusion injury. J. Mol. Cell. Cardiol. 38, 367–374.

Ashley, E.A., Niebauer, J., 2004. Coronary artery disease. In: Cardiology Explained. Remedica.

Audia, J.P., Yang, X.-M., Crockett, E.S., Housley, N., Haq, E.U., O'donnell, K., Cohen, M.V., Downey, J.M., Alvarez, D.F., 2018. Caspase-1 inhibition by VX-765 administered at reperfusion in P2Y 12 receptor antagonist-treated rats provides long-term reduction in myocardial infarct size and preservation of ventricular function. Basic Res. Cardiol. 113, 32.

Baines, C.P., Kaiser, R.A., Purcell, N.H., Blair, N.S., Osinska, H., Hambleton, M.A., Brunskill, E.W., Sayen, M.R., Gottlieb, R.A., Dorn, G.W., 2005. Loss of cyclophilin D reveals a critical role for mitochondrial permeability transition in cell death. Nature 434, 658.

Baines, C.P., Kaiser, R.A., Sheiko, T., Craigen, W.J., Molkentin, J.D., 2007. Voltage-dependent anion channels are dispensable for mitochondrial-dependent cell death. Nat. Cell Biol. 9, 550.

Balanehru, S., Nagarajan, B., 1992. Intervention of adriamycin induced free radical damage. Biochem. Int. 28, 735–744.
Benjamin, E.J., Muntner, P., Bittencourt, M.S., 2019. Heart disease and stroke statistics-2019 update: a report from the American Heart Association. Circulation 139, e56–e528.
Bergmann, O., Zdunek, S., Felker, A., Salehpour, M., Alkass, K., Bernard, S., Sjostrom, S.L., Szewczykowska, M., Jackowska, T., Dos Remedios, C., 2015. Dynamics of cell generation and turnover in the human heart. Cell 161, 1566–1575.
Bergsbaken, T., Fink, S.L., Cookson, B.T., 2009. Pyroptosis: host cell death and inflammation. Nat. Rev. Microbiol. 7, 99.
Bernales, S., Schuck, S., Walter, P., 2007. ER-phagy: selective autophagy of the endoplasmic reticulum. Autophagy 3, 285–287.
Berry, D.L., Baehrecke, E.H., 2008. Autophagy functions in programmed cell death. Autophagy 4, 359–360.
Berry, C.E., Hare, J.M., 2004. Xanthine oxidoreductase and cardiovascular disease: molecular mechanisms and pathophysiological implications. J. Physiol. 555, 589–606.
Birnbaum, Y., Hale, S.L., Kloner, R.A., 1996. The effect of coenzyme Q10 on infarct size in a rabbit model of ischemia/reperfusion. Cardiovasc. Res. 32, 861–868.
Boisguerin, P., Redt-Clouet, C., Franck-Miclo, A., Licheheb, S., Nargeot, J., Barrère-Lemaire, S., Lebleu, B., 2011. Systemic delivery of BH4 anti-apoptotic peptide using CPPs prevents cardiac ischemia–reperfusion injuries in vivo. J. Control. Release 156, 146–153.
Bonora, M., Bononi, A., De Marchi, E., Giorgi, C., Lebiedzinska, M., Marchi, S., Patergnani, S., Rimessi, A., Suski, J.M., Wojtala, A., 2013. Role of the c subunit of the FO ATP synthase in mitochondrial permeability transition. Cell Cycle 12, 674–683.
Bourdillon, P.D., Poole-Wilson, P.A., 1982. The effects of verapamil, quiescence, and cardioplegia on calcium exchange and mechanical function in ischemic rabbit myocardium. Circ. Res. 50, 360–368.
Braunersreuther, V., Montecucco, F., Ashri, M., Pelli, G., Galan, K., Frias, M., Burger, F., Quinderé, A.L.G., Montessuit, C., Krause, K.-H., 2013. Role of NADPH oxidase isoforms NOX1, NOX2 and NOX4 in myocardial ischemia/reperfusion injury. J. Mol. Cell. Cardiol. 64, 99–107.
Brigelius-Flohé, R., Maiorino, M., 2013. Glutathione peroxidases. Biochim. Biophys. Acta 1830, 3289–3303.
Brown, D.I., Willis, M.S., Berthiaume, J.M., 2016. Influence of ischemia-reperfusion injury on cardiac metabolism. In: The Scientist's Guide to Cardiac Metabolism. Elsevier.
Brustovetsky, N., Klingenberg, M., 1996. Mitochondrial ADP/ATP carrier can be reversibly converted into a large channel by Ca2+. Biochemistry 35, 8483–8488.
Brustovetsky, N., Tropschug, M., Heimpel, S., Heidkämper, D., Klingenberg, M., 2002. A large Ca2+−dependent channel formed by recombinant ADP/ATP carrier from neurospora crassa resembles the mitochondrial permeability transition pore. Biochemistry 41, 11804–11811.
Budoff, M.J., Gul, K.M., 2008. Expert review on coronary calcium. Vasc. Health Risk Manag. 4, 315.
Burman, C., Ktistakis, N.T., 2010. Regulation of autophagy by phosphatidylinositol 3-phosphate. FEBS Lett. 584, 1302–1312.
Cadenas, S., 2018. ROS and redox signaling in myocardial ischemia-reperfusion injury and cardioprotection. Free Radic. Biol. Med. 117, 76–89.
Caforio, A.L., Pankuweit, S., Arbustini, E., Basso, C., Gimeno-Blanes, J., Felix, S.B., Fu, M., Heliö, T., Heymans, S., Jahns, R., 2013. Current state of knowledge on aetiology, diagnosis, management, and therapy of myocarditis: a position statement of the European Society of Cardiology Working Group on myocardial and pericardial diseases. Eur. Heart J. 34, 2636–2648.

Cai, Z., Jitkaew, S., Zhao, J., Chiang, H.-C., Choksi, S., Liu, J., Ward, Y., Wu, L.-G., Liu, Z.-G., 2014. Plasma membrane translocation of trimerized MLKL protein is required for TNF-induced necroptosis. Nat. Cell Biol. 16, 55.

Campos, J.C., Queliconi, B.B., Bozi, L.H., Bechara, L.R., Dourado, P.M., Andres, A.M., Jannig, P.R., Gomes, K.M., Zambelli, V.O., Rocha-Resende, C., 2017. Exercise reestablishes autophagic flux and mitochondrial quality control in heart failure. Autophagy 13, 1304–1317.

Carroll, J., He, J., Ding, S., Fearnley, I.M., Walker, J.E., 2019. Persistence of the permeability transition pore in human mitochondria devoid of an assembled ATP synthase. Proc. Natl. Acad. Sci. U. S. A. 116, 12816–12821.

Chapman, J.G., Magee, W.P., Stukenbrok, H.A., Beckius, G.E., Milici, A.J., Tracey, W.R., 2002. A novel nonpeptidic caspase-3/7 inhibitor,(S)-(+)-5-[1-(2-methoxymethylpyrrolidinyl) sulfonyl] isatin reduces myocardial ischemic injury. Eur. J. Pharmacol. 456, 59–68.

Chatterjee, K., Zhang, J., Honbo, N., Karliner, J.S., 2010. Doxorubicin cardiomyopathy. Cardiology 115, 155–162.

Chau, D.H., Yuan, J., Zhang, H., Cheung, P., Lim, T., Liu, Z., Sall, A., Yang, D., 2007. Coxsackievirus B3 proteases 2A and 3C induce apoptotic cell death through mitochondrial injury and cleavage of eIF4GI but not DAP5/p97/NAT1. Apoptosis 12, 513–524.

Chen, Y., Liu, Y., Dorn, G.W., 2011. Mitochondrial fusion is essential for organelle function and cardiac homeostasis. Circ. Res. 109, 1327–1331.

Chen, X., He, W.-T., Hu, L., Li, J., Fang, Y., Wang, X., Xu, X., Wang, Z., Huang, K., Han, J., 2016. Pyroptosis is driven by non-selective gasdermin-D pore and its morphology is different from MLKL channel-mediated necroptosis. Cell Res. 26, 1007.

Cheng, K.T., Xiong, S., Ye, Z., Hong, Z., Di, A., Tsang, K.M., Gao, X., An, S., Mittal, M., Vogel, S.M., 2017. Caspase-11–mediated endothelial pyroptosis underlies endotoxemia-induced lung injury. J. Clin. Invest. 127, 4124–4135.

Chen-Scarabelli, C., Agrawal, P.R., Saravolatz, L., Abuniat, C., Scarabelli, G., Stephanou, A., Loomba, L., Narula, J., Scarabelli, T.M., Knight, R., 2014. The role and modulation of autophagy in experimental models of myocardial ischemia-reperfusion injury. J. Geriatr. Cardiol. 11, 338–348.

Chugh, A.R., Beache, G.M., Loughran, J.H., Mewton, N., Elmore, J.B., Kajstura, J., Pappas, P., Tatooles, A., Stoddard, M.F., Lima, J.A., 2012. Administration of cardiac stem cells in patients with ischemic cardiomyopathy: the SCIPIO trial: surgical aspects and interim analysis of myocardial function and viability by magnetic resonance. Circulation 126, S54–S64.

Cipolat, S., De Brito, O.M., Dal Zilio, B., Scorrano, L., 2004. OPA1 requires mitofusin 1 to promote mitochondrial fusion. Proc. Natl. Acad. Sci. U. S. A. 101, 15927–15932.

Condorelli, G., Roncarati, R., Ross, J., Pisani, A., Stassi, G., Todaro, M., Trocha, S., Drusco, A., Gu, Y., Russo, M.A., 2001. Heart-targeted overexpression of caspase3 in mice increases infarct size and depresses cardiac function. Proc. Natl. Acad. Sci. 98, 9977–9982.

Crompton, M., Ellinger, H., Costi, A., 1988. Inhibition by cyclosporin A of a Ca2+–dependent pore in heart mitochondria activated by inorganic phosphate and oxidative stress. Biochem. J. 255, 357.

Crompton, M., Virji, S., Ward, J.M., 1998. Cyclophilin-D binds strongly to complexes of the voltage-dependent anion channel and the adenine nucleotide translocase to form the permeability transition pore. Eur. J. Biochem. 258, 729–735.

Dabkowski, E.R., Williamson, C.L., Hollander, J.M., 2008. Mitochondria-specific transgenic overexpression of phospholipid hydroperoxide glutathione peroxidase (GPx4) attenuates ischemia/reperfusion-associated cardiac dysfunction. Free Radic. Biol. Med. 45, 855–865.

Damy, T., Ratajczak, P., Robidel, E., Bendall, J.K., Oliviéro, P., Boczkowski, J., Ebrahimian, T., Marotte, F., Samuel, J.-L., Heymes, C., 2003. Up-regulation of cardiac nitric oxide synthase 1-derived nitric oxide after myocardial infarction in senescent rats. FASEB J. 17, 1934–1936.

Das, D.K., Engelman, R.M., Clement, R., Otani, H., Prasad, M.R., Rao, P.S., 1987. Role of xanthine oxidase inhibitor as free radical scavenger: a novel mechanism of action of allopurinol and oxypurinol in myocardial salvage. Biochem. Biophys. Res. Commun. 148, 314–319.

De Jong, J.W., Schoemaker, R., De Jonge, R., Bernocchi, P., Keijzer, E., Harrison, R., Sharma, H., Ceconi, C., 2000. Enhanced expression and activity of xanthine oxidoreductase in the failing heart. J. Mol. Cell. Cardiol. 32, 2083–2089.

Ding, J., Wang, K., Liu, W., She, Y., Sun, Q., Shi, J., Sun, H., Wang, D.-C., Shao, F., 2016. Pore-forming activity and structural autoinhibition of the gasdermin family. Nature 535, 111.

Disatnik, M.H., Ferreira, J.C., Campos, J.C., Gomes, K.S., Dourado, P.M., Qi, X., Mochly-Rosen, D., 2013. Acute inhibition of excessive mitochondrial fission after myocardial infarction prevents long-term cardiac dysfunction. J. Am. Heart Assoc. 2, e000461.

Diwan, A., Krenz, M., Syed, F.M., Wansapura, J., Ren, X., Koesters, A.G., Li, H., Kirshenbaum, L.A., Hahn, H.S., Robbins, J., 2007. Inhibition of ischemic cardiomyocyte apoptosis through targeted ablation of Bnip3 restrains postinfarction remodeling in mice. J. Clin. Invest. 117, 2825–2833.

Diwan, A., Wansapura, J., Syed, F.M., Matkovich, S.J., Lorenz, J.N., Dorn 2nd, G.W., 2008. Nix-mediated apoptosis links myocardial fibrosis, cardiac remodeling, and hypertrophy decompensation. Circulation 117, 396–404.

Do Carmo, H., Arjun, S., Petrucci, O., Yellon, D.M., Davidson, S.M., 2018. The caspase 1 inhibitor VX-765 protects the isolated rat heart via the RISK pathway. Cardiovasc. Drugs Ther. 32, 165–168.

Donovan, M.J., Hahn, R., Tessarollo, L., Hempstead, B.L., 1996. Identification of an essential nonneuronal function of neurotrophin 3 in mammalian cardiac development. Nat. Genet. 14, 210.

Elmore, S., 2007. Apoptosis: a review of programmed cell death. Toxicol. Pathol. 35, 495–516.

Elrod, J.W., Wong, R., Mishra, S., Vagnozzi, R.J., Sakthievel, B., Goonasekera, S.A., Karch, J., Gabel, S., Farber, J., Force, T., 2010. Cyclophilin D controls mitochondrial pore–dependent Ca(2+) exchange, metabolic flexibility, and propensity for heart failure in mice. J. Clin. Invest. 120, 3680–3687.

Eltze, T., Boer, R., Wagner, T., Weinbrenner, S., Mcdonald, M.C., Thiemermann, C., Bürkle, A., Klein, T., 2008. Imidazoquinolinone, imidazopyridine, and isoquinolindione derivatives as novel and potent inhibitors of the poly (ADP-ribose) polymerase (PARP): a comparison with standard PARP inhibitors. Mol. Pharmacol. 74, 1587–1598.

Eskes, R., Antonsson, B., Osen-Sand, A., Montessuit, S., Richter, C., Sadoul, R., Mazzei, G., Nichols, A., Martinou, J.-C., 1998. Bax-induced cytochrome C release from mitochondria is independent of the permeability transition pore but highly dependent on Mg2+ ions. J. Cell Biol. 143, 217–224.

Fang, X., Wang, H., Han, D., Xie, E., Yang, X., Wei, J., Gu, S., Gao, F., Zhu, N., Yin, X., 2019. Ferroptosis as a target for protection against cardiomyopathy. Proc. Natl. Acad. Sci. U. S. A. 116, 2672–2680.

Faro, R., Toyoda, Y., Mccully, J.D., Jagtap, P., Szabo, E., Virag, L., Bianchi, C., Levitsky, S., Szabo, C., Sellke, F.W., 2002. Myocardial protection by PJ34, a novel potent poly (ADP-ribose) synthetase inhibitor. Ann. Thorac. Surg. 73, 575–581.

Fatokun, A.A., Dawson, V.L., Dawson, T.M., 2014. Parthanatos: mitochondrial-linked mechanisms and therapeutic opportunities. Br. J. Pharmacol. 171, 2000–2016.

Fauconnier, J., Roberge, S., Saint, N., Lacampagne, A., 2013. Type 2 ryanodine receptor: a novel therapeutic target in myocardial ischemia/reperfusion. Pharmacol. Ther. 138, 323–332.

Feng, Q., Lu, X., Jones, D.L., Shen, J., Arnold, J.M.O., 2001. Increased inducible nitric oxide synthase expression contributes to myocardial dysfunction and higher mortality after myocardial infarction in mice. Circulation 104, 700–704.

Fernandes-Alnemri, T., Wu, J., Yu, J., Datta, P., Miller, B., Jankowski, W., Rosenberg, S., Zhang, J., Alnemri, E., 2007. The pyroptosome: a supramolecular assembly of ASC dimers mediating inflammatory cell death via caspase-1 activation. Cell Death Differ. 14, 1590.

Filippone, S.M., Samidurai, A., Roh, S.K., Cain, C.K., He, J., Salloum, F.N., Kukreja, R.C., Das, A., 2017. Reperfusion therapy with rapamycin attenuates myocardial infarction through activation of AKT and ERK. Oxid. Med. Cell. Longev. 2017, 1–16.

Frantz, S., Ducharme, A., Sawyer, D., Rohde, L.E., Kobzik, L., Fukazawa, R., Tracey, D., Allen, H., Lee, R.T., Kelly, R.A., 2003. Targeted deletion of caspase-1 reduces early mortality and left ventricular dilatation following myocardial infarction. J. Mol. Cell. Cardiol. 35, 685–694.

Fung, G., Luo, H., Qiu, Y., Yang, D., Mcmanus, B., 2016. Myocarditis. Circ. Res. 118, 496–514.

Galluzzi, L., Vitale, I., Aaronson, S.A., Abrams, J.M., Adam, D., Agostinis, P., Alnemri, E.S., Altucci, L., Amelio, I., Andrews, D.W., Annicchiarico-Petruzzelli, M., Antonov, A.V., Arama, E., Baehrecke, E.H., Barlev, N.A., Bazan, N.G., Bernassola, F., Bertrand, M.J.M., Bianchi, K., Blagosklonny, M.V., Blomgren, K., Borner, C., Boya, P., Brenner, C., Campanella, M., Candi, E., Carmona-Gutierrez, D., Cecconi, F., Chan, F.K.M., Chandel, N.S., Cheng, E.H., Chipuk, J.E., Cidlowski, J.A., Ciechanover, A., Cohen, G.M., Conrad, M., Cubillos-Ruiz, J.R., Czabotar, P.E., D'angiolella, V., Dawson, T.M., Dawson, V.L., De Laurenzi, V., De Maria, R., Debatin, K.-M., Deberardinis, R.J., Deshmukh, M., Di Daniele, N., Di Virgilio, F., Dixit, V.M., Dixon, S.J., Duckett, C.S., Dynlacht, B.D., El-Deiry, W.S., Elrod, J.W., Fimia, G.M., Fulda, S., García-Sáez, A.J., Garg, A.D., Garrido, C., Gavathiotis, E., Golstein, P., Gottlieb, E., Green, D.R., Greene, L.A., Gronemeyer, H., Gross, A., Hajnoczky, G., Hardwick, J.M., Harris, I.S., Hengartner, M.O., Hetz, C., Ichijo, H., Jäättelä, M., Joseph, B., Jost, P.J., Juin, P.P., Kaiser, W.J., Karin, M., Kaufmann, T., Kepp, O., Kimchi, A., Kitsis, R.N., Klionsky, D.J., Knight, R.A., Kumar, S., Lee, S.W., Lemasters, J.J., Levine, B., Linkermann, A., Lipton, S.A., Lockshin, R.A., López-Otín, C., Lowe, S.W., Luedde, T., Lugli, E., Macfarlane, M., Madeo, F., Malewicz, M., Malorni, W., Manic, G., et al., 2018. Molecular mechanisms of cell death: recommendations of the nomenclature committee on cell death 2018. Cell Death Differ. 25, 486–541.

Gardner, T.J., Stewart, J.R., Casale, A.S., Downey, J.M., Chambers, D.E., 1983. Reduction of myocardial ischemic injury with oxygen-derived free radical scavengers. Surgery 94, 423–427.

Geisler, S., Holmström, K.M., Treis, A., Skujat, D., Weber, S.S., Fiesel, F.C., Kahle, P.J., Springer, W., 2010. The PINK1/Parkin-mediated mitophagy is compromised by PD-associated mutations. Autophagy 6, 871–878.

Gharanei, M., Hussain, A., Janneh, O., Maddock, H.L., 2013. Doxorubicin induced myocardial injury is exacerbated following ischaemic stress via opening of the mitochondrial permeability transition pore. Toxicol. Appl. Pharmacol. 268, 149–156.

Ghyselinck, N.B., Wendling, O., Messaddeq, N., Dierich, A., Lampron, C., Décimo, D., Viville, S., Chambon, P., Mark, M., 1998. Contribution of retinoic acid receptor β isoforms to the formation of the conotruncal septum of the embryonic heart. Dev. Biol. 198, 303–318.

Giorgio, V., Von Stockum, S., Antoniel, M., Fabbro, A., Fogolari, F., Forte, M., Glick, G.D., Petronilli, V., Zoratti, M., Szabó, I., Lippe, G., Bernardi, P., 2013. Dimers of mitochondrial ATP synthase form the permeability transition pore. Proc. Natl. Acad. Sci. U. S. A. 110, 5887–5892.

Granger, D.N., Kvietys, P.R., 2015. Reperfusion injury and reactive oxygen species: the evolution of a concept. Redox Biol. 6, 524–551.

Griffiths, E.J., Halestrap, A.P., 1993. Protection by cyclosporin A of ischemia/reperfusion-induced damage in isolated rat hearts. J. Mol. Cell. Cardiol. 25, 1461–1469.

Gross, O., Poeck, H., Bscheider, M., Dostert, C., Hannesschläger, N., Endres, S., Hartmann, G., Tardivel, A., Schweighoffer, E., Tybulewicz, V., Mocsai, A., Tschopp, J., Ruland, J., 2009. Syk kinase signalling couples to the Nlrp3 inflammasome for anti-fungal host defence. Nature 459, 433.

Grupp, I.L., Jackson, T.M., Hake, P., Grupp, G., Szabó, C., 1999. Protection against hypoxia-reoxygenation in the absence of poly (ADP-ribose) synthetase in isolated working hearts. J. Mol. Cell. Cardiol. 31, 297–303.

Guan, Y., Drake, J.C., Yan, Z., 2019. Exercise-induced mitophagy in skeletal muscle and heart. Exerc. Sport Sci. Rev. 47, 151–156.

Gutiérrez-Aguilar, M., Douglas, D.L., Gibson, A.K., Domeier, T.L., Molkentin, J.D., Baines, C.P., 2014. Genetic manipulation of the cardiac mitochondrial phosphate carrier does not affect permeability transition. J. Mol. Cell. Cardiol. 72, 316–325.

Halestrap, A.P., Brenner, C., 2003. The adenine nucleotide translocase: a central component of the mitochondrial permeability transition pore and key player in cell death. Curr. Med. Chem. 10, 1507–1525.

Halestrap, A.P., Pasdois, P., 2009. The role of the mitochondrial permeability transition pore in heart disease. Biochim. Biophys. Acta 1787, 1402–1415.

Halestrap, A., Connern, C., Griffiths, E., Kerr, P., 1997. Cyclosporin A binding to mitochondrial cyclophilin inhibits the permeability transition pore and protects hearts from ischaemia/reperfusion injury. In: Detection of Mitochondrial Diseases. Springer.

Halestrap, A.P., Kerr, P.M., Javadov, S., Woodfield, K.-Y., 1998. Elucidating the molecular mechanism of the permeability transition pore and its role in reperfusion injury of the heart. Biochim. Biophys. Acta 1366, 79–94.

Harris, K.G., Morosky, S.A., Drummond, C.G., Patel, M., Kim, C., Stolz, D.B., Bergelson, J.M., Cherry, S., Coyne, C.B., 2015. RIP3 regulates autophagy and promotes coxsackievirus B3 infection of intestinal epithelial cells. Cell Host Microbe 18, 221–232.

Harwood, S.M., Yaqoob, M.M., Allen, D.A., 2005. Caspase and calpain function in cell death: bridging the gap between apoptosis and necrosis. Ann. Clin. Biochem. 42, 415–431.

Hausenloy, D.J., Duchen, M.R., Yellon, D.M., 2003. Inhibiting mitochondrial permeability transition pore opening at reperfusion protects against ischaemia–reperfusion injury. Cardiovasc. Res. 60, 617–625.

He, J., Carroll, J., Ding, S., Fearnley, I.M., Walker, J.E., 2017a. Permeability transition in human mitochondria persists in the absence of peripheral stalk subunits of ATP synthase. Proc. Natl. Acad. Sci. U. S. A. 114, 9086–9091.

He, J., Ford, H.C., Carroll, J., Ding, S., Fearnley, I.M., Walker, J.E., 2017b. Persistence of the mitochondrial permeability transition in the absence of subunit c of human ATP synthase. Proc. Natl. Acad. Sci. U. S. A. 114, 3409–3414.

Hearse, D., Humphrey, S., Chain, E., 1973. Abrupt reoxygenation of the anoxic potassium-arrested perfused rat heart: a study of myocardial enzyme release. J. Mol. Cell. Cardiol. 5, 395–407.

Henz, K., Al-Zebeeby, A., Basoglu, M., Fulda, S., Cohen, G.M., Varadarajan, S., Vogler, M., 2019. Selective BH3-mimetics targeting BCL-2, BCL-XL or MCL-1 induce severe mitochondrial perturbations. Biol. Chem. 400, 181–185.

Hochhauser, E., Cheporko, Y., Yasovich, N., Pinchas, L., Offen, D., Barhum, Y., Pannet, H., Tobar, A., Vidne, B., Birk, E., 2007. Bax deficiency reduces infarct size and improves long-term function after myocardial infarction. Cell Biochem. Biophys. 47, 11–19.

Hoshino, A., Mita, Y., Okawa, Y., Ariyoshi, M., Iwai-Kanai, E., Ueyama, T., Ikeda, K., Ogata, T., Matoba, S., 2013. Cytosolic p53 inhibits Parkin-mediated mitophagy and promotes mitochondrial dysfunction in the mouse heart. Nat. Commun. 4, 2308.

Huang, J.-Q., Radinovic, S., Rezaiefar, P., Black, S.C., 2000. In vivo myocardial infarct size reduction by a caspase inhibitor administered after the onset of ischemia. Eur. J. Pharmacol. 402, 139–142.

Huang, J., Ito, Y., Morikawa, M., Uchida, H., Kobune, M., Sasaki, K., Abe, T., Hamada, H., 2003. Bcl-xL gene transfer protects the heart against ischemia/reperfusion injury. Biochem. Biophys. Res. Commun. 311, 64–70.

Ikeda, Y., Shirakabe, A., Maejima, Y., Zhai, P., Sciarretta, S., Toli, J., Nomura, M., Mihara, K., Egashira, K., Ohishi, M., 2015. Endogenous Drp1 mediates mitochondrial autophagy and protects the heart against energy stress. Circ. Res. 116, 264–278.

Imai, H., Nakagawa, Y., 2003. Biological significance of phospholipid hydroperoxide glutathione peroxidase (PHGPx, GPx4) in mammalian cells. Free Radic. Biol. Med. 34, 145–169.

Inserte, J., Barba, I., Hernando, V., Abellán, A., Ruiz-Meana, M., Rodríguez-Sinovas, A., Garcia-Dorado, D., 2007. Effect of acidic reperfusion on prolongation of intracellular acidosis and myocardial salvage. Cardiovasc. Res. 77, 782–790.

Inserte, J., Cardona, M., Poncelas-Nozal, M., Hernando, V., Vilardosa, Ú., Aluja, D., Parra, V.M., Sanchis, D., Garcia-Dorado, D., 2016. Studies on the role of apoptosis after transient myocardial ischemia: genetic deletion of the executioner caspases-3 and-7 does not limit infarct size and ventricular remodeling. Basic Res. Cardiol. 111, 18.

Iwamoto, H., Miura, T., Okamura, T., Shirakawa, K., Iwatate, M., Kawamura, S., Tatsuno, H., Ikeda, Y., Matsuzaki, M., 1999. Calpain inhibitor-1 reduces infarct size and DNA fragmentation of myocardium in ischemic/reperfused rat heart. J. Cardiovasc. Pharmacol. 33, 580–586.

Javadov, S.A., Clarke, S., Das, M., Griffiths, E.J., Lim, K.H.H., Halestrap, A.P., 2003. Ischaemic preconditioning inhibits opening of mitochondrial permeability transition pores in the reperfused rat heart. J. Physiol. 549, 513–524.

Jeddi, S., Ghasemi, A., Asgari, A., Nezami-Asl, A., 2018. Role of inducible nitric oxide synthase in myocardial ischemia-reperfusion injury in sleep-deprived rats. Sleep Breath. 22, 353–359.

Jennings, R., Shen, A., 1972. Calcium in experimental myocardial ischemia. Recent Adv. Stud. Cardiac Struct. Metab. 1, 639–655.

Jia, C., Chen, H., Zhang, J., Zhou, K., Zhuge, Y., Niu, C., Qiu, J., Rong, X., Shi, Z., Xiao, J., 2019. Role of pyroptosis in cardiovascular diseases. Int. Immunopharmacol. 67, 311–318.

Kabeya, Y., Mizushima, N., Ueno, T., Yamamoto, A., Kirisako, T., Noda, T., Kominami, E., Ohsumi, Y., Yoshimori, T., 2000. LC3, a mammalian homologue of yeast Apg8p, is localized in autophagosome membranes after processing. EMBO J. 19, 5720–5728.

Kaiser, W.J., Upton, J.W., Long, A.B., Livingston-Rosanoff, D., Daley-Bauer, L.P., Hakem, R., Caspary, T., Mocarski, E.S., 2011. RIP3 mediates the embryonic lethality of caspase-8-deficient mice. Nature 471, 368–372.

Kaizuka, T., Mizushima, N., 2016. Atg13 is essential for autophagy and cardiac development in mice. Mol. Cell. Biol. 36, 585–595.

Kalivendi, S.V., Konorev, E.A., Cunningham, S., Vanamala, S.K., Kaji, E.H., Joseph, J., Kalyanaraman, B., 2005. Doxorubicin activates nuclear factor of activated

T-lymphocytes and Fas ligand transcription: role of mitochondrial reactive oxygen species and calcium. Biochem. J. 389, 527–539.
Kalkavan, H., Green, D.R., 2018. MOMP, cell suicide as a BCL-2 family business. Cell Death Differ. 25, 46–55.
Kalogeris, T., Baines, C.P., Krenz, M., Korthuis, R.J., 2012. Cell biology of ischemia/reperfusion injury. Int. Rev. Cell Mol. Biol. 298, 229–317.
Karch, J., Molkentin, J.D., 2015. Regulated necrotic cell death: the passive aggressive side of Bax and Bak. Circ. Res. 116, 1800–1809.
Karch, J., Molkentin, J.D., 2018. Identity of the elusive mitochondrial permeability transition pore: what it might be, what it was, and what it still could be. Curr. Opin. Physiol. 3, 57–62.
Karch, J., Kwong, J.Q., Burr, A.R., Sargent, M.A., Elrod, J.W., Peixoto, P.M., Martinez-Caballero, S., Osinska, H., Cheng, E.H., Robbins, J., 2013. Bax and Bak function as the outer membrane component of the mitochondrial permeability pore in regulating necrotic cell death in mice. Elife 2, e00772.
Karch, J., Kanisicak, O., Brody, M.J., Sargent, M.A., Michael, D.M., Molkentin, J.D., 2015. Necroptosis interfaces with MOMP and the MPTP in mediating cell death. PLoS One 10, e0130520.
Karch, J., Schips, T.G., Maliken, B.D., Brody, M.J., Sargent, M.A., Kanisicak, O., Molkentin, J.D., 2017. Autophagic cell death is dependent on lysosomal membrane permeability through Bax and Bak. Elife 6, e30543.
Karch, J., Bround, M.J., Khalil, H., Sargent, M.A., Latchman, N., Terada, N., Peixoto, P.M., Molkentin, J.D., 2019. Inhibition of mitochondrial permeability transition by deletion of the ANT family and CypD. Sci. Adv. 5, eaaw4597.
Kawaguchi, M., Takahashi, M., Hata, T., Kashima, Y., Usui, F., Morimoto, H., Izawa, A., Takahashi, Y., Masumoto, J., Koyama, J., 2011. Inflammasome activation of cardiac fibroblasts is essential for myocardial ischemia/reperfusion injury. Circulation 123, 594–604.
Kehat, I., Molkentin, J.D., 2010. Molecular pathways underlying cardiac remodeling during pathophysiological stimulation. Circulation 122, 2727–2735.
Khan, T.A., Ruel, M., Bianchi, C., Voisine, P., Komjáti, K., Szabo, C., Sellke, F.W., 2003. Poly (ADP-ribose) polymerase inhibition improves postischemic myocardial function after cardioplegia-cardiopulmonary bypass. J. Am. Coll. Surg. 197, 270–277.
Kim, G.-T., Chun, Y.-S., Park, J.-W., Kim, M.-S., 2003. Role of apoptosis-inducing factor in myocardial cell death by ischemia–reperfusion. Biochem. Biophys. Res. Commun. 309, 619–624.
Kim, E.H., Koh, E.H., Park, J.-Y., Lee, K.-U., 2010. Adenine nucleotide translocator as a regulator of mitochondrial function: implication in the pathogenesis of metabolic syndrome. Korean Diabetes J. 34, 146–153.
Klingel, K., Sauter, M., Bock, C.T., Szalay, G., Schnorr, J.-J., Kandolf, R., 2004. Molecular pathology of inflammatory cardiomyopathy. Med. Microbiol. Immunol. 193, 101–107.
Kokoszka, J.E., Waymire, K.G., Levy, S.E., Sligh, J.E., Cai, J., Jones, D.P., Macgregor, G.R., Wallace, D.C., 2004. The ADP/ATP translocator is not essential for the mitochondrial permeability transition pore. Nature 427, 461.
Kondapalli, C., Kazlauskaite, A., Zhang, N., Woodroof, H.I., Campbell, D.G., Gourlay, R., Burchell, L., Walden, H., Macartney, T.J., Deak, M., 2012. PINK1 is activated by mitochondrial membrane potential depolarization and stimulates Parkin E3 ligase activity by phosphorylating Serine 65. Open Biol. 2, 120080.
Konstantinidis, K., Whelan, R.S., Kitsis, R.N., 2012. Mechanisms of cell death in heart disease. Arterioscler. Thromb. Vasc. Biol. 32, 1552–1562.
Koshinuma, S., Miyamae, M., Kaneda, K., Kotani, J., Figueredo, V.M., 2014. Combination of necroptosis and apoptosis inhibition enhances cardioprotection against myocardial ischemia–reperfusion injury. J. Anesth. 28, 235–241.

Koudstaal, S., Oerlemans, M.I., Van Der Spoel, T.I., Janssen, A.W., Hoefer, I.E., Doevendans, P.A., Sluijter, J.P., Chamuleau, S.A., 2015. Necrostatin-1 alleviates reperfusion injury following acute myocardial infarction in pigs. Eur. J. Clin. Invest. 45, 150–159.

Kovacs, P., Bak, I., Szendrei, L., Vecsernyes, M., Varga, E., Blasig, I.E., Tosaki, A., 2001. Non-specific caspase inhibition reduces infarct size and improves post-ischaemic recovery in isolated ischaemic/reperfused rat hearts. Naunyn Schmiedebergs Arch. Pharmacol. 364, 501–507.

Kristen, A.V., Ackermann, K., Buss, S., Lehmann, L., Schnabel, P.A., Haunstetter, A., Katus, H.A., Hardt, S.E., 2013. Inhibition of apoptosis by the intrinsic but not the extrinsic apoptotic pathway in myocardial ischemia-reperfusion. Cardiovasc. Pathol. 22, 280–286.

Kroemer, G., El-Deiry, W.S., Golstein, P., Peter, M.E., Vaux, D., Vandenabeele, P., Zhivotovsky, B., Blagosklonny, M.V., Malorni, W., Knight, R.A., Piacentini, M., Nagata, S., Melino, G., 2005. Classification of cell death: recommendations of the nomenclature committee on cell death. Cell Death Differ. 12, 1463–1467.

Kroemer, G., Galluzzi, L., Vandenabeele, P., Abrams, J., Alnemri, E.S., Baehrecke, E., Blagosklonny, M., EL-Deiry, W., Golstein, P., Green, D., 2009. Classification of cell death: recommendations of the nomenclature committee on cell death 2009. Cell Death Differ. 16, 3.

Kubli, D.A., Zhang, X., Lee, Y., Hanna, R.A., Quinsay, M.N., Nguyen, C.K., Jimenez, R., Petrosyan, S., Murphy, A.N., Gustafsson, Å.B., 2013. Parkin protein deficiency exacerbates cardiac injury and reduces survival following myocardial infarction. J. Biol. Chem. 288, 915–926.

Kuma, A., Komatsu, M., Mizushima, N., 2017. Autophagy-monitoring and autophagy-deficient mice. Autophagy 13, 1619–1628.

Kung, G., Konstantinidis, K., Kitsis, R.N., 2011. Programmed necrosis, not apoptosis, in the heart. Circ. Res. 108, 1017–1036.

Kurihara, Y., Kurihara, H., Oda, H., Maemura, K., Nagai, R., Ishikawa, T., Yazaki, Y., 1995. Aortic arch malformations and ventricular septal defect in mice deficient in endothelin-1. J. Clin. Invest. 96, 293–300.

Kwong, J., Davis, J., Baines, C., Sargent, M., Karch, J., Wang, X., Huang, T., Molkentin, J., 2014. Genetic deletion of the mitochondrial phosphate carrier desensitizes the mitochondrial permeability transition pore and causes cardiomyopathy. Cell Death Differ. 21, 1209.

Kwong, J.Q., Lu, X., Correll, R.N., Schwanekamp, J.A., Vagnozzi, R.J., Sargent, M.A., York, A.J., Zhang, J., Bers, D.M., Molkentin, J.D., 2015. The mitochondrial calcium uniporter selectively matches metabolic output to acute contractile stress in the heart. Cell Rep. 12, 15–22.

L'ecuyer, T., Sanjeev, S., Thomas, R., Novak, R., Das, L., Campbell, W., Heide, R.V., 2006. DNA damage is an early event in doxorubicin-induced cardiac myocyte death. Am. J. Physiol. Heart Circ. Physiol. 291, H1273–H1280.

Lee, J.-Y., Nagano, Y., Taylor, J.P., Lim, K.L., Yao, T.-P., 2010. Disease-causing mutations in parkin impair mitochondrial ubiquitination, aggregation, and HDAC6-dependent mitophagy. J. Cell Biol. 189, 671–679.

Leipner, C., Grün, K., Schneider, I., Glück, B., Sigusch, H.H., Stelzner, A., 2004. Coxsackievirus B3-induced myocarditis: differences in the immune response of C57BL/6 and Balb/c mice. Med. Microbiol. Immunol. 193, 141–147.

Lemasters, J.J., Nieminen, A.-L., Qian, T., Trost, L.C., Elmore, S.P., Nishimura, Y., Crowe, R.A., Cascio, W.E., Bradham, C.A., Brenner, D.A., Herman, B., 1998. The mitochondrial permeability transition in cell death: a common mechanism in necrosis, apoptosis and autophagy. Biochim. Biophys. Acta 1366, 177–196.

Leung, A.W., Varanyuwatana, P., Halestrap, A.P., 2008. The mitochondrial phosphate carrier interacts with cyclophilin D and may play a key role in the permeability transition. J. Biol. Chem. 283, 26312–26323.
Li, H., Zhu, H., Xu, C.-J., Yuan, J., 1998. Cleavage of BID by caspase 8 mediates the mitochondrial damage in the Fas pathway of apoptosis. Cell 94, 491–501.
Liesa, M., Shirihai, O.S., 2013. Mitochondrial dynamics in the regulation of nutrient utilization and energy expenditure. Cell Metab. 17, 491–506.
Lim, S., Davidson, S., Mocanu, M., Yellon, D., Smith, C., 2007. The cardioprotective effect of necrostatin requires the cyclophilin-D component of the mitochondrial permeability transition pore. Cardiovasc. Drugs Ther. 21, 467–469.
Lim, C.C., Bryan, N.S., Jain, M., Garcia-Saura, M.F., Fernandez, B.O., Sawyer, D.B., Handy, D.E., Loscalzo, J., Feelisch, M., Liao, R., 2009. Glutathione peroxidase deficiency exacerbates ischemia-reperfusion injury in male but not female myocardium: insights into antioxidant compensatory mechanisms. Am. J. Physiol. Heart Circ. Physiol. 297, H2144–H2153.
Linkermann, A., Bräsen, J.H., Darding, M., Jin, M.K., Sanz, A.B., Heller, J.-O., De Zen, F., Weinlich, R., Ortiz, A., Walczak, H., 2013. Two independent pathways of regulated necrosis mediate ischemia–reperfusion injury. Proc. Natl. Acad. Sci. U. S. A. 110, 12024–12029.
Linton, P.-J., Gurney, M., Sengstock, D., Mentzer Jr., R.M., Gottlieb, R.A., 2015. This old heart: cardiac aging and autophagy. J. Mol. Cell. Cardiol. 83, 44–54.
Liu, P.P., Mason, J.W., 2001. Advances in the understanding of myocarditis. Circulation 104, 1076–1082.
Liu, N., Su, H., Zhang, Y., Liu, Z., Kong, J., 2017. Cholecalciterol cholesterol emulsion attenuates experimental autoimmune myocarditis in mice via inhibition of the pyroptosis signaling pathway. Biochem. Biophys. Res. Commun. 493, 422–428.
Lockshin, R.A., Beaulaton, J., 1974. Programmed cell death. Life Sci. 15, 1549–1565.
Lum, J.J., Bauer, D.E., Kong, M., Harris, M.H., Li, C., Lindsten, T., Thompson, C.B., 2005. Growth factor regulation of autophagy and cell survival in the absence of apoptosis. Cell 120, 237–248.
Ma, X., Liu, H., Foyil, S.R., Godar, R.J., Weinheimer, C.J., Hill, J.A., Diwan, A., 2012. Impaired autophagosome clearance contributes to cardiomyocyte death in ischemia/reperfusion injury. Circulation 125, 3170–3181.
Ma, S., Wang, Y., Chen, Y., Cao, F., 2015. The role of the autophagy in myocardial ischemia/reperfusion injury. Biochim. Biophys. Acta 1852, 271–276.
Maekawa, N., Wada, H., Kanda, T., Niwa, T., Yamada, Y., Saito, K., Fujiwara, H., Sekikawa, K., Seishima, M., 2002. Improved myocardial ischemia/reperfusion injury in mice lacking tumor necrosis factor-α. J. Am. Coll. Cardiol. 39, 1229–1235.
Maliken, B.D., Molkentin, J.D., 2018. Undeniable evidence that the adult mammalian heart lacks an endogenous regenerative stem cell. Circulation 138, 806–808.
Manasek, F.J., 1969. Myocardial cell death in the embryonic chick ventricle. Development 21, 271–284.
Mandl, A., Huong Pham, L., Toth, K., Zambetti, G., Erhardt, P., 2011. Puma deletion delays cardiac dysfunction in murine heart failure models through attenuation of apoptosis. Circulation 124, 31–39.
Marchetti, C., Chojnacki, J., Toldo, S., Mezzaroma, E., Tranchida, N., Rose, S.W., Federici, M., Van Tassell, B.W., Zhang, S., Abbate, A., 2014. A novel pharmacologic inhibitor of the NLRP3 inflammasome limits myocardial injury following ischemia-reperfusion in the mouse. J. Cardiovasc. Pharmacol. 63, 316.
Marchetti, C., Toldo, S., Chojnacki, J., Mezzaroma, E., Liu, K., Salloum, F.N., Nordio, A., Carbone, S., Mauro, A.G., Das, A., 2015. Pharmacologic inhibition of the NLRP3 inflammasome preserves cardiac function after ischemic and non-ischemic injury in the mouse. J. Cardiovasc. Pharmacol. 66, 1.

Marquez, R.T., Xu, L., 2012. Bcl-2:Beclin 1 complex: multiple, mechanisms regulating autophagy/apoptosis toggle switch. Am. J. Cancer Res. 2, 214.

Marzo, I., Brenner, C., Zamzami, N., Jürgensmeier, J.M., Susin, S.A., Vieira, H.L.A., Prévost, M.-C., Xie, Z., Matsuyama, S., Reed, J.C., Kroemer, G., 1998. Bax and adenine nucleotide translocator cooperate in the mitochondrial control of apoptosis. Science 281, 2027–2031.

Matsuda, N., Sato, S., Shiba, K., Okatsu, K., Saisho, K., Gautier, C.A., Sou, Y.-S., Saiki, S., Kawajiri, S., Sato, F., 2010. PINK1 stabilized by mitochondrial depolarization recruits Parkin to damaged mitochondria and activates latent Parkin for mitophagy. J. Cell Biol. 189, 211–221.

McArthur, K., Whitehead, L.W., Heddleston, J.M., Li, L., Padman, B.S., Oorschot, V., Geoghegan, N.D., Chappaz, S., Davidson, S., San Chin, H., Lane, R.M., Dramicanin, M., Saunders, T.L., Sugiana, C., Lessene, R., Osellame, L.D., Chew, T.L., Dewson, G., Lazarou, M., Ramm, G., Lessene, G., Ryan, M.T., Rogers, K.L., Van Delft, M.F., Kile, B.T., 2018. BAK/BAX macropores facilitate mitochondrial herniation and mtDNA efflux during apoptosis. Science 359, 1–12.

Merkle, S., Frantz, S., Schön Michael, P., Bauersachs, J., Buitrago, M., Frost Robert, J.A., Schmitteckert Eva, M., Lohse Martin, J., Engelhardt, S., 2007. A role for caspase-1 in heart failure. Circ. Res. 100, 645–653.

Miyamoto, M.I., Del Monte, F., Schmidt, U., Disalvo, T.S., Kang, Z.B., Matsui, T., Guerrero, J.L., Gwathmey, J.K., Rosenzweig, A., Hajjar, R.J., 2000. Adenoviral gene transfer of SERCA2a improves left-ventricular function in aortic-banded rats in transition to heart failure. Proc. Natl. Acad. Sci. U. S. A. 97, 793–798.

Mizushima, N., Komatsu, M., 2011. Autophagy: renovation of cells and tissues. Cell 147, 728–741.

Mocanu, M.M., Baxter, G.F., Yellon, D.M., 2000. Caspase inhibition and limitation of myocardial infarct size: protection against lethal reperfusion injury. Br. J. Pharmacol. 130, 197–200.

Mollova, M., Bersell, K., Walsh, S., Savla, J., Das, L.T., Park, S.-Y., Silberstein, L.E., Dos Remedios, C.G., Graham, D., Colan, S., 2013. Cardiomyocyte proliferation contributes to heart growth in young humans. Proc. Natl. Acad. Sci. U. S. A. 110, 1446–1451.

Murphy, M.P., 2009. How mitochondria produce reactive oxygen species. Biochem. J. 417, 1–13.

Murry, C.E., Jennings, R.B., Reimer, K.A., 1986. Preconditioning with ischemia: a delay of lethal cell injury in ischemic myocardium. Circulation 74, 1124–1136.

Nakagawa, T., Shimizu, S., Watanabe, T., Yamaguchi, O., Otsu, K., Yamagata, H., Inohara, H., Kubo, T., Tsujimoto, Y., 2005. Cyclophilin D-dependent mitochondrial permeability transition regulates some necrotic but not apoptotic cell death. Nature 434, 652.

Nakai, A., Yamaguchi, O., Takeda, T., Higuchi, Y., Hikoso, S., Taniike, M., Omiya, S., Mizote, I., Matsumura, Y., Asahi, M., 2007. The role of autophagy in cardiomyocytes in the basal state and in response to hemodynamic stress. Nat. Med. 13, 619.

Nakamura, T., Ueda, Y., Juan, Y., Katsuda, S., Takahashi, H., Koh, E., 2000. Fas-mediated apoptosis in adriamycin-induced cardiomyopathy in rats: in vivo study. Circulation 102, 572–578.

Narendra, D., Tanaka, A., Suen, D.-F., Youle, R.J., 2008. Parkin is recruited selectively to impaired mitochondria and promotes their autophagy. J. Cell Biol. 183, 795–803.

Narendra, D.P., Jin, S.M., Tanaka, A., Suen, D.-F., Gautier, C.A., Shen, J., Cookson, M.R., Youle, R.J., 2010. PINK1 is selectively stabilized on impaired mitochondria to activate Parkin. PLoS Biol. 8, e1000298.

Narita, M., Shimizu, S., Ito, T., Chittenden, T., Lutz, R.J., Matsuda, H., Tsujimoto, Y., 1998. Bax interacts with the permeability transition pore to induce permeability

transition and cytochrome c release in isolated mitochondria. Proc. Natl. Acad. Sci. U. S. A. 95, 14681–14686.
Narula, J., Pandey, P., Arbustini, E., Haider, N., Narula, N., Kolodgie, F.D., Dal Bello, B., Semigran, M.J., Bielsa-Masdeu, A., Dec, G.W., 1999. Apoptosis in heart failure: release of cytochrome c from mitochondria and activation of caspase-3 in human cardiomyopathy. Proc. Natl. Acad. Sci. U. S. A. 96, 8144–8149.
Neuhof, C., Neuhof, H., 2014. Calpain system and its involvement in myocardial ischemia and reperfusion injury. World J. Cardiol. 6, 638.
Niemann, B., Chen, Y., Issa, H., Silber, R.-E., Rohrbach, S., 2010. Caloric restriction delays cardiac ageing in rats: role of mitochondria. Cardiovasc. Res. 88, 267–276.
Nishida, K., Kyoi, S., Yamaguchi, O., Sadoshima, J., Otsu, K., 2009. The role of autophagy in the heart. Cell Death Differ. 16, 31.
Nitobe, J., Yamaguchi, S., Okuyama, M., Nozaki, N., Sata, M., Miyamoto, T., Takeishi, Y., Kubota, I., Tomoike, H., 2003. Reactive oxygen species regulate FLICE inhibitory protein (FLIP) and susceptibility to Fas-mediated apoptosis in cardiac myocytes. Cardiovasc. Res. 57, 119–128.
Niu, J., Azfer, A., Wang, K., Wang, X., Kolattukudy, P.E., 2009. Cardiac-targeted expression of soluble fas attenuates doxorubicin-induced cardiotoxicity in mice. J. Pharmacol. Exp. Ther. 328, 740–748.
Niwano, K., Arai, M., Koitabashi, N., Watanabe, A., Ikeda, Y., Miyoshi, H., Kurabayashi, M., 2008. Lentiviral vector–mediated SERCA2 gene transfer protects against heart failure and left ventricular remodeling after myocardial infarction in rats. Mol. Ther. 16, 1026–1032.
Oberst, A., Green, D.R., 2011. It cuts both ways: reconciling the dual roles of caspase 8 in cell death and survival. Nat. Rev. Mol. Cell Biol. 12, 757.
Oberst, A., Dillon, C.P., Weinlich, R., Mccormick, L.L., Fitzgerald, P., Pop, C., Hakem, R., Salvesen, G.S., Green, D.R., 2011. Catalytic activity of the caspase-8–FLIP L complex inhibits RIPK3-dependent necrosis. Nature 471, 363.
Oerlemans, M.I., Liu, J., Arslan, F., Den Ouden, K., Van Middelaar, B.J., Doevendans, P.A., Sluijter, J.P., 2012. Inhibition of RIP1-dependent necrosis prevents adverse cardiac remodeling after myocardial ischemia–reperfusion in vivo. Basic Res. Cardiol. 107 270.
Ohtsuka, M., Takano, H., Suzuki, M., Zou, Y., Akazawa, H., Tamagawa, M., Wakimoto, K., Nakaya, H., Komuro, I., 2004. Role of Na+–Ca2+ exchanger in myocardial ischemia/reperfusion injury: evaluation using a heterozygous Na+–Ca2+ exchanger knockout mouse model. Biochem. Biophys. Res. Commun. 314, 849–853.
Okamoto, N., Satow, Y., 1975. Cell death in bulbar cushion of normal and abnormal developing heart. In: Liberman, M., Sano, T. (Eds.), Developmental and Physiological Correlates of Cardiac Muscle.(Perspectives in Cardiovascular Research, Vol. 1). Raven Press, New York, pp. 51–65.
Olivetti, G., Melissari, M., Capasso, J., Anversa, P., 1991. Cardiomyopathy of the aging human heart. Myocyte loss and reactive cellular hypertrophy. Circ. Res. 68, 1560–1568.
Ong, S.-B., Subrayan, S., Lim, S.Y., Yellon, D.M., Davidson, S.M., Hausenloy, D.J., 2010. Inhibiting mitochondrial fission protects the heart against ischemia/reperfusion injury. Circulation 121, 2012.
Pacher, P., Szabó, C., 2007. Role of poly (ADP-ribose) polymerase 1 (PARP-1) in cardiovascular diseases: the therapeutic potential of PARP inhibitors. Cardiovasc. Drug Rev. 25, 235–260.
Papanicolaou, K.N., Khairallah, R.J., Ngoh, G.A., Chikando, A., Luptak, I., O'shea, K.M., Riley, D.D., Lugus, J.J., Colucci, W.S., Lederer, W.J., 2011. Mitofusin-2 maintains mitochondrial structure and contributes to stress-induced permeability transition in cardiac myocytes. Mol. Cell. Biol. 31, 1309–1328.

Papanicolaou, K.N., Kikuchi, R., Ngoh, G.A., Coughlan, K.A., Dominguez, I., Stanley, W.C., Walsh, K., 2012. Mitofusins 1 and 2 are essential for postnatal metabolic remodeling in heart. Circ. Res. 111, 1012–1026.

Paravicini, T.M., Touyz, R.M., 2008. NADPH oxidases, reactive oxygen species, and hypertension: clinical implications and therapeutic possibilities. Diabetes Care 31, S170–S180.

Pattingre, S., Tassa, A., Qu, X., Garuti, R., Liang, X.H., Mizushima, N., Packer, M., Schneider, M.D., Levine, B., 2005. Bcl-2 antiapoptotic proteins inhibit Beclin 1-dependent autophagy. Cell 122, 927–939.

Peart, J., Headrick, J.P., 2003. Adenosine-mediated early preconditioning in mouse: protective signaling and concentration dependent effects. Cardiovasc. Res. 58, 589–601.

Peart, J., Willems, L., Headrick, J.P., 2003. Receptor and non-receptor-dependent mechanisms of cardioprotection with adenosine. Am. J. Physiol. Heart Circ. Physiol. 284, H519–H527.

Petronilli, V., Nicolli, A., Costantini, P., Colonna, R., Bernardi, P., 1994. Regulation of the permeability transition pore, a voltage-dependent mitochondrial channel inhibited by cyclosporin A. Biochim. Biophys. Acta 1187, 255–259.

Phillips, L., Toledo, A.H., Lopez-Neblina, F., Anaya-Prado, R., Toledo-Pereyra, L.H., 2009. Nitric oxide mechanism of protection in ischemia and reperfusion injury. J. Invest. Surg. 22, 46–55.

Pieper, A.A., Walles, T., Wei, G., Clements, E.E., Verma, A., Snyder, S.H., Zweier, J.L., 2000. Myocardial postischemic injury is reduced by polyADPribose polymerase-1 gene disruption. Mol. Med. 6, 271.

Piper, H.M., Balser, C., Ladilov, Y.V., Schäfer, M., Siegmund, B., Ruiz-Meana, M., Garcia-Dorado, D., 1996. The role of Na+/H+ exchange in ischemia-reperfusion. Basic Res. Cardiol. 91, 191–202.

Piquereau, J., Caffin, F., Novotova, M., Prola, A., Garnier, A., Mateo, P., Fortin, D., Huynh, L.H., Nicolas, V., Alavi, M.V., 2012. Down-regulation of OPA1 alters mouse mitochondrial morphology, PTP function, and cardiac adaptation to pressure overload. Cardiovasc. Res. 94, 408–417.

Piquereau, J., Godin, R., Deschênes, S., Bessi, V.L., Mofarrahi, M., Hussain, S.N., Burelle, Y., 2013. Protective role of PARK2/Parkin in sepsis-induced cardiac contractile and mitochondrial dysfunction. Autophagy 9, 1837–1851.

Potts, M.B., Vaughn, A.E., Mcdonough, H., Patterson, C., Deshmukh, M., 2005. Reduced Apaf-1 levels in cardiomyocytes engage strict regulation of apoptosis by endogenous XIAP. J. Cell Biol. 171, 925–930.

Qin, X., Peterson, M.R., Haller, S.E., Cao, L., Thomas, D.P., He, G., 2018. Caspase recruitment domain-containing protein 9 (CARD9) knockout reduces regional ischemia/reperfusion injury through an attenuated inflammatory response. PLoS One 13, e0199711.

Rafieian-Kopaei, M., Setorki, M., Doudi, M., Baradaran, A., Nasri, H., 2014. Atherosclerosis: process, indicators, risk factors and new hopes. Int. J. Prev. Med. 5, 927.

Rhee, S.G., Kang, S.W., Chang, T.S., Jeong, W., Kim, K., 2001. Peroxiredoxin, a novel family of peroxidases. IUBMB Life 52, 35–41.

Rhee, S.G., Yang, K.-S., Kang, S.W., Woo, H.A., Chang, T.-S., 2005. Controlled elimination of intracellular H_2O_2: regulation of peroxiredoxin, catalase, and glutathione peroxidase via post-translational modification. Antioxid. Redox Signal. 7, 619–626.

Riedl, S.J., Salvesen, G.S., 2007. The apoptosome: signalling platform of cell death. Nat. Rev. Mol. Cell Biol. 8, 405.

Riley, J.S., Quarato, G., Cloix, C., Lopez, J., O'prey, J., Pearson, M., Chapman, J., Sesaki, H., Carlin, L.M., Passos, J.F., 2018. Mitochondrial inner membrane permeabilisation enables mtDNA release during apoptosis. EMBO J. 37, 1–16.

Rinkenberger, J.L., Horning, S., Klocke, B., Roth, K., Korsmeyer, S.J., 2000. Mcl-1 deficiency results in peri-implantation embryonic lethality. Genes Dev. 14, 23–27.

Roberta, A., 2011. Cell death pathways in acute I/R injury. J. Cardiovasc. Pharmacol. Ther. 16, 233–238.

Rodriguez-Enriquez, S., Kim, I., Currin, R.T., Lemasters, J.J., 2006. Tracker dyes to probe mitochondrial autophagy (mitophagy) in rat hepatocytes. Autophagy 2, 39–46.

Roy, S., Nicholson, D.W., 2000. Cross-talk in cell death signaling. J. Exp. Med. 192, F21–F26.

Ruck, A., Dolder, M., Wallimann, T., Brdiczka, D., 1998. Reconstituted adenine nucleotide translocase forms a channel for small molecules comparable to the mitochondrial permeability transition pore. FEBS Lett. 426, 97–101.

Sadat, U., 2009. Signaling pathways of cardioprotective ischemic preconditioning. Int. J. Surg. 7, 490–498.

Saito, T., Sadoshima, J., 2015. Molecular mechanisms of mitochondrial autophagy/mitophagy in the heart. Circ. Res. 116, 1477–1490.

Sallinen, P., Mänttäri, S., Leskinen, H., Ilves, M., Ruskoaho, H., Saarela, S., 2007. Time course of changes in the expression of DHPR, RyR 2, and SERCA2 after myocardial infarction in the rat left ventricle. Mol. Cell. Biochem. 303, 97–103.

Sandanger, Ø., Ranheim, T., Vinge, L.E., Bliksøen, M., Alfsnes, K., Finsen, A.V., Dahl, C.P., Askevold, E.T., Florholmen, G., Christensen, G., 2013. The NLRP3 inflammasome is up-regulated in cardiac fibroblasts and mediates myocardial ischaemia–reperfusion injury. Cardiovasc. Res. 99, 164–174.

Sanford, L.P., Ormsby, I., Gittenberger-De Groot, A.C., Sariola, H., Friedman, R., Boivin, G.P., Cardell, E.L., Doetschman, T., 1997. TGFbeta2 knockout mice have multiple developmental defects that are non-overlapping with other TGFbeta knockout phenotypes. Development 124, 2659–2670.

Scherz-Shouval, R., Shvets, E., Fass, E., Shorer, H., Gil, L., Elazar, Z., 2007. Reactive oxygen species are essential for autophagy and specifically regulate the activity of Atg4. EMBO J. 26, 1749–1760.

Schieber, M., Chandel, N.S., 2014. ROS function in redox signaling and oxidative stress. Curr. Biol. 24, R453–R462.

Sciarretta, S., Maejima, Y., Zablocki, D., Sadoshima, J., 2018. The role of autophagy in the heart. Annu. Rev. Physiol. 80, 1–26.

Seko, Y., Kayagaki, N., Seino, K.-I., Yagita, H., Okumura, K., Nagai, R., 2002. Role of Fas/FasL pathway in the activation of infiltrating cells in murine acute myocarditis caused by Coxsackievirus B3. J. Am. Coll. Cardiol. 39, 1399–1403.

Shen, Q., Yamano, K., Head, B.P., Kawajiri, S., Cheung, J.T., Wang, C., Cho, J.-H., Hattori, N., Youle, R.J., Van Der Bliek, A.M., 2014. Mutations in Fis1 disrupt orderly disposal of defective mitochondria. Mol. Biol. Cell 25, 145–159.

Sheth, S., Brito, R., Mukherjea, D., Rybak, L., Ramkumar, V., 2014. Adenosine receptors: expression, function and regulation. Int. J. Mol. Sci. 15, 2024–2052.

Shi, Y., Chen, C., Lisewski, U., Wrackmeyer, U., Radke, M., Westermann, D., Sauter, M., Tschöpe, C., Poller, W., Klingel, K., 2009. Cardiac deletion of the Coxsackievirus-adenovirus receptor abolishes Coxsackievirus B3 infection and prevents myocarditis in vivo. J. Am. Coll. Cardiol. 53, 1219–1226.

Shi, J., Abdelwahid, E., Wei, L., 2011. Apoptosis in anthracycline cardiomyopathy. Curr. Pediatr. Rev. 7, 329–336.

Shi, J., Zhao, Y., Wang, Y., Gao, W., Ding, J., Li, P., Hu, L., Shao, F., 2014. Inflammatory caspases are innate immune receptors for intracellular LPS. Nature 514, 187.

Shimizu, S., Narita, M., Tsujimoto, Y., Tsujimoto, Y., 1999. Bcl-2 family proteins regulate the release of apoptogenic cytochrome c by the mitochondrial channel VDAC. Nature 399, 483–487.

Shinmura, K., Tamaki, K., Sano, M., Murata, M., Yamakawa, H., Ishida, H., Fukuda, K., 2011. Impact of long-term caloric restriction on cardiac senescence: caloric restriction ameliorates cardiac diastolic dysfunction associated with aging. J. Mol. Cell. Cardiol. 50, 117–127.

Shiomi, T., Tsutsui, H., Matsusaka, H., Murakami, K., Hayashidani, S., Ikeuchi, M., Wen, J., Kubota, T., Utsumi, H., Takeshita, A., 2004. Overexpression of glutathione peroxidase prevents left ventricular remodeling and failure after myocardial infarction in mice. Circulation 109, 544–549.

Shirakabe, A., Ikeda, Y., Sciarretta, S., Zablocki, D.K., Sadoshima, J., 2016. Aging and autophagy in the heart. Circ. Res. 118, 1563–1576.

Shizukuda, Y., Matoba, S., Mian, O.Y., Nguyen, T., Hwang, P.M., 2005. Targeted disruption of p53 attenuates doxorubicin-induced cardiac toxicity in mice. Mol. Cell. Biochem. 273, 25–32.

Siddall, H.K., Yellon, D.M., Ong, S.-B., Mukherjee, U.A., Burke, N., Hall, A.R., Angelova, P.R., Ludtmann, M.H., Deas, E., Davidson, S.M., 2013. Loss of PINK1 increases the heart's vulnerability to ischemia-reperfusion injury. PLoS One 8, e62400.

Smith, C.C.T., Davidson, S.M., Lim, S.Y., Simpkin, J.C., Hothersall, J.S., Yellon, D.M., 2007. Necrostatin: a potentially novel cardioprotective agent? Cardiovasc. Drugs Ther. 21, 227–233.

Soriano, P., 1997. The PDGF alpha receptor is required for neural crest cell development and for normal patterning of the somites. Development 124, 2691–2700.

Stamenkovic, A., Pierce, G.N., Ravandi, A., 2019. Phospholipid oxidation products in ferroptotic myocardial cell death. Am. J. Physiol. Heart Circ. Physiol. 317, H156–H163.

Strait, J.B., Lakatta, E.G., 2012. Aging-associated cardiovascular changes and their relationship to heart failure. Heart Fail. Clin. 8, 143–164.

Sun, J., Picht, E., Ginsburg, K.S., Bers, D.M., Steenbergen, C., Murphy, E., 2006. Hypercontractile female hearts exhibit increased S-nitrosylation of the L-type Ca2+ channel α1 subunit and reduced ischemia/reperfusion injury. Circ. Res. 98, 403–411.

Sun, L., Wang, H., Wang, Z., He, S., Chen, S., Liao, D., Wang, L., Yan, J., Liu, W., Lei, X., Wang, X., 2012. Mixed lineage kinase domain-like protein mediates necrosis signaling downstream of RIP3 kinase. Cell 148, 213–227.

Swain, S.M., Whaley, F.S., Ewer, M.S., 2003. Congestive heart failure in patients treated with doxorubicin: a retrospective analysis of three trials. Cancer 97, 2869–2879.

Taffet, G.E., Pham, T.T., Hartley, C.J., 1997. The age-associated alterations in late diastolic function in mice are improved by caloric restriction. J. Gerontol. A Biol. Sci. Med. Sci. 52, B285–B290.

Tait, S.W., Green, D.R., 2010. Mitochondria and cell death: outer membrane permeabilization and beyond. Nat. Rev. Mol. Cell Biol. 11, 621.

Takemura, G., Fujiwara, H., 2007. Doxorubicin-induced cardiomyopathy: from the cardiotoxic mechanisms to management. Prog. Cardiovasc. Dis. 49, 330–352.

Talukder, M.H., Zweier, J.L., Periasamy, M., 2009. Targeting calcium transport in ischaemic heart disease. Cardiovasc. Res. 84, 345–352.

Tanaka, M., Nakae, S., Terry, R.D., Mokhtari, G.K., Gunawan, F., Balsam, L.B., Kaneda, H., Kofidis, T., Tsao, P.S., Robbins, R.C., 2004. Cardiomyocyte-specific Bcl-2 overexpression attenuates ischemia-reperfusion injury, immune response during acute rejection, and graft coronary artery disease. Blood 104, 3789–3796.

Taneike, M., Yamaguchi, O., Nakai, A., Hikoso, S., Takeda, T., Mizote, I., Oka, T., Tamai, T., Oyabu, J., Murakawa, T., 2010. Inhibition of autophagy in the heart induces age-related cardiomyopathy. Autophagy 6, 600–606.

Tang, J.N., Cores, J., Huang, K., Cui, X.L., Luo, L., Zhang, J.Y., Li, T.S., Qian, L., Cheng, K., 2018. Concise review: is cardiac cell therapy dead? Embarrassing trial outcomes and new directions for the future. Stem Cells Transl. Med. 7, 354–359.

Tang, D., Kang, R., Berghe, T.V., Vandenabeele, P., Kroemer, G., 2019. The molecular machinery of regulated cell death. Cell Res. 29, 347–364.

Tani, M., Neely, J.R., 1989. Role of intracellular Na+ in Ca2+ overload and depressed recovery of ventricular function of reperfused ischemic rat hearts. Possible involvement of H+-Na+ and Na+-Ca2+ exchange. Circ. Res. 65, 1045–1056.

Thomas, R.L., Roberts, D.J., Kubli, D.A., Lee, Y., Quinsay, M.N., Owens, J.B., Fischer, K.M., Sussman, M.A., Miyamoto, S., Gustafsson, Å.B., 2013. Loss of MCL-1 leads to impaired autophagy and rapid development of heart failure. Genes Dev. 27, 1365–1377.

Toldo, S., Mauro, A.G., Cutter, Z., Van Tassell, B.W., Mezzaroma, E., Del Buono, M.G., Prestamburgo, A., Potere, N., Abbate, A., 2019. The NLRP3 inflammasome inhibitor, OLT1177 (dapansutrile), reduces infarct size and preserves contractile function after ischemia reperfusion injury in the mouse. J. Cardiovasc. Pharmacol. 73, 215–222.

Tong, H., Imahashi, K., Steenbergen, C., Murphy, E., 2002. Phosphorylation of glycogen synthase kinase-3β during preconditioning through a phosphatidylinositol-3-kinase–dependent pathway is cardioprotective. Circ. Res. 90, 377–379.

Valentim, L., Laurence, K.M., Townsend, P.A., Carroll, C.J., Soond, S., Scarabelli, T.M., Knight, R.A., Latchman, D.S., Stephanou, A., 2006. Urocortin inhibits Beclin1-mediated autophagic cell death in cardiac myocytes exposed to ischaemia/reperfusion injury. J. Mol. Cell. Cardiol. 40, 846–852.

Van Berlo, J.H., Molkentin, J.D., 2014. An emerging consensus on cardiac regeneration. Nat. Med. 20, 1386.

Van Den Hoff, M.J., Van Den Eijnde, S.M., Viragh, S., Moorman, A.F., 2000. Programmed cell death in the developing heart. Cardiovasc. Res. 45, 603–620.

Vandenabeele, P., Galluzzi, L., Berghe, T.V., Kroemer, G., 2010. Molecular mechanisms of necroptosis: an ordered cellular explosion. Nat. Rev. Mol. Cell Biol. 11, 700.

Vander Heide, R.S., Steenbergen, C., 2013. Cardioprotection and myocardial reperfusion: pitfalls to clinical application. Circ. Res. 113, 464–477.

Varfolomeev, E.E., Schuchmann, M., Luria, V., Chiannilkulchai, N., Beckmann, J.S., Mett, I.L., Rebrikov, D., Brodianski, V.M., Kemper, O.C., Kollet, O., 1998. Targeted disruption of the mouse caspase 8 gene ablates cell death induction by the TNF receptors, Fas/Apo1, and DR3 and is lethal prenatally. Immunity 9, 267–276.

Vaux, D.L., Korsmeyer, S.J., 1999. Cell death in development. Cell 96, 245–254.

Vives-Bauza, C., Zhou, C., Huang, Y., Cui, M., De Vries, R.L., Kim, J., May, J., Tocilescu, M.A., Liu, W., Ko, H.S., 2010. PINK1-dependent recruitment of Parkin to mitochondria in mitophagy. Proc. Natl. Acad. Sci. U. S. A. 107, 378–383.

Wai, T., García-Prieto, J., Baker, M.J., Merkwirth, C., Benit, P., Rustin, P., Rupérez, F.J., Barbas, C., Ibañez, B., Langer, T., 2015. Imbalanced OPA1 processing and mitochondrial fragmentation cause heart failure in mice. Science 350, aad0116.

Wang, Y., Guo, Y., Zhang, S.X., Wu, W.-J., Wang, J., Bao, W., Bolli, R., 2002. Ischemic preconditioning upregulates inducible nitric oxide synthase in cardiac myocyte. J. Mol. Cell. Cardiol. 34, 5–15.

Wang, Y., Kim, N.S., Haince, J.-F., Kang, H.C., David, K.K., Andrabi, S.A., Poirier, G.G., Dawson, V.L., Dawson, T.M., 2011. Poly(ADP-ribose) (PAR) binding to apoptosis-inducing factor is critical for PAR polymerase-1-dependent cell death (parthanatos). Sci. Signal. 4, ra20.

Wang, X., Bathina, M., Lynch, J., Koss, B., Calabrese, C., Frase, S., Schuetz, J.D., Rehg, J.E., Opferman, J.T., 2013. Deletion of MCL-1 causes lethal cardiac failure and mitochondrial dysfunction. Genes Dev. 27, 1351–1364.

Wang, Y., Gao, B., Xiong, S., 2014. Involvement of NLRP3 inflammasome in CVB3-induced viral myocarditis. Am. J. Physiol. Heart Circ. Physiol. 307, H1438–H1447.

Waterhouse, N.J., Finucane, D.M., Green, D.R., Elce, J.S., Kumar, S., Alnemri, E.S., Litwack, G., Khanna, K., Lavin, M.F., Watters, D.J., 1998. Calpain activation is upstream of caspases in radiation-induced apoptosis. Cell Death Differ. 5, 1051–1061.

Weeber, E.J., Levy, M., Sampson, M.J., Anflous, K., Armstrong, D.L., Brown, S.E., Sweatt, J.D., Craigen, W.J., 2002. The role of mitochondrial porins and the permeability transition pore in learning and synaptic plasticity. J. Biol. Chem. 277, 18891–18897.

Wei, M.C., Zong, W.-X., Cheng, E.H.-Y., Lindsten, T., Panoutsakopoulou, V., Ross, A.J., Roth, K.A., Macgregor, G.R., Thompson, C.B., Korsmeyer, S.J., 2001. Proapoptotic BAX and BAK: a requisite gateway to mitochondrial dysfunction and death. Science 292, 727–730.

Weisiger, R.A., Fridovich, I., 1973. Superoxide dismutase organelle specificity. J. Biol. Chem. 248, 3582–3592.

Westphal, D., Dewson, G., Czabotar, P.E., Kluck, R.M., 2011. Molecular biology of Bax and Bak activation and action. Biochim. Biophys. Acta 1813, 521–531.

Whelan, R.S., Konstantinidis, K., Wei, A.-C., Chen, Y., Reyna, D.E., Jha, S., Yang, Y., Calvert, J.W., Lindsten, T., Thompson, C.B., Crow, M.T., Gavathiotis, E., Dorn, G.W., O'rourke, B., Kitsis, R.N., 2012. Bax regulates primary necrosis through mitochondrial dynamics. Proc. Natl. Acad. Sci. U. S. A. 109, 6566–6571.

Williams, R.E., Zweier, J.L., Flaherty, J.T., 1991. Treatment with deferoxamine during ischemia improves functional and metabolic recovery and reduces reperfusion-induced oxygen radical generation in rabbit hearts. Circulation 83, 1006–1014.

Wohlgemuth, S.E., Julian, D., Akin, D.E., Fried, J., Toscano, K., Leeuwenburgh, C., Dunn, J., William, A., 2007. Autophagy in the heart and liver during normal aging and calorie restriction. Rejuvenation Res. 10, 281–292.

Woodfield, K., Rück, A., Brdiczka, D., Halestrap, A.P., 1998. Direct demonstration of a specific interaction between cyclophilin-D and the adenine nucleotide translocase confirms their role in the mitochondrial permeability transition. Biochem. J. 336 (Pt. 2), 287–290.

Xiao, C.-Y., Chen, M., Zsengellér, Z., Szabó, C., 2004. Poly(ADP-ribose) polymerase contributes to the development of myocardial infarction in diabetic rats and regulates the nuclear translocation of apoptosis-inducing factor. J. Pharmacol. Exp. Ther. 310, 498–504.

Xin, M., Olson, E.N., Bassel-Duby, R., 2013. Mending broken hearts: cardiac development as a basis for adult heart regeneration and repair. Nat. Rev. Mol. Cell Biol. 14, 529.

Xu, Y.-J., Zheng, L., Hu, Y.-W., Wang, Q., 2018. Pyroptosis and its relationship to atherosclerosis. Clin. Chim. Acta 476, 28–37.

Xu, T., Ding, W., Ao, X., Chu, X., Wan, Q., Wang, Y., Xiao, D., Yu, W., Li, M., Yu, F., 2019. ARC regulates programmed necrosis and myocardial ischemia/reperfusion injury through the inhibition of mPTP opening. Redox Biol. 20, 414–426.

Yang, W.S., Stockwell, B.R., 2016. Ferroptosis: death by lipid peroxidation. Trends Cell Biol. 26, 165–176.

Yaoita, H., Ogawa, K., Maehara, K., Maruyama, Y., 1998. Attenuation of ischemia/reperfusion injury in rats by a caspase inhibitor. Circulation 97, 276–281.

Yoshida, K., Taga, T., Saito, M., Suematsu, S., Kumanogoh, A., Tanaka, T., Fujiwara, H., Hirata, M., Yamagami, T., Nakahata, T., 1996. Targeted disruption of gp130, a common signal transducer for the interleukin 6 family of cytokines, leads to myocardial and hematological disorders. Proc. Natl. Acad. Sci. U. S. A. 93, 407–411.

Youle, R.J., Narendra, D.P., 2011. Mechanisms of mitophagy. Nat. Rev. Mol. Cell Biol. 12, 9.

Youle, R.J., Van Der Bliek, A.M., 2012. Mitochondrial fission, fusion, and stress. Science 337, 1062–1065.

Yu, X., Ge, L., Niu, L., Lian, X., Ma, H., Pang, L., 2018. The dual role of inducible nitric oxide synthase in myocardial ischemia/reperfusion injury: friend or foe? Oxid. Med. Cell. Longev. 2018, 1–7.

Zepeda, R., Kuzmicic, J., Parra, V., Troncoso, R., Pennanen, C., Riquelme, J.A., Pedrozo, Z., Chiong, M., Sánchez, G., Lavandero, S., 2014. Drp1 loss-of-function reduces cardiomyocyte oxygen dependence protecting the heart from ischemia-reperfusion injury. J. Cardiovasc. Pharmacol. 63, 477–487.

Zhai, P., Sciarretta, S., Galeotti, J., Volpe, M., Sadoshima, J., 2011. Differential roles of GSK-3β during myocardial ischemia and ischemia/reperfusion. Circ. Res. 109, 502–511.

Zhang, T., Zhang, Y., Cui, M., Jin, L., Wang, Y., Lv, F., Liu, Y., Zheng, W., Shang, H., Zhang, J., 2016. CaMKII is a RIP3 substrate mediating ischemia- and oxidative stress–induced myocardial necroptosis. Nat. Med. 22, 175.

Zhe-Wei, S., Li-Sha, G., Yue-Chun, L., 2018. The role of necroptosis in cardiovascular disease. Front. Pharmacol. 9, 1–9.

Zhou, W., Marinelli, F., Nief, C., Faraldo-Gómez, J.D., 2017. Atomistic simulations indicate the c-subunit ring of the F1Fo ATP synthase is not the mitochondrial permeability transition pore. Elife 6, e23781.

Zhou, F., Jiang, X., Teng, L., Yang, J., Ding, J., He, C., 2018a. Necroptosis may be a novel mechanism for cardiomyocyte death in acute myocarditis. Mol. Cell. Biochem. 442, 11–18.

Zhou, H., Wang, S., Hu, S., Chen, Y., Ren, J., 2018b. ER–mitochondria microdomains in cardiac ischemia–reperfusion injury: a fresh perspective. Front. Physiol. 9, 1–13.

Zhu, H., Tannous, P., Johnstone, J.L., Kong, Y., Shelton, J.M., Richardson, J.A., Le, V., Levine, B., Rothermel, B.A., Hill, J.A., 2007. Cardiac autophagy is a maladaptive response to hemodynamic stress. J. Clin. Invest. 117, 1782–1793.

Zhu, P., Hu, S., Jin, Q., Li, D., Tian, F., Toan, S., Li, Y., Zhou, H., Chen, Y., 2018. Ripk3 promotes ER stress-induced necroptosis in cardiac IR injury: a mechanism involving calcium overload/XO/ROS/mPTP pathway. Redox Biol. 16, 157–168.

CHAPTER FIVE

The involvement of regulated cell death forms in modulating the bacterial and viral pathogenesis

Gergely Imre*
Institute of General Pharmacology and Toxicology, University Hospital Frankfurt, Goethe University, Frankfurt am Main, Germany
*Corresponding author: e-mail address: Imre@med.uni-frankfurt.de

Contents

Abstract

Apoptosis, necroptosis and pyroptosis represent three distinct types of regulated cell death forms, which play significant roles in response to viral and bacterial infections. Whereas apoptosis is characterized by cell shrinkage, nuclear condensation, bleb formation and retained membrane integrity, necroptosis and pyroptosis exhibit osmotic imbalance driven cytoplasmic swelling and early membrane damage. These three cell death forms exert distinct immune stimulatory potential. The caspase driven apoptotic cell demise is considered in many circumstances as anti-inflammatory, whereas the two lytic cell death modalities can efficiently trigger immune response by releasing damage associated molecular patterns to the extracellular space. The relevance of these cell death modalities in infections can be best demonstrated by the presence of viral proteins that directly interfere with cell death pathways. Conversely, some pathogens

International Review of Cell and Molecular Biology, Volume 353
ISSN 1937-6448
https://doi.org/10.1016/bs.ircmb.2019.12.008

hijack the cell death signaling routes to initiate a targeted attack against the immune cells of the host, and extracellular bacteria can benefit from the destruction of intact extracellular barriers upon cell death induction. The complexity and the crosstalk between these cell death modalities reflect a continuous evolutionary race between pathogens and host. This chapter discusses the current advances in the research of cell death signaling with regard to viral and bacterial infections and describes the network of the cell death initiating molecular mechanisms that selectively recognize pathogen associated molecular patterns.

1. Introduction

In the late 1960s, it was recognized that cell death could be considered as a regulated process. For the genetically programmed, regulated cell demise the term apoptosis (from Greek: "*falling off*") was coined, which exhibited distinctive morphological and molecular patterns (Kerr et al., 1972). That time necrosis was interpreted as a total counterpart of apoptosis, a non-physiological and non-regulated, lytic cell death modality. In the recent decades, additional newly discovered regulated cell death modalities have been added to the list of cell death types (Galluzzi et al., 2018), and it became evident that necrotic-like, lytic cell death can also take place in a regulated manner. Apoptosis is a non-lytic form of cell death characterized by nuclear condensation, shrinkage of the cell, formation of apoptotic bodies and preservation of the cytoplasm membrane integrity until the late stages of the process (Kerr et al., 1972). Importantly, apoptosis actively labels itself by translocating phosphatidylserine (PS) to the outer leaflet of the membrane lipid bilayer. The PS exposure on the outer surface provides a so-called "eat me" signal for the neighboring monocytes and macrophages and ensuring the clearance of the apoptotic debris (Fadok et al., 1992). These two main attributes qualify apoptosis as a non-inflammatory cell death. In contrast, necroptosis and pyroptosis are lytic cell death modalities, accompanied by early membrane rupture and osmotic imbalance, which provoke inflammatory signals by releasing damage associated molecular patterns (DAMPs) to the extracellular space (Galluzzi et al., 2018). It is important to note that the immunogenicity of the cell death is a more complex process than above outlined. The lytic property of a cell demise does not necessarily result in an immune stimulation, while the caspase-8 driven apoptosis may lead to immunogenic processes (Yatim et al., 2015). This also clearly indicates that the release of DAMPs is a signaling-dependent, actively regulated process.

The human body developed elaborate strategies to efficiently avoid and eliminate infectious pathogens. Two corner stones of this response are the innate and the adaptive immune signaling, which also include controlled killing/self-killing mechanisms of the infected host cells (Jorgensen et al., 2017). This review focuses on the infection related aspects of regulated cell death forms, which in many instances provide benefit for the hosts by destroying the intracellular niche of the pathogen. Nevertheless, it is evident that some extracellular pathogens—and also intracellular ones—can benefit from the death of the host cells. In this chapter, based on the current advances in the research, we give a detailed description about the key cell death modalities, including apoptosis, necroptosis and pyroptosis emerging in response to pathogenic insults, and we discuss how bacterial and viral infections can modulate these signaling pathways.

2. Role of apoptosis in bacterial and viral infection

2.1 The apoptotic core machinery

Apoptosis is conducted by caspase activation. Caspases are cysteine-driven and aspartate-directed peptidases, which present in inactive forms, and become cleaved and activated upon selective apoptotic stimulation (Cohen, 1997). The initiator caspases (for instance caspase-8 and -9) are the upstream components of the caspase-cascade. Their structurally unique feature is the large pro-domain, that enables the recruitment to multimeric protein complexes (Mace and Riedl, 2010). Extrinsic and intrinsic pathways represent the two major routes of initiator caspase activation. In the extrinsic pathway, extracellular ligands from the tumor necrosis factor (TNF) superfamily interact with transmembrane death receptors (DR) of the TNF superfamily. Consequently, by the help of an adaptor protein, caspase-8 and -10 are recruited into the DR containing cell death inducing signaling complex (DISC), which leads to proximity induced cleavage and activation of these initiator caspases (Wilson et al., 2009). The intrinsic pathway can be triggered by intracellular perturbations, for instance by genotoxic stress, starvation and ER-stress. Upon stimulus, two pro-apoptotic members of the B cell lymphoma 2 (BCL-2) protein family BCL-2-associated X protein (BAX) and BCL-2-antagonistic killer (BAK) go through conformational changes and translocate into the mitochondria (Tait and Green, 2010). In turn, these proteins oligomerize and form large pores throughout the mitochondrial membrane (Große et al., 2016). This event initiates the mitochondrial outer membrane permeabilization (MOMP). Caspase-9 activation takes place in the apoptosome protein complex. Formation of this complex is dependent on the MOMP-driven release of mitochondrial pro-apoptotic proteins, such as cytochrome-C (Riedl and Salvesen, 2007). The activation of initiator caspases in both pathways culminates in the cleavage of executioner caspase-3, -6 and -7. In turn, the effector caspases cleave selective substrates, including poly ADP-ribose polymerase (PARP), inhibitor of caspase activated DNAse (ICAD) and lamins calling forth the biochemical and morphological features of apoptosis (Cohen, 1997; Tewari et al., 1995).

2.2 Apoptosis in virus infection

Viruses are obligatory intracellular pathogens and their propagation is fully dependent on the intracellular milieu of the host cell. The host organisms developed various strategies to detect and destroy the invading viral pathogens. The immunological responses, involving the transcription of anti-viral genes (e.g., interferons and cytokines) given to a specific virus infection are well characterized. It is widely accepted view that apoptosis can be beneficial for the host by destroying the intracellular niche of the pathogen (Evavold et al., 2018). The most striking evidence for the anti-viral advantage of apoptosis is the observation that many viruses produce anti-apoptotic/necroptotic proteins that block cell death signals (Thome et al., 1997; Zhou et al., 1997). However, based on recent studies, it is obvious that the picture is more complex than originally expected, and whether the cell death is beneficial or detrimental for the host seems to be dependent on multiple factors, including timing, cell type and the effectiveness of the cell death modality to trigger or suppress immune response.

2.2.1 The role of death receptors and the extrinsic apoptotic pathway

The TNF ligand superfamily acts via transmembrane TNF receptors. Specific members of these receptor family, also called death receptors (DR), possess a cytoplasmic 80 amino acid motif, termed death effector domain. These receptors, including TNFR1, first apoptosis signal (FAS, also called CD95) and the TNF-related apoptosis-inducing ligand receptor-1/2 (TRAILR1/2) can unleash various signaling pathways, such as proliferation, survival, immune response and apoptosis (Wilson et al., 2009). In response to ligand binding, the trimerized DRs can recruit one of the two death domain (DD) containing adaptor proteins: TNFR associated DD (TRADD) or FAS associated DD (FADD). Recruitment of TRADD leads to the assembly of a complex containing several proteins, including the DD containing receptor interacting protein kinase-1 (RIPK1) (Hsu et al., 1996), the cellular FLICE/Casp8 inhibitory protein (cFLIP), the TNFR associated factor-2 (TRAF2) and the E3 ubiquitin ligase cellular inhibitor of apoptosis (cIAP) (Varfolomeev and Vucic, 2008). This protein complex conveys a non-apoptotic signal via the activation of the NF-κB signaling. In contrast, the binding of FADD triggers the interaction of caspase-8 and -10, leading to apoptotic signaling either via the mitochondrial amplification loop or via direct caspase-3/-7 cleavage in immune cells. TNFR stimulation primarily results in the recruitment of TRADD, leading to non-apoptotic signals via NF-κB activation and expression of pro-survival genes such as cFLIP. Once this primary signal is compromised by the inhibition of the expression of anti-apoptotic cFLIP, the apoptotic signal can proceed. Conversely, activation of FAS and TRAILR predominately transduces FADD dependent apoptotic signaling (Wilson et al., 2009). Signaling through DRs represents one of the major forms of the immune cell directed host cell killing in infection.

Cytotoxic T cells (CTL) are key components of the adaptive immune response and natural killer (NK) cells play analogous role in the innate immune response. CTLs express FAS ligands (FASL) on their surface (Fig. 1) (Halle et al., 2017). The trimerized ligands bind the DRs expressed on the cell surface, leading to DR clustering and binding of the adaptor protein FADD, which consequently results in the recruitment of initiator caspase-8/-10. This cell killing mechanism can reduce the spreading of the virus by eliminating the intracellular niche of the pathogen (Halle et al., 2017). This occurs in human vascular endothelial cells in response to Dengue virus (DV), where the cells exhibit increased FASL and FAS expression and increased number of FAS on the cell surface. This phenomenon is accompanied by elevated rate of apoptosis (Liao et al., 2010). Similarly, influenza A virus (IAV) infection leads to the expression of FASL, FADD and caspase-8, and downregulates the expression of cFLIP in human lung epithelial cell lines. However, it is not entirely clear if the host or the virus profits from the apoptosis induction. For instance, the replication of the IAV can be amplified by overexpression of pro-apoptotic genes and is blocked by the upregulation of anti-apoptotic genes (Wang et al., 2014). This study suggests that an initially protective apoptotic over-activation can turn against the host in some instances. TNFα, the ligand of TNFR1 can also be expressed paracrine or in autocrine manner as the part of the anti-viral immune response. Hepatitis B virus x protein (HBx) sensitizes host cells to TNF driven apoptosis by interfering the anti-apoptotic function of cFLIP, thereby resulting in caspase-8 over-activation upon stimuli (Kim and Seong, 2003).

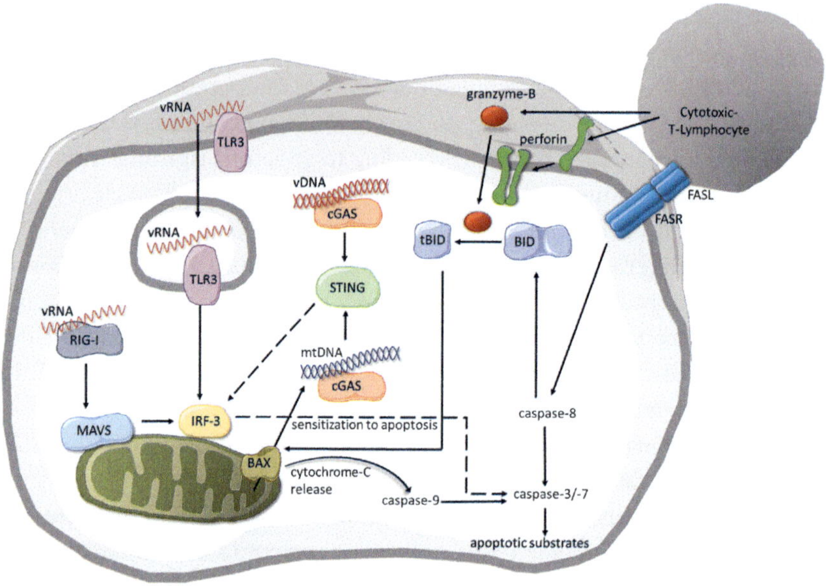

Fig. 1 Apoptosis signaling in response to infection. Extracellular and intracellular Viral RNA (vRNA) fragments (left) activate TLR3 and RIG-I, respectively, resulting in IRF-3 activation. Cytoplasmic viral, bacterial or mitochondrial DNA (mtDNA) is detected by cGAS, which in turn activates the STING pathway, leading to apoptotic signaling via a putative IRF-3 driven mechanism. Apoptosis can be triggered by effector cells of the immune system. Ligation of the FASR (right) leads to the assembly of the adaptor FADD and caspase-8 containing DISC, resulting in the activation of caspase-8. Activation of caspase-8 either results in the direct cleavage of caspase-3/7 or leads to the BID cleavage (tBID) driven amplification loop of the intrinsic pathway. The secretion of the pro-apoptotic perforin and granzyme-B (middle) triggers the activation of the intrinsic apoptotic pathway.

2.2.2 Intrinsic apoptosis triggered by immune cells: Granzyme-B

In addition, CTLs and NK cells induce apoptosis in virus infected target cells through the action of secreted effector molecules. Granzyme-B enters the cells along with another secreted protein called perforin (Fig. 1) (Lowin et al., 1994; Shresta et al., 1995). Granzyme-B is a serine protease that has been shown to induce mitochondrial apoptosis by inducing Bid cleavage and subsequent mitochondrial cytochrome-C release, leading to caspase-9 activation (Pinkoski et al., 2001).

2.2.3 Virus induced direct self-destruction in infected cells

Apart from the significant role of immune system driven apoptotic responses, a plethora of evidence demonstrates the relevance of apoptosis, unleashed directly by the viral pathogens intracellularly. Enterovirus 71 (EV71), a virus in the family of Picornaviridae, is one of the main causative agents of hand, foot and mouth disease in infected infants and young children. EV71 can directly induce the intrinsic apoptotic pathway by activating BAX and the subsequent MOMP, leading to the caspase-9 driven apoptotic signal. Importantly, depletion of caspase-8 has no influence on the cell death initiated by the virus, excluding the role of a DR driven external stimulation (Han and Cong, 2017). Analogously, Coxsackievirus A16 (CA16) can trigger both the extrinsic and the intrinsic way of apoptosis in human neuronal and muscle cell lines accompanied by the typical apoptotic features: caspase cleavage, DNA fragmentation, and PS translocation (Li et al., 2014). The West Nile virus (WNV) is a mosquito born single-stranded RNA virus, causing West Nile encephalitis in humans. The non-structural proteins 2B and 3 (NS2B and NS3) conduct the proteolytic cleavage and the processing of the large viral polyprotein, that is formed directly from the RNA of the virus. Besides, NS3 initiates apoptosis in a caspase-8 dependent way, which then results in typical apoptotic features in multiple human cell lines (Ramanathan et al., 2006).

The virus-host co-evolution is well demonstrated by the examples, where the pathogen has developed strategies to hijack the originally protective apoptotic signaling of the host. Chikungunya virus (CHIKV) was reported to infect macrophages and induce apoptosis by triggering the intrinsic apoptotic pathway via the caspase-9/-3 axis (Nayak et al., 2017). Intriguingly, CHIKV virus exploits the formation of the apoptotic blebs to transport the virus particles into neighboring macrophages (Krejbich-Trotot et al., 2011). The apoptosis can be beneficial for the pathogen, if the defensive immune cells are selectively deleted by the virus-driven cell death signals. The human immunodeficiency virus (HIV) protease cleaves procaspase-8 to a 41 kDa fragment, called casp8p41, which lacks the catalytic site of the caspase. Interestingly, the catalytically inactive fragment can facilitate BAX oligomerization and apoptosis in the host CD4 cells (Sainski et al., 2014).

2.2.4 Viral RNA sensors involved in apoptosis

Probably one of the most thrilling question, emerging from the studies on the host-pathogen interactions, is how the host organisms differentiate between friend and foe. Detection of intracellular pathogens in mammalian cells is a multifaceted mechanism, which involves the recognition of various signals deriving from the pathogens, collectively termed pathogen associated molecular patterns (PAMPs). For instance, the viral RNA and DNA fragments represent a prominent group of PAMPs. PAMPs are detected by PAMP recognition receptors (PRR) (Chow et al., 2015), which mediate the expression and activation of immunomodulatory cytokines, chemokines, and interferons. Furthermore, the signaling through PRRs leads to cell death signals primarily via the transcriptional activation of apoptotic genes.

Depending on the spatial localization of the viral RNA fragments, RNA viruses can either stimulate Toll-like receptors (TLRs) or activate the retinoic acid inducible gene-I (RIG-I) pathway. TLR3, which resides in the cytoplasmic and endosomal membranes (Fig. 1), recognizes extracellular double-stranded viral RNA (vRNA) fragments (Alexopoulou et al., 2001). TLR3 utilizes the adaptor protein Toll/interleukin-1 receptor (TIR) domain containing adaptor inducing interferon (TRIF) (Yamamoto et al., 2003) to recruit the inhibitor of kappa B kinase (IKK), subsequently leading to the activation of transcription factor nuclear factor kappa-light-chain-enhancer of activated B cells (NF-κB) and the TANK binding kinase 1 (TBK1) (Fitzgerald et al., 2003), which in turn induces the interferon regulatory factor-3 (IRF3).

Intracellular dsRNA, produced by replicating viruses is detected by cytoplasmic RNA helicases, RIG-I and melanoma differentiation-associated gene 5 (MDA5) (Fig. 1). RIG-I plays important role in recognizing RNAs from RNA viruses, including Newcastle disease virus (NDV), Sendai virus (SeV) and Vesicular stomatitis virus (VSV) (Kato et al., 2006). They recruit members of the TRAF family via the mitochondrial anti-viral-signaling protein (MAVS). In turn, MAVS, analogously to the TLR3 pathway activates the transcription factors IRF3 and NF-κB. Importantly, cells lacking the IRF3 gene are not susceptible to apoptosis upon SeV infection (Peters et al., 2008). Conversely, MAVS is shown to induce apoptosis via caspase-8 but not via IRF3 upon Semliki Forest virus infection (El Maadidi et al., 2014). Infection of a genetically engineered Vaccinia virus strain, lacking the viral BCL-2 analog F1L protein induces apoptosis via the activation of IRF3, which in turn mediates the expression of the pro-apoptotic protein Noxa (from Latin: "*damage*") in monocytes and lymphocytes (Eitz Ferrer et al., 2011), whereas the wild type strain, possessing the F1L protein cannot stimulate the same response, indicating that the Vaccinia virus successfully adapted in an evolutionary race to overcome the apoptotic inducing capacity of the host.

It was recognized that TLR3 stimulation was able to trigger apoptosis via the adaptor protein TRIF (Kaiser and Offermann, 2005) and that TLR3 mediated direct caspase-8 activation, primarily depending on the presence of cFLIP isoforms (Feoktistova et al., 2011; Weber et al., 2010). Along with this, immunoprecipitation experiments of TLR3 demonstrate the formation of a TLR3 associated DISC-like complex, including TRIF, RIPK1, FADD, cFLIP and caspase-8 in response to IAV infection. The administration of interleukin-24 (IL-24) facilitates the caspase-8 activation by decreasing the presence of cFLIP in the complex and renders the cells to apoptosis (Weiss et al., 2013). The relevance of TLR-3/caspase-8 dependent apoptosis is further confirmed in mouse intestinal epithelial cells, where the apoptosis is independent from the TNFR signaling and from the effect of external cytokines (McAllister et al., 2013).

2.2.5 Viral DNA sensors inducing apoptosis

The cyclic GMP-AMP (cGAMP) synthase (cGAS) has been shown to interact with viral DNA (vDNA) molecules. Upon interaction, cGAS produces its second messenger cGAMP and this in turn activates the stimulator of interferon gene (STING) protein (Fig. 1), which is located in the endoplasmic reticulum (ER). Consequently, STING can activate IRF3 (Ishikawa and Barber, 2008; Sun et al., 2013) and result in mitochondrial apoptosis (Diner et al., 2016). Intriguingly, mitochondrial DNA (mtDNA) that is released upon the apoptotic mitochondrial membrane permeabilization can also activate cGAS. This can trigger IRF3 activation and consequently result in anti-viral responses even in circumstances where viruses efficiently block apoptosis downstream of MOMP, for example at the level of caspase activation. It has been demonstrated that apoptotic caspases not only mediate an anti-inflammatory cell death pathway, but also directly block the mtDNA triggered STING-mediated IFN production in mice (White et al., 2014). A recent study provides evidence that apoptotic caspases cleave the cytoplasmic anti-viral sensors cGAS, MAVS and IRF3 to limit cytokine production. In this study caspase-3 deficient mice exerts increased resistance to virus infection, suggesting that caspase activation may contribute to virus propagation by cutting off the pro-inflammatory signals deriving from the infected cells (Ning et al., 2019).

2.2.6 When apoptosis does not work: Anti-apoptotic proteins encoded by viruses

The viruses developed elaborate countermeasures to avoid or delay a premature apoptotic demise, which consequently would eliminate the essential environment of the virus replication. These strategies include the expression of apoptosis inhibitor proteins. The majority of the anti-apoptotic factors are encoded by large DNA viruses, which generally replicate with slower kinetics than RNA viruses. A broad group of inhibitors attack the extrinsic pathway of apoptosis, which is triggered by an anti-viral immune response. Viral FLICE-like inhibitory protein (vFLIP) is encoded by the Molluscum contagiosum virus and several gamma Herpes viruses, including Kaposi's sarcoma-associated herpes virus (KSHV). The vFLIP exerts its inhibitory activity by binding the adaptor protein FADD and forming inactive caspase-8/vFLIP heterodimers (Thome et al., 1997). Human cytomegalovirus (CMV) is a common infection in humans, which does not lead to disease in healthy individuals, whereas it causes severe infections or death in patients with immunodeficiencies. The mouse CMV (MCMCV) is the experimental model of human CMV infection. The M36 protein of the virus blocks caspase-8 driven apoptosis and thus contributing to the pathogenicity of the microbe (Ebermann et al., 2012; Skaletskaya et al., 2001). The large subunit of the herpes simplex virus (HSV) ribonucleotide reductase ICP10 demonstrates caspase-8 inhibitory potential by directly interacting with the initiator caspase (Dufour et al., 2011; Langelier et al., 2002). The cytokine response modifier A (CrmA), a product of the Cowpox virus, prevents apoptosis by blocking the activity of caspase-8 (Zhou et al., 1997). Other group of viral anti-apoptotic proteins modulates the expression of cellular anti-apoptotic proteins. The human T-cell leukemia virus type-1 (HTLV-1)-associated adult T-cell leukemia/lymphoma (ATL) is a malignancy of infected mature CD4 T cells in humans (Matsuoka and Jeang, 2007). The transactivator protein TAX of HTLV-1 upregulates the expression of cFLIP by binding to the inhibitor of κB kinase (IKK) and inducing the NF-κB pathway in the host cells, and thereby the virus efficiently amplifies the apoptosis resistance in infected cells (Okamoto et al., 2006). A further group of virally encoded anti-apoptotic factors target the intrinsic apoptosis signaling. The Vaccinia virus encoded F1L protein inhibits the mitochondrial apoptotic signaling in two distinct ways. F1L functions both as an inhibitor of pro-apoptotic BCL-2 family proteins and as a direct caspase-9 inhibitor (Zhai et al., 2010). Human papillomavirus (HPV), a double-stranded DNA virus has been associated with cervical cancer developments in humans (Dürst et al., 1983). The early gene-coding region protein 6 (E6) of HPV is reported to interfere with apoptosis signaling at the level of p53 and by interacting with the pro-apoptotic BAK protein.

2.3 Bacterial infection driven apoptosis

2.3.1 Canonical caspase-8 activation in response to bacterial infection

In contrast to viruses, bacteria are not essentially dependent on the intracellular environment. Moreover, intact cellular layers exhibit physical barriers against bacterial penetration. Therefore, extracellular bacteria developed strategies to efficiently destroy these barrier functions. A novel study presents the bacterial qourum-sensing autoinducer of the *Pseudomonas aeruginosa* N-(3-oxo-dodecanoyl) homoserine lactone (3oc), a small chemical released to control microbial communication, as an inducer of TNFR1. It is demonstrated that 3oc directly disrupts the lipid domain structures, containing sphingolipids and cholesterol, and induces the translocation of TNFR1 into the disordered lipid phase of the membrane, which in turn triggers the trimerization of the TNFR and leads to subsequent apoptosis in human and mice monocytes (Song et al., 2019). This type of TNFR activation ensures a ligand independent induction of the extrinsic pathway. Another type of cell killing involves neighboring immune cells. The Gram-positive bacteria *Listeria monocytogenes* conducts the sorting of the bacterial DNA into extracellular vesicles, which in turn are delivered to bystander T cells, where they trigger the DNA sensor cGAS-STING pathway mediated apoptosis (Nandakumar et al., 2019).

2.3.2 TLR mediated caspase activation upon bacterial infection

Yersinia pestis, the causative agent of pest, induces significant cytotoxicity. The type three secretion system (T3SS) is a host sensor and effector system, which can be found in several Gram-negative bacteria. This system promotes the virulence of the bacteria by injecting effector proteins into the cytosol (Ashida et al., 2011). Structurally, TT3S consists of a base embedded in the bacterial membrane, the needle, which directly interacts with the host membrane, and the rod, which connects the base to the needle. Yersinia outer protein J (YopJ), an effector protein of the T3SS, rapidly kills macrophages in a caspase-8 dependent way, which occurs independent of the canonical inducers of extrinsic apoptosis. In this model, TLR4 stimulation by the pathogen activates the TRIF/RIPK1/caspase-8 axis, which leads to caspase-1 cleavage and cell death. Consistently, mice lacking caspase-8 and RIPK3 are more susceptible to the infection, pinpointing the importance of caspase-8 and RIPK driven signaling in tackling this disease (Gröbner et al., 2007; Weng et al., 2014). *Mycobacterium (M.) tuberculosis* triggers caspase-8 dependent apoptosis in human monocytes, which is induced by extracellular bacterial RNA fragments detected by TLR3 (Obregón-Henao et al., 2012). The polymorphic GC-rich repetitive sequence containing PE_PGRS33, a surface exposed protein and the 19-kDa glycolipoprotein (p19) of the *M. tuberculosis* can both engage TLR2 and initiate apoptosis signal-regulating kinase-1 (ASK1) driven TNF and TNFR expression in mouse macrophages (Basu et al., 2007; López et al., 2003). Finally, group B streptococcus, a pathogen causing neonatal meningitis, induces apoptosis via the activation of TLR2, which engages its adaptor protein myeloma differentiation primary response-88 (MyD88), and consequently results in caspase-8 activation in microglial cells (Lehnardt et al., 2007).

2.3.3 Role of caspase-2 in bacterial infection

Caspase-2 is unique among caspases. It exhibits initiator characteristics structurally, since it has an N-terminal caspase activation and recruitment domain (CARD). In addition, caspase-2 represents executioner features functionally, because it can selectively cleave substrates similar to that of caspase-3 or -7 (Olsson et al., 2015). Caspase-2, along with caspase-1 plays a significant role in *Brucella abortus* and *Brucella suis* induced cell death (Bronner et al., 2013; Chen et al., 2011). One possible mechanism of activation is derived from the studies on bacterial pore forming toxins (PFT), alpha toxin and aerolysin secreted by the human pathogens *Staphylococcus aureus* and *Pseudomonas aeruginosa*, respectively. Based on this study, caspase-2 activation can be triggered by the PFT driven potassium efflux in epithelial cells, which facilitates the assembly of a caspase-2 containing high molecular weight protein complex (Imre et al., 2012). Importantly, the low potassium concentration triggered activation of caspases is not without precedent in the literature. Though, the role of potassium efflux has been demonstrated in terms of the pro-inflammatory caspase-1 activation in immune cells (see pyroptosis), it is not clear whether caspase-2 activation takes place in an inflammasome-like complex in response to pathogen insults in epithelial cells.

2.3.4 Anti-apoptotic strategies of the bacteria

Several pathogenic bacteria developed strategies to evade the apoptotic demolition. *Francisella tularensis*, a highly virulent Gram-negative pathogen inhibits apoptosis by retaining mitochondrial integrity via the inhibition of BID processing and BAX translocation, and successfully extends the lifespan of the infected neutrophils (McCracken et al., 2016). *Helicobacter pylori* leads to a persistent, yet asymptomatic infection, which, however, can increase the incidence of the chronic gastritis and the gastric adenocarcinoma. This pathogen inhibits the anti-bacterial and cell death responses simultaneously by enhancing the level of ubiquitin-editing enzyme A20, which in turn bi-functionally blocks both the NF-κB activity and the caspase-8 activity (Lim et al., 2017). Enteropathogenic *Escherichia coli* (EPEC) transports effector proteins directly into the cytosol of infected cells by utilizing the T3SS. One of these effector proteins is the non-locus of enterocyte effacement encoded effector B1 (NLeB1). NleB1 transfers an *N*-acetyl-glucosamine residue to Arg117 in the death domain of FADD and inhibits the FADD dependent caspase-8 activation (Pearson et al., 2013). *Chlamydia trachomatis* is an obligate intracellular pathogen with medical relevance, which interferes with apoptosis by blocking the internalization of the cell death receptor TNFR1. This mechanism renders the specific blockage of the apoptotic signaling but preserves the function of the NF-κB signaling of the TNFR1 (Waguia Kontchou et al., 2016).

3. Role of necroptosis in bacterial and viral infection

3.1 Necroptosis signaling

In the previous chapter, we have seen several examples for apoptosis being induced in response to PAMPs. Nevertheless, apoptosis in many circumstances does not support the immune stimulation, which on the long run assists the pathogen to evade the counterattacks of the immune system. Necroptosis represents an immunogenic cell death modality. This non-apoptotic programmed cell death form, that morphologically resembles necrosis, has gained attention in the past 2 decades (Degterev et al., 2005; Vercammen et al., 1998), and its relevance in infection has been recognized only recently (Pan et al., 2014). In contrast to apoptosis, necroptosis is characterized by swelling of the cytoplasm, osmotic perturbations and the early rupture of the cytoplasm membrane, resulting in the release of DAMPs into the extracellular space. Receptor interacting protein kinase-1 and -3 (RIPK1 and RIPK3) play indispensable role in conducting necroptosis signaling. RIPKs interact via their RIPK homotypic interaction motif (RHIM) domains. The most studied pathway, by which RIPK dependent necroptosis can be triggered is the TNFR1 induced pathway. TNFR1 activation primarily results in NF-κB dependent pro-proliferative response. Upon the ligation of TNFR1 the formation of the so-called complex-I takes place, consisting of the adaptor proteins TRADD, TRAF2, cIAP and ubiquitinated RIPK1 (Fig. 2) (Wilson et al., 2009). Once, the pro-proliferative signal is blocked,

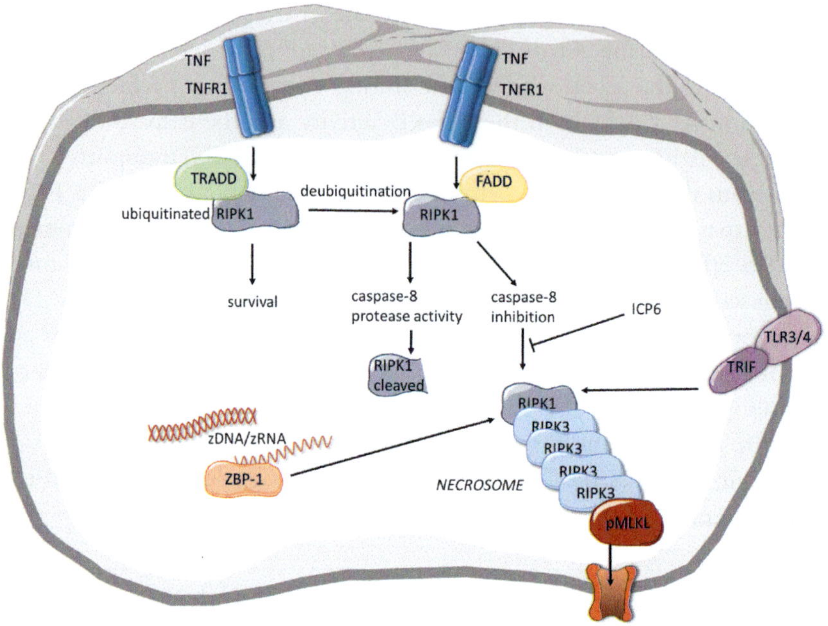

Fig. 2 Necroptosis signaling pathways. Necroptosis can be triggered by TNF ligation (top). RIPK1 deubiquitination promotes the cytotoxic function of RIPK1. Active caspase-8 cleaves RIPK1. Inhibition of caspase-8 leads to RIPK1 dependent assembly of the RIPK3 and MLKL containing necrosome. Phosphorylation of the MLKL in the necrosome results in MLKL oligomerization and its translocation into the cytoplasm membrane. TLR3 and 4 ligation (right) recruits the adaptor protein TRIF, which in turn activates RIPK3 and triggers the oligomerization of the MLKL. ZBP1 detects z-DNA and z-RNA fragments, subsequently leading to the binding of the RIPK3 and to the oligomerization of the MLKL.

the FADD containing apoptotic signal takes over, leading to caspase-8 activation. Surprisingly, the inhibition of the TNFR induced apoptosis by caspase inhibitors results in a necrotic-like cell demise. This programmed necrotic cell death, also termed necroptosis can be inhibited by RIPK1 inhibitor necrostatin-1 (Xie et al., 2013). It is also acknowledged that the proteolytic activity of caspase-8 acts against the necroptotic activity of three key components of the pathway: RIPK1, RIPK3 and cylindromatosis (CYLD) (O'Donnell et al., 2011; Tenev et al., 2011), hence inhibition of caspase-8 activity is a prerequisite for the induction of the TNFR1 driven necroptosis (Fig. 2). Once caspase-8 activity is blocked, RIPK1 is enabled to recruit RIPK3. Consequently, RIPK3 activation can lead to the phosphorylation of the necroptotic effector mixed lineage kinase-like protein (MLKL) (Fig. 2) (Sun et al., 2012). Upon the phosphorylation driven conformational changes, MLKL oligomerizes and forms pores throughout the cytoplasm membrane, resulting in membrane integrity loss. In the past decade, it became clear that RIPK1 is a multifaceted scaffold protein. If ubiquitinated, RIPK1 inhibits the caspase-8 driven apoptosis and the RIPK3/MLKL dependent necroptosis. However, as a deubiquitinated protein, RIPK1 can promote the stimulation of both cell death signaling pathways (Varfolomeev and Vucic, 2018). Apart from the decisive role of ubiquitination, phosphorylation at Ser89 by protein kinase A, C, and phosphorylation at several sites by IKKs have been found to be important to attenuate the cytotoxic potential of this signaling protein (Dondelinger et al., 2019; McQuade et al., 2013).

3.2 Necroptosis modulation in response to viral infections

3.2.1 Virus induced necroptosis via extracellular factors

The current observations indicate that the majority of the human pathogen viruses developed strategies to avoid the necroptotic signaling of the host. Conversely, human immunodeficiency virus type 1 (HIV-1) induces TNFα dependent necroptotic cell death in CD4+ T lymphocytes, which can be inhibited by necrostatin-1. Unlike apoptosis, this cell death occurs in infected cells and spares the bystander cells (Pan et al., 2014). Nevertheless, the HIV triggered necroptosis is considered detrimental for the host by targeting the defensive mechanisms of the immune system. Another example, where necroptosis is triggered by viruses, takes places in response to human coronavirus (HCoV) infection. HCoV is a neuroinvasive respiratory pathogen, which induces a cell death pathway dependent on RIPK1 and MLKL in human neuroblastoma cells (Meessen-Pinard et al., 2017).

3.2.2 Necroptosis via ZBP1

The protein z-DNA/RNA binding protein-1 (ZBP1; also known as DAI) recognizes z-RNA or z-DNA structures, which are common as a consequence of a rapid viral RNA and DNA synthesis, but distinct from the B structure of the eukaryotic RNA or DNA. ZBP1 can activate RIPK3 (Upton et al., 2012), since it is capable to interact the RIPKs via its RHIM (Fig. 2). Intriguingly, MEFs, liver cells, lymphoid cells and skin cells isolated from mice exhibiting catalytically inactive RIPK1 die early and exert RIPK3 auto-phosphorylation, a hallmark of the necroptotic activation, whereas RIPK3 and MLKL deficient mice are protected from the cell death ex vivo and in vivo. Furthermore, the ZBP1-RIPK3 interaction can be blocked in macrophages from wild type mice, whereas the interaction takes place in mouse macrophages that express a RHIM deficient RIPK1.

These results indicate that RIPK1 acts as a negative regulator of ZPB1-RIPK3 interaction during development in order to prevent premature necroptosis (Newton et al., 2016). To date, IAV infection is the only example, where ZBP1 dependent necroptosis occurs in response to viral infections employing non-engineered (wild type) virus strains. IAV induces RIPK3 and MLKL dependent, but TNFα and TLR independent necroptosis (Nogusa et al., 2016). This pathway is mediated through ZBP1 by sensing viral genomic RNA segments (Thapa et al., 2016). A recent study (Wang et al., 2019b) performed in mice further confirms the role of ZBP-1 as the key necroptosis modulator protein in response to IAV infection in vivo. Aside from the necroptosis stimulatory effects of the viral RNA fragments, the NS1 protein of the IAV is suggested as a direct interactor of MLKL (Gaba et al., 2019). Finally, the transcriptional regulation of ZBP1 seems to be dependent on the interferon regulatory factor 1 (IRF1), which is activated in mouse bone marrow-derived macrophages in response to IAV infection (Kuriakose et al., 2018). Consequently, depletion of IRF1 can reduce the level of ZBP1 and diminish the IAV induced necroptotic features, including MLKL phosphorylation and membrane integrity loss.

Murine cytomegalovirus (MCMV) lacking the viral M45-encoded inhibitor of RIPK (vIRA) triggers premature cell death in mice. Importantly, RIPK3 knockout can rescue the premature death phenotype, indicating that the MCMV infection induces a RIPK3 dependent necroptotic cell death (Upton et al., 2010). Similar to IAV infection, the RIPK3 driven necroptosis is independent of the effect of RIPK1 and the TNF stimulation, but ZBP1 is activated via its putative RNA binding domain Zα (Sridharan et al., 2017). Analogously, the Vaccinia virus (VACV) encoded innate immune evasion protein-3 (E3) can directly interact with the Zα domain of the ZBP1. Along with this, E3 deleted VACV strain triggers immediate necroptosis in interferon treated mouse L292 and human embryonal kidney (HEK293T) cells. The cytotoxic effect of the engineered VACV strain is attributed to necroptotic signaling, because the pathogenicity of the virus can be restored in vivo by the depletion of either RIPK3 or ZBP1 in mice (Koehler et al., 2017).

3.2.3 Species-specific dual role of the viral ICP6 in necroptosis regulation

The large subunit of the viral ribonucleotide reductase ICP6 (Fig. 2) is capable to engage both RIPK1 and RIPK3 through its RHIM-like domain upon herpes simplex virus type 1 (HSV-1) infection and to result in necroptosis, which also contributes to the limited viral propagation in mice. Intriguingly, ICP6 inhibits the necroptosis signaling in its natural host, in human cells. This observation points out the species-specific differences in cell death signaling pathways and indicates that viruses in their natural hosts evolved to overcome host defense mechanisms (Huang et al., 2015). HSV-1 induces necroptosis not only via its ICP6 protein, but also by triggering the ZBP1 pathway in mice. Conversely, in order to induce the necroptotic pathway in human cells, the expression of ZBP1 has to be first enhanced (Guo et al., 2018). Moreover, ICP10, the virally encoded large subunit of the ribonucleotide reductase of HSV-2, can efficiently prevent necroptosis that is induced by TNFα (Guo et al., 2015). One possible explanation for the difference between mice and human responses might be provided by a recent study on HSV-1 demonstrating that ICP6 blocks TNFR1 induced MLKL necrosome assembly in membrane vesicles (Ali et al., 2019). This compartmentalization process of the necrosome represents a critical step for the progression of necroptosis in human cells. Consistent with this, RHIM domain deficient RIPK3 can efficiently phosphorylate MLKL but does not lead to necroptosis in human cells. Thus, one can speculate that ICP6 can trigger necrosome formation in mice, because the compartmentalization step may not be required for the effector mechanisms of necroptosis. Additionally, inositol phosphate (IP) kinases largely contribute to the oligomerization of MLKL. IP kinase mutants are unable to achieve the MLKL oligomerization and membrane localization despite a functioning RIPK3 in response to a genetically engineered ICP-6 mutant HSV-1 infection (Dovey et al., 2018). This study suggests that the RIPK3 dependent phosphorylation at the C-terminal Ser358 and the subsequent events, such as oligomerization and membrane localization are separately regulated in human cells. Accordingly, in order to accomplish necroptosis, MLKL activation requires a highly phosphorylated IP product to bind to the N-terminal domain. It is currently not clear what the exact role of the highly phosphorylated product can be, nevertheless the stabilization of the oligomerized MLKL complex can serve as one plausible explanation.

3.2.4 Viral inhibitors of necroptosis

The latent membrane protein-1 (LMP1) of the γ-herpes virus Epstein-Barr virus (EBV) is considered as the major transforming factor of the pathogen. Initially, this transmembrane protein has been reported to exert anti-apoptotic effects in B cells (Henderson et al., 1991). Later, LMP1 has been associated with necroptosis inhibition in EBV-infected human nasopharyngeal epithelial cells and nasopharyngeal carcinoma cells. LMP1 can interact both RIPKs and can successfully mediate the polyubiquitination of them, which in turn leads to the inhibition of the kinase function of the RIPK3 and activates the pro-survival pathway (Liu et al., 2018). As previously discussed, the vIRA protein of MCMV can efficiently block RIPKs and hamper the necroptotic signaling. Though, distinct from the necroptosis inhibitory mechanism of the vIRA protein, the human CMV has also developed a strategy to suppress necroptosis upstream of MLKL phosphorylation (Omoto et al., 2015). Furthermore, Rhinoviruses, the causative agents of common cold can actively suppress both apoptosis and necroptosis. The viral 3C protease directly interacts and cleaves RIPK1 and blocks necroptosis stimulated by Poly(I:C) in human airway epithelial cells (Lötzerich et al., 2018). It is therefore somewhat surprising that the viral protease still triggers a lytic cell death pathway, which is distinct from necroptosis and apoptosis.

3.3 The role of necroptosis in bacterial infection

Whereas the relevance of necroptosis initiation in viral infection has been proven only in few cases, for instance in response to IAV, the connection between necroptosis and bacterial infection seems to be more obvious. The significance of type I interferons in initiating necroptosis is demonstrated during *Salmonella typhimurium* infection. Mice deficient in the receptor for type I interferons (Ifnar1) exhibit improved survival, and macrophages derived from these mice exert resistance to the bacteria triggered cell death. Consistently, RIPK3 deficient macrophages can similarly evade the cell death upon infection. This study gives an example for a necroptosis induction, which is detrimental for the host and beneficial for the bacteria (Robinson et al., 2012). Similarly, *Staphylococcus (S.) aureus* infection induced cytotoxicity can be prevented by RIPK3 depletion or inhibition of RIPK1 and MLKL in primary human macrophages. Surprisingly, necroptosis induction by pore forming toxins (PFT) of *S. aureus* occurs without employing any external caspase inhibition, which suggests that the caspase-8 signal remains inactive throughout the whole infection process (Kitur et al., 2015). In concordance with this, several PFT producing bacterial strains, including *Streptococcus pneumoniae* and *Staphylococcus marcescens* induce necroptotic cell death in mouse macrophages, which can be arrested by necroptosis inhibitors (e.g., necrostatins) but not by the depletion/inhibition of caspases (González-Juarbe et al., 2015). Surprisingly, a PFT secreted by *S. marcescens* can trigger MLKL dependent necroptosis in mouse lung epithelial cells in the absence of TNFR1/2 or

TLR4 signaling (González-Juarbe et al., 2017). Potassium efflux, as a key event, has been previously studied in necroptosis initiation. However, how this osmotic imbalance orchestrates the necroptotic process still needs to be elucidated. An in vivo study describes the PFT driven activation of caspase-2, -4 and -10, which facilitates the release of alarmins to the extracellular space (Gonzalez-Juarbe et al., 2018). *Mycobacterium tuberculosis* infection induces NAD+ depletion triggered necroptosis via the RIPK3-MLKL axis, which is initiated by the tuberculosis necrotizing toxin (TNT) in human macrophages (Pajuelo et al., 2018). In contrast to the observation in human macrophages, the pathogenesis of *Mycobacterium tuberculosis* infection is independent from the role of RIPK3 and necroptosis in mice (Stutz et al., 2018).

Yersinia pestis outer protein J (YopJ) is required for the lymphatic spread of the bacteria, which leads to the formation of swollen and necrotic structures called buboes. YopJ suppresses the downregulators of the RIPK1 driven necroptosis, resulting in necroptotic spread of the pathogen. In concordance with this, necrostatin treatment efficiently protects the mice from lethal infections of the bacteria (Arifuzzaman et al., 2018). It is demonstrated that YopJ interferes with the pro-survival activity of the FLIPs in the DISC, thereby promoting the necroptotic feature of the RIPKs.

In contrast to the previous examples, some bacteria have developed countermeasures to inhibit the necroptotic signals of the host. The enterophatogenic *E. coli* (EPEC) directly interferes with the elements of the necroptotic signal transduction. Espl, an effector component of the EPEC T3SS is identified as a protease, that can target and cleave host proteins containing a RHIM domain, including RIPK1/3, TRIF and ZBP1 (Pearson et al., 2017). Further example for necroptosis inhibition is demonstrated in a study employing the Salmonella induced colitis model. The salmonella outer protein B (SopB) attenuates the necroptotic cell death by decreasing the level of MLKL phosphorylation in the cells (Hu et al., 2019).

3.3.1 Necroptosis via TLR signaling

TLR-3 and TLR-4 are PAMP recognition receptors and engage dsRNA and bacterial lipopolysaccharide (LPS), respectively, which in turn initiate the recruitment of Toll/IL-1R (TIR) domain containing adapters to activate gene expression via transcription factors, such as NF-κB and IRF3/7. Over the last years, the role of TLR receptors in mediating cell death pathways has been intensively studied. Induction of TLRs can induce caspase-8 activation (Kaiser and Offermann, 2005) and the inhibition of caspase-8 activity leads to a RIPK3 dependent necrotic-like cell death in mouse macrophages (He et al., 2011). The adaptor protein TRIF is necessary to transduce signals from TLR3 stimulation, whereas activation of TLR4 can engage both TRIF and MyD88. These pathways proceed independent of RIPK1 in mouse fibroblasts, but can efficiently trigger MLKL activation downstream of RIPK3 (Kaiser et al., 2013) (Fig. 2). In response to TNFR ligation, the ubiquitination status of RIPK1 is defined as the key factor in the decision making between survival and death. Upon TNF stimulation, the activated caspase-8 cleaves CYLD, the deubiquitinase enzyme of RIPK1 (O'Donnell et al., 2011). The LPS induced TLR-4 ligation recruits FADD and caspase-8 in a complex and initiates the cleavage of CYLD (Legarda et al., 2016). The non-histone chromatin binding protein high mobility group box-1 (HMGB1) is released into the extracellular space (Lu et al., 2012) during necroptosis and functions as a pro-inflammatory DAMP (Scaffidi et al., 2002). Interestingly, HMGB1 can form a complex with LPS upon infection, which leads to an amplified activation of TRIF dependent immune responses (Meng et al., 2019).

4. Role of pyroptosis in bacterial and viral infections

4.1 The pyroptotic machinery

Pyroptosis is an inflammatory and regulated cell death modality (Fink and Cookson, 2006) that partially shares features with apoptosis, because it requires caspase activation. On the other hand, the pyroptotic membrane rupture occurs rapidly in the affected cells, which is a common feature shared with necroptosis. Pyroptosis has been observed in monocytes, macrophages, dendritic cells, lymphocytes, epithelial cells, and endothelial cells (Man et al., 2017). Contrary to apoptosis, induction of pyroptosis activates inflammatory caspases, which, similarly to apoptotic initiator caspases, possess a large prodomain structure. The inflammatory caspase-1, caspase-4/5 in humans and caspase-11 in mice are being activated in large molecular weight complexes, termed inflammasomes (van Opdenbosch and Lamkanfi, 2019). In general, inflammasomes consist of sensors that can recognize pathogenic insults, also termed PAMPs. Depending on the nature of a PAMP signal, different sensors are triggered. The nucleotide binding oligomerization domain (NOD)-like receptors represent the best studied inflammasome components, that recognize PAMPs (Fig. 3), but various other PAMP receptors have also been identified in the last years. These pathogen sensors can recruit caspases either directly or via an adaptor molecule. Once, the inflammatory caspase-1 is activated, it can process interleukin (IL) 1β and 18 into their mature forms.

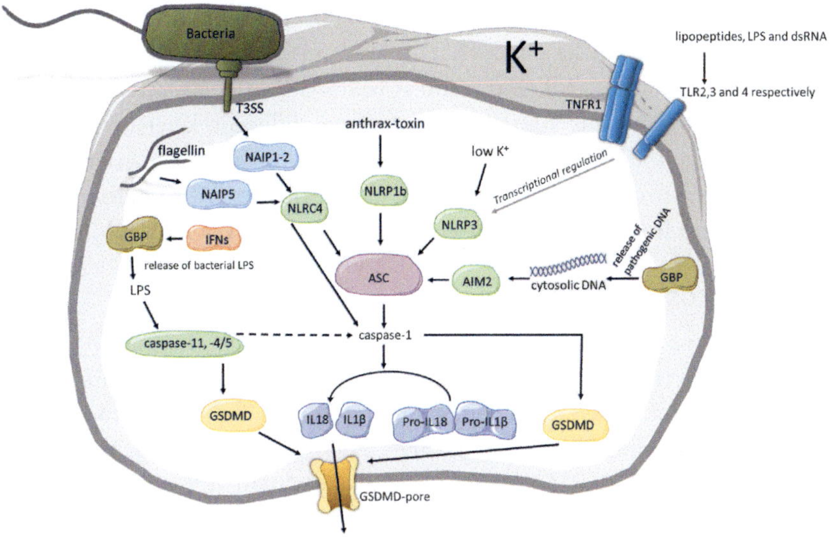

Fig. 3 Inflammasome activation and pyroptosis signaling. Components of the bacterial T3SS and bacterial flagellins can activate the NAIPs. In turn, NAIPs stimulate the NLRC-4 inflammasome (top left), leading to caspase-1 activation. Anthrax lethal toxin (top middle) directly interacts with NLRP1b and triggers the recruitment of ASC. Intracellular perturbations culminating in potassium efflux stimulate the assembly of the NLRP3 inflammasome (top right), leading to the recruitment of the adaptor protein ASC. NLRP3 requires priming signals via the ligation of TNF or TLRs. The IFN-stimulated GBPs facilitate the release of the pathogenic DNA fragments into the cytosol, which in turn can stimulate the AIM2 inflammasome. Caspase-11 and caspase-4/5 (left) can directly detect LPS, released by the GBPs into the cytosol. Active caspase-1 can process pro-IL1β/IL18 and cleave GSDMD, subsequently leading to GSDMD pore formation and release of the ILs into the extracellular space.

Furthermore, both caspase-1 and caspase-11 can process gasdermin-D (GSDMD), a member of the less studied gasdermin protein family (Kayagaki et al., 2015). The N-terminal leaflet of GSDMD inserts into the lipid bilayer and forms oligomer pores with an inner diameter of approximately 20 nm (Sborgi et al., 2016), leading to IL-1β/-18 release (He et al., 2015) and to osmotic imbalance driven swelling of the cells (Fig. 3). Several studies conclude the role of GSDMD as a key factor for IL secretion. In this setup, the pyroptotic membrane integrity loss, the swelling and the cell death are interpreted as the collateral damage of the pore formation. In fact, GSDMD driven IL release can proceed without any cell destruction (Evavold et al., 2018). In contrast, a recent study on mice infected with *Burkholderia thailandensis* leads to the conclusion that the action of GSDMD on bacterial killing occurs in a pyroptosis dependent manner and is entirely independent from an indirect effect of IL production (Wang et al., 2019a). Interestingly, GSDMD is also capable to insert into the bacterial membrane and to kill extracellular bacteria (Liu et al., 2016). Thus, it seems that GSDMD is represented at three different layers of defense depending on the localization of the pathogen. Besides, gasdermin E, another member of the gasdermin family, is also reported to participate in pyroptosis and in post-apoptotic secondary necrosis (Rogers et al., 2017).

4.2 Pyroptosis in virus infection

Whereas, virus triggered inflammasome activation is a common event, inflammasome induced pyroptosis cannot be observed in all circumstances. Nevertheless, inflammasome dependent pyroptosis has been reported in response to infection by several viral pathogens, including HIV, DV (Tan and Chu, 2013), IAV (Kuriakose et al., 2016), Coxsacivirus (CV) (Wang et al., 2018), Parvovirus (Deng et al., 2017), hepatitis C virus (Kofahi et al., 2016), and Enterovirus 71 (EV71) (Huang et al., 2015; Zhu et al., 2018). An in vivo study on attenuated Rabies strain provides evidence for the beneficial role of pyroptosis against viral infections. The infection with the attenuated Rabies virus strain, which normally leads to the onset of mild symptoms, can be aggravated by caspase-1/-11 depletion, but not by caspase-3/7 or by IL-1β/IL-18 depletion (Kip et al., 2017). These observations suggest that the pyroptosis can be necessary to limit the Rabies infection. Further argument for the anti-viral nature of pyroptosis is provided by the observation that some viruses developed strategies in order to inhibit the effector mechanisms of this lytic cell death modality. EV71 triggers GSDMD by the viral protease 3C, and the cleaved fragments fail to unleash pyroptosis in THP-1 macrophages, HEK293T embryonal kidney and rhabdomyosarcoma cell lines (Lei et al., 2017).

The nature of the viral PAMPs determines which inflammasome mediated pyroptotic pathway is activated in response to a specific stimulus. In accordance with this, viral nucleotide patterns can lead to the activation of absent in melanoma-2 (AIM2) and interferon-γ-inducible protein 16 (IFI16) inflammasome activation, whereas various virally triggered stress signals can result in the assembly of the NOD-like receptor (NLR) family pyrin domain containing 3 (NLRP3) inflammasome (Fig. 3).

4.2.1 The role of NLRP3 inflammasome

NLRP3 protein is the major component of the NLRP3 inflammasome, which is the most studied inflammasome to date. NLRP3 consist of a pyrin domain, a nucleotide binding site, and a leucine-rich repeat motif. Upon activation by PAMPs, NLRP3 interacts with the apoptosis-associated speck-like protein (ASC) via its pyrin domain (Fig. 3). This in turn induces ASC polymerization and the assembly of the ASC specks. The adaptor protein ASC possesses a caspase recruitment CARD domain and is therefore capable to recruit pro-inflammatory caspases, such as caspase-1 (Martinon, 2008). Consequently, caspase-1 is dimerized and auto-processed based on the principle of the induced proximity.

Several PAMP signals have been demonstrated to culminate in NLRP3 activation. For instance, lytic cell death induced potassium efflux is proposed as the major mechanism of the NLRP3 inflammasome activation following Encephalomyocarditis virus (EMCV) and VSV infection (da Costa et al., 2019). In these infections, the lytic cell death precedes the inflammasome formation, whereas the infection with IAV results in rapid NLRP3 dependent pyroptosis in dendritic cells (Fernandez et al., 2016; Kuriakose et al., 2016), indicating that other mechanism can also trigger NLRP3 assembly. It is of note, that IAV infection results in pyroptosis with a delayed kinetic in respiratory epithelial cells (Lee et al., 2018). In the activation of NLRP3 inflammasome in DV infection, the cytoplasmic membrane surface C-type lectin 5A (CLEC5A) is found to be critical, which interacts directly the DV virion and serves as a pathogen sensing receptor in human inflammatory macrophages (Wu et al., 2013). A highly pathogenic avian influenza A (H5N1) results in pneumonia and acute respiratory distress syndrome in humans. The depletion of galectin-B, a b-galactosidase binding protein leads to a striking reduction of inflammation in lungs of the infected mice compared to that of the wild type mice. Consequently, co-immunoprecipitation experiments demonstrate the interaction of galectin-B, NLRP3 and ASC in bone marrow-derived macrophages upon infection (Chen et al., 2018). The gp120 envelop protein of the HIV is the major contributing factor of the HIV-associated neurocognitive disorder accompanied by neuroinflammation. The gp120 triggered neuropathy is NLRP3 dependent and the administration of a selective NLRP3 inhibitor can alleviate the neuronal death and the impaired cognitive functions in gp120 transgenic mice. In a recent study, the role of the Purinergic 2X4 (P2X4), a cationic ion channel involved in neuropathic pain, is shown to mediate the gp120 dependent inflammasome activation and the caspase-1 cleavage in glial cells (Zhao et al., 2019). Most of these studies indicate the role of osmotic imbalances, such as potassium efflux as one of the major causative of NLRP3 inflammasome stimulation, however, how the low potassium concentration renders changes in the NLRP3 structure and whether this is a direct or an indirect effect still needs to be clarified.

4.2.2 The role of IFI16 in the HIV induced T lymphocyte pyroptosis

HIV triggers bystander pyroptosis in non-infected T lymphocytes (Doitsh et al., 2014). The bystander CD4 T cells, that are not permissive to HIV infection, play a crucial role in the acquired immunodeficiency syndrome (AIDS). These cells go through an incomplete virus replication cycle, resulting in aberrant transcripts of the viral genome. These transcripts have been demonstrated to stimulate the DNA sensor interferon-γ-inducible protein 16 (IFI16), leading to caspase-1 activation mediated pyroptosis (Monroe et al., 2014) (Fig. 3). The immune stimulatory dying of the T lymphocytes initiates a pathogenic positive feedback loop by recruiting more immune cells to the site of infections, which leads to even more accelerated propagation of the virus. Cell free virions are unable to trigger pyroptosis. Thus, a cell to cell transmission of the virions is required for the induction of this lytic inflammatory cell death (Galloway et al., 2015). A clinical antiretroviral therapy study in humans demonstrates that the non-responding individuals have increased level of NLRP3 and caspase-1. In addition, the non-responding group exhibits significantly lower levels of CD4 T lymphocytes in comparison to that of the responder patients. Though, this patient data supports the previous in vitro and mouse experiments, which suggest the relevance of CD4 lymphocyte loss in the pathogenesis of HIV infection (Bandera et al., 2018), it does not provide direct evidence for an IFI16 driven pyroptotic cell death.

4.2.3 AIM2 inflammasome

AIM2 has been shown to bind DNA and engage the caspase-1-activating adaptor protein ASC to form an inflammasome. AIM2 conducts the maturation of IL-1β/IL-18 and induces pyroptosis in response to exposure of synthetic double-stranded DNA (Fig. 3). In natural situations, AIM2 dependent pyroptosis is observed in response to MCMV, Vaccinia virus (Rathinam et al., 2010) and EV71 infections (Yogarajah et al., 2017). Zika virus (ZIKV), an RNA virus, has gained increased attention over the last years due to its association with severe birth defects, including microcephaly. Infection with this flavivirus triggers higher expression levels of NLRP1, 3, AIM2 and caspase-1 in brain tissues collected from fatal human cases (de Sousa et al., 2018). It is, however, not clear whether the accompanied brain tissue damage is the direct consequence of the GSDMD driven pyroptosis or the detected higher inflammasome expression levels represent a completely unrelated phenomenon.

4.2.4 Alternative inflammasomes in virus infection

Virus induced pyroptosis can be triggered cell type specifically for instance in Rotavirus infection, a leading cause of gastroenteritis and diarrhea in small children. NLRP9 is a member of the NLR family that is specifically expressed in intestinal epithelial cells in the site of the Rotavirus infection. NLRP9 can recognize dsRNA fragments and form inflammasome with ASC to initiate caspase-1 cleavage and GSDMD driven pyroptosis (Zhu et al., 2017). Moreover, Caspase-4, the human homolog of the mouse caspase-11 has been reported to function as a PAMP receptor. In response to DV infection, caspase-4 is activated and can process caspase-1 in order to accomplish pyroptosis (Cheung et al., 2018).

4.3 Pyroptosis in response to bacterial infection

The first report describing infection driven pyroptosis was performed in *Salmonella enterica serovar Typhimurium* (also called *Salmonella typhimurium*) infected mouse macrophages (Fink and Cookson, 2006), although caspase-1 dependent cell death in response to bacterial infection was mentioned in the literature earlier (Chen et al., 1996; Monack et al., 2001). The detection of PAMPs in bacterial infections includes the recognition of bacterial LPS, flagellins, components of the bacterial T3SS and the detection of cytoplasmic bacterial DNA.

4.3.1 Detection of bacterial proteins by the NLRC4 inflammasome

The components of the bacterial T3SS trigger inflammasome formation and pyroptotic cell death in Shigella infected macrophages through the activation of the NLR family CARD domain containing protein 4 (NLRC4) (Fig. 3). NLRC4 possess a CARD domain, therefore this NLR family member can directly engage and process caspase-1 (Suzuki et al., 2007). Besides, flagellins of various bacterial strains, including *Listeria (L.) monocytogenes* (Cervantes et al., 2008; Warren et al., 2008), *Legionella pneumophila* and *Burkholderia thailandensis* (Miao et al., 2010) can lead to the assembly of the NLRC4 inflammasome (Case et al., 2009). The activation of NLRC4 inflammasome has been shown to lead to bacterial clearance independent from the effect of IL-1 and IL-18 production, since this process functioned efficiently in IL-1/ IL-18 knockout mice (Miao et al., 2010), thus implicating the primary role of pyroptosis in the bacterial clearance. This process involves the pyroptotic release of the intracellular pathogens from the macrophages and leads to the uptake of the bacteria by neutrophil granulocytes. In contrast, pyroptosis can be also detrimental for the host organisms. The extracellular opportunistic pathogen *Pseudomonas aeruginosa* infection in mice leads to NLRC4 inflammasome activation, pyroptosis and increased mortality, whereas depletion of NLRC4 results in improved bacterial clearance and increased survival (Cohen and Prince, 2013).

The NLRC4 inflammasome utilizes various sensors in order to recognize PAMPs. The human NLR family apoptosis inhibitory protein 2 (NAIP2) and the mouse counterpart Naip5 prevent apoptosis by inhibiting caspases. The flagellin of the *Legionella pneumophila* can result in caspase-1 activation and cell death in a NAIP2 dependent manner in macrophages (Fig. 3) (Katagiri et al., 2012). Importantly, not all bacterial strains express flagellins constantly. Nevertheless, the inner rod of the T3SS needle complex of the non-flagellated Shigella bacteria results in the engagement of mouse Naip2 and leads to the NLRC4 inflammasome mediated pyroptosis in primary mouse macrophages (Suzuki et al., 2014). The hNAIP and the mouse counterpart Naip1 (Fig. 3) can recognize the needle component of the Salmonella T3SS in dendritic cells, and to less extent in macrophages (Yang et al., 2013). NLRC4 activation is mediated by NAIP independent phosphorylation events, which suggest a biphasic model for the stimulation of this intracellular pathogen sensor (Matusiak et al., 2015; Qu et al., 2012). The existence of a second stimulatory signal is further supported by the observations on the T3SS effector protein YopJ of the *Yersinia pestis*, which translocates into the cytoplasm upon infection and induces a rapid apoptosis in macrophages. In contrast, a preceding activation by LPS or by the ligation of the PRRs TLR2, 3 and 4 leads to a YopJ independent pyroptosis (Bergsbaken and Cookson, 2007). Furthermore, interferon regulatory factor 8 (IRF8) is required for the activation of the NLRC4 inflammasome in bone marrow-derived macrophages infected with *Salmonella typhimurium*, *Burkholderia thailandensis* and *Pseudomonas aeruginosa*. In these infections, IRF8 conducts the transcription of the NAIPs (Karki et al., 2018). These results implicate that the activation of NLRC4 relies on a second PAMP signal, which then initiate transcriptional priming and can mediate further stimulatory phosphorylation steps.

4.3.2 The NLRP3 inflammasome

NLRP3 indirectly senses pathogen associated signals such as extracellular ATP and uric acid crystals (Martinon et al., 2006), which then leads to a conformational change, resulting in the binding and the oligomerization of the adaptor molecule ASC. Bacterial PFTs, including the hemolysin of the bacteria *Vibrio cholerae*, or the M surface protein of the group A Streptococcus can efficiently induce potassium efflux, which has been defined as the unified trigger of the NLRP3 inflammasome activation (Muñoz-Planillo et al., 2013; Queen et al., 2015; Valderrama et al., 2017). Alternatively, disruption of the glycolysis, leading to mitochondrial ROS production is also proposed as the intracellular mechanism of the NLRP3 activation (Sanman et al., 2016). Besides, a TLR triggered second priming event by ubiquitination is necessary to stimulate the NLRP3 inflammasome in mice myeloid cells (Humphries et al., 2018).

4.3.3 NLRP1B and the anthrax lethal toxin

Anthrax lethal toxin (LT), the critical virulence factor of the *Bacillus anthracis* induces caspase-1 activation and lytic cell death through the cytosolic NLR family member NLRP1B in macrophages (Boyden and Dietrich, 2006) and dendritic cells (Fink et al., 2008) (Fig. 3). In humans, NLRP1B possess a pyrin domain and lacks the CARD domain, therefore ASC is required as an adaptor protein to activate caspase-1. It has been demonstrated that LT was transported to the cytosol, where it could interact with additional factors, however the question whether NLRP1B directly interacted the LT remained elusive. This question was answered few years later in a study where the direct proteolytic cleavage of NLRP1B could activate NLRP1B driven pyroptosis in mice and rat cells (Chavarría-Smith and Vance, 2013).

4.3.4 The role of caspase-11 and caspase-4/-5

The mouse caspase-11 and its duplicated human homologs caspase-4 and -5 are often referred as non-canonical inflammasome, since they function as direct LPS sensors in the cytosol (Fig. 3). The CARD domain of the caspase-4 is capable to bind LPS molecules and, surprisingly, can directly cleave LPS micelles (An et al., 2019). The advantage of the caspase-11 and caspase-4/5 inflammasome can be understood in situations in which the pathogen attempts to evade caspase-1 activation, for instance by blocking the NLRC-4 inflammasome activation. It has been recently shown that the Shiga toxin from the pathogenic enterohemorrhagic *Escherichia coli* could lead to the activation of caspase-4 and result in the cleavage of GSDMD upstream of the NLRP3 activation in human THP-1 cells (Platnich et al., 2018). Furthermore, the Gram-negative *Burkholderia thailandensis* infection triggers the activation of caspase-11 in mice (Aachoui et al., 2013) and

Legionella (L.) pneumophila similarly elicits the activation of caspase-11, which in this scenario is dependent on the IFN signaling. Interestingly, loss of IFNs does not lead to the inhibition of caspase-11 expression, but it hampers the caspase-11 dependent pyroptosis in macrophages (Case et al., 2013). The requirement of IFNs leads us to the problem of the cytosolic LPS detection. The Gram-negative bacteria *Salmonella typhimurium* and *L. pneumophila* are normally located in vacuoles, hence they are protected from cytoplasmic pathogen sensors as for instance caspase-11. The guanylate-binding proteins (GBPs) represent a family of IFN-inducible GTPases. It has been demonstrated that IFN-stimulated GBP members were capable to bind the vacuoles and initiate the rupture of the membrane, which in turn resulted in the release of the bacterial LPS, exposing it to the cytoplasmic caspase-11 (Meunier et al., 2014; Rupper and Cardelli, 2008). Conversely, IFNs can also directly regulate the transcription of the pro-inflammatory caspases. Based on a new genome-wide screen, the interferon regulatory factor 2 (IRF2) seems to act as the main transcription factor directly regulating caspase-4 expression upon LPS stimulus in human monocytes (Benaoudia et al., 2019).

4.3.5 AIM2 inflammasome in bacterial infection: DNA sensing

The role of the AIM2 inflammasome in the recognition of the intracellular bacteria has been demonstrated in response to *Listeria monocytogenes*, *Francisella (F.) tularensis* (Kim et al., 2010; Rathinam et al., 2010) and *Brucella abortus* infection (Costa Franco et al., 2019). Since the recognition by AIM2 is fully dependent on the availability of cytosolic DNA, strains engineered to lyse in the cytosol can induce pyroptosis more efficiently (Sauer et al., 2010). Conversely, in response to a naturally-occurring infection by non-lysing strains, including *F. novicida* (Man et al., 2016), *F. tularensis* (Man et al., 2015) and *Chlamydia trachomatis* (Finethy et al., 2015) the cytosolic release of the bacterial DNA is facilitated by the TLR mediated GBPs, hence supporting the relevance of the IFN dependent priming events to activate this inflammasome.

4.3.6 Cross talk between inflammatory and apoptotic cell death pathways

A study on caspase-1 deficient murine macrophages implicates the existence of the cross talk between pyroptosis and apoptotic caspases. In the absence of caspase-1, the stimulation of the AIM2/ASC axis by *F. tularensis* leads to caspase-8 cleavage and apoptosis, which can restrict bacterial replication in vitro (Pierini et al., 2012). The mechanistic details of the activation, however, still need to be elucidated. Surprisingly, the loss of caspase-1 results in faster kinetics of the AIM-2 speck formation, indicating the presence of a caspase-1 dependent negative feedback loop (Juruj et al., 2013). We therefore speculate that an over-activation of the AIM2 speck formation may result in the loss of specificity, consequently leading to the recruitment of other initiator caspases into the complex. In fact, the recruitment of caspase-8 into the NLRP3-ASC inflammasome has been demonstrated in LPS and nigericin stimulated murine bone marrow-derived dendritic cells in the absence of caspase-1 and -11 (Antonopoulos et al., 2015). Furthermore, the Yersinia effector protein YopJ has been shown to inhibit TAK1 and IKK, which in turn led to RIPK1 and caspase-8 activation in macrophages. This and another group simultaneously published the finding that caspase-8 could directly cleave GSDMD, leading to pyroptotic cell death (Orning et al., 2018; Sarhan et al., 2018). Finally, in a third study the NLRC4 dependent assembly of the ASC caspase-8 inflammasome has been demonstrated in *Salmonella typhimurium* infected cells (Man et al., 2013).

5. Concluding remarks

In this review, we discussed the role of three major cell death pathways: apoptosis, necroptosis, and pyroptosis in the context of viral and bacterial infections. The past few years of research in the field of host-pathogen interaction demonstrates us that the connection between the pathogens and the host cell death signaling is more complex than previously expected. The complexity of the apoptotic, necroptotic, and pyroptotic signaling pathways, the presence of the cross talks and the feedback mechanisms reflect the significance of a continuously running evolutionary race between the pathogens and their natural hosts. This is best proven by the observations that successful human pathogens developed counter measures to attenuate the cell death induction potential of the host, whereas other pathogens developed strategies to hijack the cell death machinery in order to accomplish the targeted destruction of the defense mechanism of the host. Thus, looking it from an evolutionary perspective, the different layers of cell death pathways can be interpreted as different evolutionary steps built upon each other in response to a selective pathogen insult. The better understanding of these mechanisms can provide us novel strategies to more efficiently treat infectious diseases in the future.

Acknowledgments

The author is grateful to Madeleine Eichler for the graphical design of the figures presented in the manuscript and to Usman Nasrullah for the critical reading of the manuscript.

References

Aachoui, Y., Leaf, I.A., Hagar, J.A., Fontana, M.F., Campos, C.G., Zak, D.E., Tan, M.H., Cotter, P.A., Vance, R.E., Aderem, A., Miao, E.A., 2013. Caspase-11 protects against bacteria that escape the vacuole. Science (New York, NY) 339 (6122), 975–978.

Alexopoulou, L., Holt, A.C., Medzhitov, R., Flavell, R.A., 2001. Recognition of double-stranded RNA and activation of NF-kappaB by Toll-like receptor 3. Nature 413 (6857), 732–738.

Ali, M., Roback, L., Mocarski, E.S., 2019. Herpes simplex virus 1 ICP6 impedes TNF receptor 1-induced necrosome assembly during compartmentalization to detergent-resistant membrane vesicles. J. Biol. Chem. 294 (3), 991–1004.

An, J., Kim, S.H., Hwang, D., Lee, K.E., Kim, M.J., Yang, E.G., Kim, S.Y., Chung, H.S., 2019. Caspase-4 disaggregates lipopolysaccharide micelles via LPS-CARD interaction. Sci. Rep. 9 (1), 826.

Antonopoulos, C., Russo, H.M., El Sanadi, C., Martin, B.N., Li, X., Kaiser, W.J., Mocarski, E.S., Dubyak, G.R., 2015. Caspase-8 as an effector and regulator of NLRP3 inflammasome signaling. J. Biol. Chem. 290 (33), 20167–20184.

Arifuzzaman, M., Ang, W.X.G., Choi, H.W., Nilles, M.L., St John, A.L., Abraham, S.N., 2018. Necroptosis of infiltrated macrophages drives Yersinia pestis dispersal within buboes. JCI insight 3, 18.

Ashida, H., Ogawa, M., Kim, M., Mimuro, H., Sasakawa, C., 2011. Bacteria and host interactions in the gut epithelial barrier. Nat. Chem. Biol. 8 (1), 36–45.

Bandera, A., Masetti, M., Fabbiani, M., Biasin, M., Muscatello, A., Squillace, N., Clerici, M., Gori, A., Trabattoni, D., 2018. The NLRP3 inflammasome is upregulated in HIV-infected antiretroviral therapy-treated individuals with defective immune recovery. Front. Immunol. 9, 214.

Basu, S., Pathak, S.K., Banerjee, A., Pathak, S., Bhattacharyya, A., Yang, Z., Talarico, S., Kundu, M., Basu, J., 2007. Execution of macrophage apoptosis by PE_PGRS33 of *Mycobacterium tuberculosis* is mediated by Toll-like receptor 2-dependent release of tumor necrosis factor-alpha. J. Biol. Chem. 282 (2), 1039–1050.

Benaoudia, S., Martin, A., Puig Gamez, M., Gay, G., Lagrange, B., Cornut, M., Krasnykov, K., Claude, J.-B., Bourgeois, C.F., Hughes, S., Gillet, B., Allatif, O., Corbin, A., Ricci, R., Henry, T., 2019. A genome-wide screen identifies IRF2 as a key regulator of caspase-4 in human cells. EMBO Rep. 20 (9), e48235.

Bergsbaken, T., Cookson, B.T., 2007. Macrophage activation redirects yersinia-infected host cell death from apoptosis to caspase-1-dependent pyroptosis. PLoS Pathog. 3 (11), e161.

Boyden, E.D., Dietrich, W.F., 2006. Nalp1b controls mouse macrophage susceptibility to anthrax lethal toxin. Nat. Genet. 38 (2), 240–244.

Bronner, D.N., O'Riordan, M.X.D., He, Y., 2013. Caspase-2 mediates a *Brucella abortus* RB51-induced hybrid cell death having features of apoptosis and pyroptosis. Front. Cell. Infect. Microbiol. 3, 83.

Case, C.L., Shin, S., Roy, C.R., 2009. Asc and Ipaf inflammasomes direct distinct pathways for caspase-1 activation in response to *Legionella pneumophila*. Infect. Immun. 77 (5), 1981–1991.

Case, C.L., Kohler, L.J., Lima, J.B., Strowig, T., de Zoete, M.R., Flavell, R.A., Zamboni, D.S., Roy, C.R., 2013. Caspase-11 stimulates rapid flagellin-independent pyroptosis in response to *Legionella pneumophila*. Proc. Natl. Acad. Sci. U. S. A. 110 (5), 1851–1856.

Cervantes, J., Nagata, T., Uchijima, M., Shibata, K., Koide, Y., 2008. Intracytosolic Listeria monocytogenes induces cell death through caspase-1 activation in murine macrophages. Cell Microbiol. 10 (1), 41–52.

Chavarría-Smith, J., Vance, R.E., 2013. Direct proteolytic cleavage of NLRP1B is necessary and sufficient for inflammasome activation by anthrax lethal factor. PLoS Pathog. 9 (6), e1003452.

Chen, Y., Smith, M.R., Thirumalai, K., Zychlinsky, A., 1996. A bacterial invasin induces macrophage apoptosis by binding directly to ICE. EMBO J. 15 (15), 3853–3860.

Chen, F., Ding, X., Ding, Y., Xiang, Z., Li, X., Ghosh, D., Schurig, G.G., Sriranganathan, N., Boyle, S.M., He, Y., 2011. Proinflammatory caspase-2-mediated macrophage cell death induced by a rough attenuated Brucella suis strain. Infect. Immun. 79 (6), 2460–2469.

Chen, Y.-J., Wang, S.-F., Weng, I.-C., Hong, M.-H., Lo, T.-H., Jan, J.-T., Hsu, L.-C., Chen, H.-Y., Liu, F.-T., 2018. Galectin-3 enhances avian H5N1 influenza A virus-induced pulmonary inflammation by promoting NLRP3 inflammasome activation. Am. J. Pathol. 188 (4), 1031–1042.

Cheung, K.T., Sze, D.M., Chan, K.H., Leung, P.H., 2018. Involvement of caspase-4 in IL-1 beta production and pyroptosis in human macrophages during dengue virus infection. Immunobiology 223 (4–5), 356–364.

Chow, J., Franz, K.M., Kagan, J.C., 2015. PRRs are watching you: localization of innate sensing and signaling regulators. Virology 479–480, 104–109.

Cohen, G.M., 1997. Caspases: the executioners of apoptosis. Biochem. J. 326 (Pt 1), 1–16.

Cohen, T.S., Prince, A.S., 2013. Activation of inflammasome signaling mediates pathology of acute *P. aeruginosa* pneumonia. J. Clin. Invest. 123 (4), 1630–1637.

Costa Franco, M.M.S., Marim, F.M., Alves-Silva, J., Cerqueira, D., Rungue, M., Tavares, I.P., Oliveira, S.C., 2019. AIM2 senses *Brucella abortus* DNA in dendritic cells to induce IL-1β secretion, pyroptosis and resistance to bacterial infection in mice. Microbes Infect. 21 (2), 85–93.

da Costa, L.S., Outlioua, A., Anginot, A., Akarid, K., Arnoult, D., 2019. RNA viruses promote activation of the NLRP3 inflammasome through cytopathogenic effect-induced potassium efflux. Cell Death Dis. 10 (5), 346.

de Sousa, J.R., Azevedo, R.D.S.D.S., Martins Filho, A.J., de Araujo, M.T.F., Cruz, E.D.R.M., Vasconcelos, B.C.B., Cruz, A.C.R., de Oliveira, C.S., Martins, L.C., Vasconcelos, B.H.B., Casseb, L.M.N., Chiang, J.O., Quaresma, J.A.S., Vasconcelos, P.F.D.C., 2018. In situ inflammasome activation results in severe damage to the central nervous system in fatal Zika virus microcephaly cases. Cytokine 111, 255–264.

Degterev, A., Huang, Z., Boyce, M., Li, Y., Jagtap, P., Mizushima, N., Cuny, G.D., Mitchison, T.J., Moskowitz, M.A., Yuan, J., 2005. Chemical inhibitor of nonapoptotic cell death with therapeutic potential for ischemic brain injury. Nat. Chem. Biol. 1 (2), 112–119.

Deng, X., Zou, W., Xiong, M., Wang, Z., Engelhardt, J.F., Ye, S.Q., Yan, Z., Qiu, J., 2017. Human parvovirus infection of human airway epithelia induces pyroptotic cell death by inhibiting apoptosis. J. Virol. 91 (24), e01533–17.

Diner, B.A., Lum, K.K., Toettcher, J.E., Cristea, I.M., 2016. Viral DNA sensors IFI16 and cyclic GMP-AMP synthase possess distinct functions in regulating viral gene expression, immune defenses, and apoptotic responses during herpesvirus infection. mBio 7 (6), e01553-16.

Doitsh, G., Galloway, N.L.K., Geng, X., Yang, Z., Monroe, K.M., Zepeda, O., Hunt, P.W., Hatano, H., Sowinski, S., Muñoz-Arias, I., Greene, W.C., 2014. Cell death by pyroptosis drives CD4 T-cell depletion in HIV-1 infection. Nature 505 (7484), 509–514.

Dondelinger, Y., Delanghe, T., Priem, D., Wynosky-Dolfi, M.A., Sorobetea, D., Rojas-Rivera, D., Giansanti, P., Roelandt, R., Gropengiesser, J., Ruckdeschel, K., Savvides, S.N., Heck, A.J.R., Vandenabeele, P., Brodsky, I.E., Bertrand, M.J.M., 2019. Serine 25 phosphorylation inhibits RIPK1 kinase-dependent cell death in models of infection and inflammation. Nat. Commun. 10 (1), 1729.

Dovey, C.M., Diep, J., Clarke, B.P., Hale, A.T., McNamara, D.E., Guo, H., Brown, N.W., Cao, J.Y., Grace, C.R., Gough, P.J., Bertin, J., Dixon, S.J., Fiedler, D., Mocarski, E.S., Kaiser, W.J., Moldoveanu, T., York, J.D., Carette, J.E., 2018. MLKL requires the inositol phosphate code to execute necroptosis. Mol. Cell 70 (5), 936–948.

Dufour, F., Sasseville, A.M.-J., Chabaud, S., Massie, B., Siegel, R.M., Langelier, Y., 2011. The ribonucleotide reductase R1 subunits of herpes simplex virus types 1 and 2 protect cells against TNFα- and FasL-induced apoptosis by interacting with caspase-8. Apoptosis 16 (3), 256–271.

Dürst, M., Gissmann, L., Ikenberg, H., zur Hausen, H., 1983. A papillomavirus DNA from a cervical carcinoma and its prevalence in cancer biopsy samples from different geographic regions. Proc. Natl. Acad. Sci. U. S. A. 80 (12), 3812–3815.

Ebermann, L., Ruzsics, Z., Guzmán, C.A., van Rooijen, N., Casalegno-Garduño, R., Koszinowski, U., Čičin-Šain, L., 2012. Block of death-receptor apoptosis protects mouse cytomegalovirus from macrophages and is a determinant of virulence in immunodeficient hosts. PLoS Pathog. 8 (12), e1003062.

Eitz Ferrer, P., Potthoff, S., Kirschnek, S., Gasteiger, G., Kastenmüller, W., Ludwig, H., Paschen, S.A., Villunger, A., Sutter, G., Drexler, I., Häcker, G., 2011. Induction of Noxa-mediated apoptosis by modified vaccinia virus Ankara depends on viral recognition by cytosolic helicases, leading to IRF-3/IFN-β-dependent induction of pro-apoptotic Noxa. PLoS Pathog. 7 (6), e1002083.

El Maadidi, S., Faletti, L., Berg, B., Wenzl, C., Wieland, K., Chen, Z.J., Maurer, U., Borner, C., 2014. A novel mitochondrial MAVS/Caspase-8 platform links RNA virus-induced innate antiviral signaling to Bax/Bak-independent apoptosis. J. Immunol. 192 (3), 1171–1183.

Evavold, C.L., Ruan, J., Tan, Y., Xia, S., Wu, H., Kagan, J.C., 2018. The pore-forming protein gasdermin D regulates interleukin-1 secretion from living macrophages. Immunity 48 (1), 35–44.

Fadok, V.A., Voelker, D.R., Campbell, P.A., Cohen, J.J., Bratton, D.L., Henson, P.M., 1992. Exposure of phosphatidylserine on the surface of apoptotic lymphocytes triggers specific recognition and removal by macrophages. J. Immunol. 7 (148), 2207–2216.

Feoktistova, M., Geserick, P., Kellert, B., Dimitrova, D.P., Langlais, C., Hupe, M., Cain, K., MacFarlane, M., Häcker, G., Leverkus, M., 2011. cIAPs block Ripoptosome formation, a RIP1/caspase-8 containing intracellular cell death complex differentially regulated by cFLIP isoforms. Mol. Cell 43 (3), 449–463.

Fernandez, M.V., Miller, E., Krammer, F., Gopal, R., Greenbaum, B.D., Bhardwaj, N., 2016. Ion efflux and influenza infection trigger NLRP3 inflammasome signaling in human dendritic cells. J. Leukoc. Biol. 99 (5), 723–734.

Finethy, R., Jorgensen, I., Haldar, A.K., de Zoete, M.R., Strowig, T., Flavell, R.A., Yamamoto, M., Nagarajan, U.M., Miao, E.A., Coers, J., 2015. Guanylate binding proteins enable rapid activation of canonical and noncanonical inflammasomes in Chlamydia-infected macrophages. Infect. Immun. 83 (12), 4740–4749.

Fink, S.L., Cookson, B.T., 2006. Caspase-1-dependent pore formation during pyroptosis leads to osmotic lysis of infected host macrophages. Cell. Microbiol. 8 (11), 1812–1825.

Fink, S.L., Bergsbaken, T., Cookson, B.T., 2008. Anthrax lethal toxin and Salmonella elicit the common cell death pathway of caspase-1-dependent pyroptosis via distinct mechanisms. Proc. Natl. Acad. Sci. U. S. A. 105 (11), 4312–4317.

Fitzgerald, K.A., McWhirter, S.M., Faia, K.L., Rowe, D.C., Latz, E., Golenbock, D.T., Coyle, A.J., Liao, S.-M., Maniatis, T., 2003. IKKepsilon and TBK1 are essential components of the IRF3 signaling pathway. Nat. Immunol. 4 (5), 491–496.

Gaba, A., Xu, F., Lu, Y., Park, H.-S., Liu, G., Zhou, Y., 2019. The NS1 protein of influenza A virus participates in necroptosis by interacting with MLKL and increasing its oligomerization and membrane translocation. J. Virol. 93 (2), e01835–18.

Galloway, N.L., Doitsh, G., Monroe, K.M., Yang, Z., Muñoz-Arias, I., Levy, D.N., Greene, W.C., 2015. Cell-to-cell transmission of HIV-1 is required to trigger pyroptotic death of lymphoid-tissue-derived CD4 T cells. Cell Rep. 12 (10), 1555–1563.

Galluzzi, L., Vitale, I., Aaronson, S.A., Abrams, J.M., Adam, D., Agostinis, P., Alnemri, E.S., Altucci, L., Amelio, I., Andrews, D.W., Annicchiarico-Petruzzelli, M., Antonov, A.V., Arama, E., Baehrecke, E.H., Barlev, N.A., Bazan, N.G., Bernassola, F., Bertrand, M.J.M., Bianchi, K., Blagosklonny, M.V., Blomgren, K., Borner, C., Boya, P., Brenner, C., Campanella, M., Candi, E., Carmona-Gutierrez, D., Cecconi, F., Chan, F.K.-M., Chandel, N.S., Cheng, E.H., Chipuk, J.E., Cidlowski, J.A., Ciechanover, A., Cohen, G.M., Conrad, M., Cubillos-Ruiz, J.R., Czabotar, P.E., D'Angiolella, V., Dawson, T.M., Dawson, V.L., de Laurenzi, V., de Maria, R., Debatin, K.-M., DeBerardinis, R.J., Deshmukh, M., Di Daniele, N., Di Virgilio, F., Dixit, V.M., Dixon, S.J., Duckett, C.S., Dynlacht, B.D., El-Deiry, W.S., Elrod, J.W., Fimia, G.M., Fulda, S., García-Sáez, A.J., Garg, A.D., Garrido, C., Gavathiotis, E., Golstein, P., Gottlieb, E., Green, D.R., Greene, L.A., Gronemeyer, H., Gross, A., Hajnoczky, G., Hardwick, J.M., Harris, I.S., Hengartner, M.O., Hetz, C., Ichijo, H., Jäättelä, M., Joseph, B., Jost, P.J., Juin, P.P., Kaiser, W.J., Karin, M., Kaufmann, T., Kepp, O., Kimchi, A., Kitsis, R.N., Klionsky, D.J., Knight, R.A., Kumar, S., Lee, S.W., Lemasters, J.J., Levine, B., Linkermann, A., Lipton, S.A., Lockshin, R.A., López-Otín, C., Lowe, S.W., Luedde, T., Lugli, E., MacFarlane, M., Madeo, F., Malewicz, M., Malorni, W., Manic, G., Marine, J.-C., Martin, S.J., Martinou, J.-C., Medema, J.P., Mehlen, P., Meier, P., Melino, S., Miao, E.A., Molkentin, J.D., Moll, U.M., Muñoz-Pinedo, C., Nagata, S., Nuñez, G., Oberst, A., Oren, M., Overholtzer, M., Pagano, M., Panaretakis, T., Pasparakis, M., Penninger, J.M., Pereira, D.M., Pervaiz, S., Peter, M.E., Piacentini, M., Pinton, P., Prehn, J.H.M., Puthalakath, H., Rabinovich, G.A., Rehm, M., Rizzuto, R., Rodrigues, C.M.P., Rubinsztein, D.C., Rudel, T., Ryan, K.M., Sayan, E., Scorrano, L., Shao, F., Shi, Y., Silke, J., Simon, H.-U., Sistigu, A., Stockwell, B.R., Strasser, A., Szabadkai, G., Tait, S.W.G., Tang, D., Tavernarakis, N., Thorburn, A., Tsujimoto, Y., Turk, B., Vanden Berghe, T., Vandenabeele, P., Vander Heiden, M.G., Villunger, A., Virgin, H.W., Vousden, K.H., Vucic, D., Wagner, E.F., Walczak, H., Wallach, D., Wang, Y., Wells, J.A., Wood, W., Yuan, J., Zakeri, Z., Zhivotovsky, B., Zitvogel, L., Melino, G., Kroemer, G., 2018. Molecular mechanisms of cell death: recommendations of the nomenclature committee on cell death 2018. Cell Death Differ. 25 (3), 486–541.

González-Juarbe, N., Gilley, R.P., Hinojosa, C.A., Bradley, K.M., Kamei, A., Gao, G., Dube, P.H., Bergman, M.A., Orihuela, C.J., 2015. Pore-forming toxins induce macrophage necroptosis during acute bacterial pneumonia. PLoS Pathog. 11 (12), e1005337.

González-Juarbe, N., Bradley, K.M., Shenoy, A.T., Gilley, R.P., Reyes, L.F., Hinojosa, C.A., Restrepo, M.I., Dube, P.H., Bergman, M.A., Orihuela, C.J., 2017. Pore-forming toxin-mediated ion dysregulation leads to death receptor-independent necroptosis of lung epithelial cells during bacterial pneumonia. Cell Death Differ. 24 (5), 917–928.

Gonzalez-Juarbe, N., Bradley, K.M., Riegler, A.N., Reyes, L.F., Brissac, T., Park, S.-S., Restrepo, M.I., Orihuela, C.J., 2018. Bacterial pore-forming toxins promote the activation of caspases in parallel to necroptosis to enhance alarmin release and inflammation during pneumonia. Sci. Rep. 8 (1), 5846.

Gröbner, S., Schulz, S., Soldanova, I., Gunst, D.S.J., Waibel, M., Wesselborg, S., Borgmann, S., Autenrieth, I.B., 2007. Absence of Toll-like receptor 4 signaling results in delayed Yersinia enterocolitica YopP-induced cell death of dendritic cells. Infect. Immun. 75 (1), 512–517.

Große, L., Wurm, C.A., Brüser, C., Neumann, D., Jans, D.C., Jakobs, S., 2016. Bax assembles into large ring-like structures remodeling the mitochondrial outer membrane in apoptosis. EMBO J. 35 (4), 402–413.

Guo, H., Omoto, S., Harris, P.A., Finger, J.N., Bertin, J., Gough, P.J., Kaiser, W.J., Mocarski, E.S., 2015. Herpes simplex virus suppresses necroptosis in human cells. Cell Host Microbe 17 (2), 243–251.

Guo, H., Gilley, R.P., Fisher, A., Lane, R., Landsteiner, V.J., Ragan, K.B., Dovey, C.M., Carette, J.E., Upton, J.W., Mocarski, E.S., Kaiser, W.J., 2018. Species-independent contribution of ZBP1/DAI/DLM-1-triggered necroptosis in host defense against HSV1. Cell Death Dis. 9 (8), 816.

Halle, S., Halle, O., Förster, R., 2017. Mechanisms and dynamics of T cell-mediated cytotoxicity in vivo. Trends Immunol. 38 (6), 432–443.

Han, X., Cong, H., 2017. Enterovirus 71 induces apoptosis by directly modulating the conformational activation of pro-apoptotic protein Bax. J. Gen. Virol. 98 (3), 422–434.

He, S., Liang, Y., Shao, F., Wang, X., 2011. Toll-like receptors activate programmed necrosis in macrophages through a receptor-interacting kinase-3-mediated pathway. Proc. Natl. Acad. Sci. U. S. A. 108 (50), 20054–20059.

He, W.-T., Wan, H., Hu, L., Chen, P., Wang, X., Huang, Z., Yang, Z.-H., Zhong, C.-Q., Han, J., 2015. Gasdermin D is an executor of pyroptosis and required for interleukin-1β secretion. Cell Res. 25 (12), 1285–1298.

Henderson, S., Rowe, M., Gregory, C., Croom-Carter, D., Wang, F., Longnecker, R., Kieff, E., Rickinson, A., 1991. Induction of bcl-2 expression by epstein-barr virus latent membrane protein 1 protects infected B cells from programmed cell death. Cell 65 (7), 1107–1115.

Hsu, H., Huang, J., Shu, H.-B., Baichwal, V., Goeddel, D.V., 1996. TNF-dependent recruitment of the protein kinase RIP to the TNF receptor-1 signaling complex. Immunity 4 (4), 387–396.

Hu, G.-Q., Yang, Y.-J., Qin, X.-X., Qi, S., Zhang, J., Yu, S.-X., Du, C.-T., Chen, W., 2019. Salmonella outer protein b suppresses colitis development via protecting cell from necroptosis. Front. Cell. Infect. Microbiol. 9, 87.

Huang, Z., Wu, S.-Q., Liang, Y., Zhou, X., Chen, W., Li, L., Wu, J., Zhuang, Q., Chen, C.A., Li, J., Zhong, C.-Q., Xia, W., Zhou, R., Zheng, C., Han, J., 2015. RIP1/RIP3 binding to HSV-1 ICP6 initiates necroptosis to restrict virus propagation in mice. Cell Host Microbe 17 (2), 229–242.

Humphries, F., Bergin, R., Jackson, R., Delagic, N., Wang, B., Yang, S., Dubois, A.V., Ingram, R.J., Moynagh, P.N., 2018. The E3 ubiquitin ligase Pellino2 mediates priming of the NLRP3 inflammasome. Nat. Commun. 9 (1), 1560.

Imre, G., Heering, J., Takeda, A.-N., Husmann, M., Thiede, B., zu Heringdorf, D.M., Green, D.R., van der Goot, F.G., Sinha, B., Dötsch, V., Rajalingam, K., 2012. Caspase-2 is an initiator caspase responsible for pore-forming toxin-mediated apoptosis. EMBO J. 31 (11), 2615–2628.

Ishikawa, H., Barber, G.N., 2008. STING is an endoplasmic reticulum adaptor that facilitates innate immune signalling. Nature 455 (7213), 674–678.

Jorgensen, I., Rayamajhi, M., Miao, E.A., 2017. Programmed cell death as a defence against infection. Nat. Rev. Immunol. 17 (3), 151–164.

Juruj, C., Lelogeais, V., Pierini, R., Perret, M., Py, B.F., Jamilloux, Y., Broz, P., Ader, F., Faure, M., Henry, T., 2013. Caspase-1 activity affects AIM2 speck formation/stability through a negative feedback loop. Front. Cell. Infect. Microbiol. 3, 14.

Kaiser, W.J., Offermann, M.K., 2005. Apoptosis induced by the Toll-like receptor adaptor TRIF is dependent on its receptor interacting protein homotypic interaction motif. J. Immunol. 174 (8), 4942–4952.

Kaiser, W.J., Sridharan, H., Huang, C., Mandal, P., Upton, J.W., Gough, P.J., Sehon, C.A., Marquis, R.W., Bertin, J., Mocarski, E.S., 2013. Toll-like receptor 3-mediated necrosis via TRIF, RIP3, and MLKL. J. Biol. Chem. 288 (43), 31268–31279.

Karki, R., Lee, E., Place, D., Samir, P., Mavuluri, J., Sharma, B.R., Balakrishnan, A., Malireddi, R.K.S., Geiger, R., Zhu, Q., Neale, G., Kanneganti, T.-D., 2018. IRF8 regulates transcription of Naips for NLRC4 inflammasome activation. Cell 173 (4), 920–933.e13.

Katagiri, N., Shobuike, T., Chang, B., Kukita, A., Miyamoto, H., 2012. The human apoptosis inhibitor NAIP induces pyroptosis in macrophages infected with *Legionella pneumophila*. Microbes Infect. 14 (13), 1123–1132.

Kato, H., Takeuchi, O., Sato, S., Yoneyama, M., Yamamoto, M., Matsui, K., Uematsu, S., Jung, A., Kawai, T., Ishii, K.J., Yamaguchi, O., Otsu, K., Tsujimura, T., Koh, C.-S., Reis e Sousa, C., Matsuura, Y., Fujita, T., Akira, S., 2006. Differential roles of MDA5 and RIG-I helicases in the recognition of RNA viruses. Nature 441 (7089), 101–105.

Kayagaki, N., Stowe, I.B., Lee, B.L., O'Rourke, K., Anderson, K., Warming, S., Cuellar, T., Haley, B., Roose-Girma, M., Phung, Q.T., Liu, P.S., Lill, J.R., Li, H., Wu, J., Kummerfeld, S., Zhang, J., Lee, W.P., Snipas, S.J., Salvesen, G.S., Morris, L.X., Fitzgerald, L., Zhang, Y., Bertram, E.M., Goodnow, C.C., Dixit, V.M., 2015. Caspase-11 cleaves gasdermin D for non-canonical inflammasome signalling. Nature 526 (7575), 666–671.

Kerr, J.F., Wyllie, A.H., Currie, A.R., 1972. Apoptosis: a basic biological phenomenon with wide-ranging implications in tissue kinetics. Br. J. Cancer 26 (4), 239–257.

Kim, S., Bauernfeind, F., Ablasser, A., Hartmann, G., Fitzgerald, K.A., Latz, E., Hornung, V., 2010. *Listeria monocytogenes* is sensed by the NLRP3 and AIM2 inflammasome. Eur. J. Immunol. 40 (6), 1545–1551.

Kim, K.-H., Seong, B.L., 2003. Pro-apoptotic function of HBV X protein is mediated by interaction with c-FLIP and enhancement of death-inducing signal. EMBO J. 22 (9), 2104–2116.

Kip, E., Nazé, F., Suin, V., Vanden Berghe, T., Francart, A., Lamoral, S., Vandenabeele, P., Beyaert, R., van Gucht, S., Kalai, M., 2017. Impact of caspase-1/11, -3, -7, or IL-1β/IL-18 deficiency on rabies virus-induced macrophage cell death and onset of disease. Cell Death Discov. 3, 17012.

Kitur, K., Parker, D., Nieto, P., Ahn, D.S., Cohen, T.S., Chung, S., Wachtel, S., Bueno, S., Prince, A., 2015. Toxin-induced necroptosis is a major mechanism of *Staphylococcus aureus* lung damage. PLoS Pathog. 11 (4), e1004820.

Koehler, H., Cotsmire, S., Langland, J., Kibler, K.V., Kalman, D., Upton, J.W., Mocarski, E.S., Jacobs, B.L., 2017. Inhibition of DAI-dependent necroptosis by the Z-DNA binding domain of the vaccinia virus innate immune evasion protein, E3. Proc. Natl. Acad. Sci. U. S. A. 114 (43), 11506–11511.

Kofahi, H.M., Taylor, N.G.A., Hirasawa, K., Grant, M.D., Russell, R.S., 2016. Hepatitis C virus infection of cultured human hepatoma cells causes apoptosis and pyroptosis in both infected and bystander cells. Sci. Rep. 6, 37433.

Krejbich-Trotot, P., Denizot, M., Hoarau, J.-J., Jaffar-Bandjee, M.-C., Das, T., Gasque, P., 2011. Chikungunya virus mobilizes the apoptotic machinery to invade host cell defenses. FASEB J. 25 (1), 314–325.

Kuriakose, T., Man, S.M., Malireddi, R.K.S., Karki, R., Kesavardhana, S., Place, D.E., Neale, G., Vogel, P., Kanneganti, T.-D., 2016. ZBP1/DAI is an innate sensor of influenza virus triggering the NLRP3 inflammasome and programmed cell death pathways. Sci. Immunol. 1 (2), aag2045.

Kuriakose, T., Zheng, M., Neale, G., Kanneganti, T.-D., 2018. IRF1 is a transcriptional regulator of ZBP1 promoting NLRP3 inflammasome activation and cell death during influenza virus infection. J. Immunol. 200 (4), 1489–1495.

Langelier, Y., Bergeron, S., Chabaud, S., Lippens, J., Guilbault, C., Sasseville, A.M.-J., Denis, S., Mosser, D.D., Massie, B., 2002. The R1 subunit of herpes simplex virus ribonucleotide reductase protects cells against apoptosis at, or upstream of, caspase-8 activation. J. Gen. Virol. 83 (Pt 11), 2779–2789.

Lee, S., Hirohama, M., Noguchi, M., Nagata, K., Kawaguchi, A., 2018. Influenza A virus infection triggers pyroptosis and apoptosis of respiratory epithelial cells through the type I interferon signaling pathway in a mutually exclusive manner. J. Virol. 92, 14.

Legarda, D., Justus, S.J., Ang, R.L., Rikhi, N., Li, W., Moran, T.M., Zhang, J., Mizoguchi, E., Zelic, M., Kelliher, M.A., Blander, J.M., Ting, A.T., 2016. CYLD proteolysis protects macrophages from TNF-mediated auto-necroptosis induced by LPS and licensed by type I IFN. Cell Rep. 15 (11), 2449–2461.

Lehnardt, S., Wennekamp, J., Freyer, D., Liedtke, C., Krueger, C., Nitsch, R., Bechmann, I., Weber, J.R., Henneke, P., 2007. TLR2 and caspase-8 are essential for group B Streptococcus-induced apoptosis in microglia. J. Immunol. 179 (9), 6134–6143.

Lei, X., Zhang, Z., Xiao, X., Qi, J., He, B., Wang, J., 2017. Enterovirus 71 inhibits pyroptosis through cleavage of gasdermin D. J. Virol. 91 (18), pii: e01069-17.

Li, Z., Yu, J., Liu, L., Wei, Z., Ehrlich, E.S., Liu, G., Li, J., Liu, X., Wang, H., Yu, X.-F., Zhang, W., 2014. Coxsackievirus A16 infection induces neural cell and non-neural cell apoptosis in vitro. PLoS One 9 (10), e111174.

Liao, H., Xu, J., Huang, J., 2010. FasL/Fas pathway is involved in dengue virus induced apoptosis of the vascular endothelial cells. J. Med. Virol. 82 (8), 1392–1399.

Lim, M.C.C., Maubach, G., Sokolova, O., Feige, M.H., Diezko, R., Buchbinder, J., Backert, S., Schlüter, D., Lavrik, I.N., Naumann, M., 2017. Pathogen-induced ubiquitin-editing enzyme A20 bifunctionally shuts off NF-κB and caspase-8-dependent apoptotic cell death. Cell Death Differ. 24 (9), 1621–1631.

Liu, X., Zhang, Z., Ruan, J., Pan, Y., Magupalli, V.G., Wu, H., Lieberman, J., 2016. Inflammasome-activated gasdermin D causes pyroptosis by forming membrane pores. Nature 535 (7610), 153–158.

Liu, X., Li, Y., Peng, S., Yu, X., Li, W., Shi, F., Luo, X., Tang, M., Tan, Z., Bode, A.M., Cao, Y., 2018. Epstein-Barr virus encoded latent membrane protein 1 suppresses necroptosis through targeting RIPK1/3 ubiquitination. Cell Death Dis. 9 (2), 53.

López, M., Sly, L.M., Luu, Y., Young, D., Cooper, H., Reiner, N.E., 2003. The 19-kDa *Mycobacterium tuberculosis* protein induces macrophage apoptosis through Toll-like receptor-2. J. Immunol. 170 (5), 2409–2416.

Lötzerich, M., Roulin, P.S., Boucke, K., Witte, R., Georgiev, O., Greber, U.F., 2018. Rhinovirus 3C protease suppresses apoptosis and triggers caspase-independent cell death. Cell Death Dis. 9 (3), 272.

Lowin, B., Beermann, F., Schmidt, A., Tschopp, J., 1994. A null mutation in the perforin gene impairs cytolytic T lymphocyte- and natural killer cell-mediated cytotoxicity. Proc. Natl. Acad. Sci. U. S. A. 91 (24), 11571–11575.

Lu, B., Nakamura, T., Inouye, K., Li, J., Tang, Y., Lundbäck, P., Valdes-Ferrer, S.I., Olofsson, P.S., Kalb, T., Roth, J., Zou, Y., Erlandsson-Harris, H., Yang, H.,

Ting, J.P.-Y., Wang, H., Andersson, U., Antoine, D.J., Chavan, S.S., Hotamisligil, G.S., Tracey, K.J., 2012. Novel role of PKR in inflammasome activation and HMGB1 release. Nature 488 (7413), 670–674.

Mace, P.D., Riedl, S.J., 2010. Molecular cell death platforms and assemblies. Curr. Opin. Cell Biol. 22 (6), 828–836.

Man, S.M., Tourlomousis, P., Hopkins, L., Monie, T.P., Fitzgerald, K.A., Bryant, C.E., 2013. Salmonella infection induces recruitment of Caspase-8 to the inflammasome to modulate IL-1β production. J. Immunol. 191 (10), 5239–5246.

Man, S.M., Karki, R., Malireddi, R.K.S., Neale, G., Vogel, P., Yamamoto, M., Lamkanfi, M., Kanneganti, T.-D., 2015. The transcription factor IRF1 and guanylate-binding proteins target activation of the AIM2 inflammasome by Francisella infection. Nat. Immunol. 16 (5), 467–475.

Man, S.M., Karki, R., Sasai, M., Place, D.E., Kesavardhana, S., Temirov, J., Frase, S., Zhu, Q., Malireddi, R.K.S., Kuriakose, T., Peters, J.L., Neale, G., Brown, S.A., Yamamoto, M., Kanneganti, T.-D., 2016. IRGB10 liberates bacterial ligands for sensing by the AIM2 and caspase-11-NLRP3 inflammasomes. Cell 167 (2), 382–396.e17.

Man, S.M., Karki, R., Kanneganti, T.-D., 2017. Molecular mechanisms and functions of pyroptosis, inflammatory caspases and inflammasomes in infectious diseases. Immunol. Rev. 277 (1), 61–75.

Martinon, F., 2008. Detection of immune danger signals by NALP3. J. Leukoc. Biol. 83 (3), 507–511.

Martinon, F., Pétrilli, V., Mayor, A., Tardivel, A., Tschopp, J., 2006. Gout-associated uric acid crystals activate the NALP3 inflammasome. Nature 440 (7081), 237–241.

Matsuoka, M., Jeang, K.-T., 2007. Human T-cell leukaemia virus type 1 (HTLV-1) infectivity and cellular transformation. Nat. Rev. Cancer 7 (4), 270–280.

Matusiak, M., van Opdenbosch, N., Vande Walle, L., Sirard, J.-C., Kanneganti, T.-D., Lamkanfi, M., 2015. Flagellin-induced NLRC4 phosphorylation primes the inflammasome for activation by NAIP5. Proc. Natl. Acad. Sci. U. S. A. 112 (5), 1541–1546.

McAllister, C.S., Lakhdari, O., Pineton de Chambrun, G., Gareau, M.G., Broquet, A., Lee, G.H., Shenouda, S., Eckmann, L., Kagnoff, M.F., 2013. TLR3, TRIF, and caspase 8 determine double-stranded RNA-induced epithelial cell death and survival in vivo. J. Immunol. 190 (1), 418–427.

McCracken, J.M., Kinkead, L.C., McCaffrey, R.L., Allen, L.-A.H., 2016. Francisella tularensis modulates a distinct subset of regulatory factors and sustains mitochondrial integrity to impair human neutrophil apoptosis. J. Innate Immun. 8 (3), 299–313.

McQuade, T., Cho, Y., Chan, F.K.-M., 2013. Positive and negative phosphorylation regulates RIP1- and RIP3-induced programmed necrosis. Biochem. J. 456 (3), 409–415.

Meessen-Pinard, M., Le Coupanec, A., Desforges, M., Talbot, P.J., 2017. Pivotal role of receptor-interacting protein kinase 1 and mixed lineage kinase domain-like in neuronal cell death induced by the human neuroinvasive coronavirus OC43. J. Virol. 91, 1.

Meng, R., Gu, L., Lu, Y., Zhao, K., Wu, J., Wang, H., Han, J., Tang, Y., Lu, B., 2019. High mobility group box 1 enables bacterial lipids to trigger receptor-interacting protein kinase 3 (RIPK3)-mediated necroptosis and apoptosis in mice. J. Biol. Chem. 294 (22), 8872–8884.

Meunier, E., Dick, M.S., Dreier, R.F., Schürmann, N., Kenzelmann Broz, D., Warming, S., Roose-Girma, M., Bumann, D., Kayagaki, N., Takeda, K., Yamamoto, M., Broz, P., 2014. Caspase-11 activation requires lysis of pathogen-containing vacuoles by IFN-induced GTPases. Nature 509 (7500), 366–370.

Miao, E.A., Leaf, I.A., Treuting, P.M., Mao, D.P., Dors, M., Sarkar, A., Warren, S.E., Wewers, M.D., Aderem, A., 2010. Caspase-1-induced pyroptosis is an innate immune effector mechanism against intracellular bacteria. Nat. Immunol. 11 (12), 1136–1142.

Monack, D.M., Navarre, W.W., Falkow, S., 2001. Salmonella-induced macrophage death: the role of caspase-1 in death and inflammation. Microbes Infect. 3 (14–15), 1201–1212.

Monroe, K.M., Yang, Z., Johnson, J.R., Geng, X., Doitsh, G., Krogan, N.J., Greene, W.C., 2014. IFI16 DNA sensor is required for death of lymphoid CD4 T cells abortively infected with HIV. Science (New York, NY) 343 (6169), 428–432.

Muñoz-Planillo, R., Kuffa, P., Martínez-Colón, G., Smith, B.L., Rajendiran, T.M., Núñez, G., 2013. K^+ efflux is the common trigger of NLRP3 inflammasome activation by bacterial toxins and particulate matter. Immunity 38 (6), 1142–1153.

Nandakumar, R., Tschismarov, R., Meissner, F., Prabakaran, T., Krissanaprasit, A., Farahani, E., Zhang, B.-C., Assil, S., Martin, A., Bertrams, W., Holm, C.K., Ablasser, A., Klause, T., Thomsen, M.K., Schmeck, B., Howard, K.A., Henry, T., Gothelf, K.V., Decker, T., Paludan, S.R., 2019. Intracellular bacteria engage a STING-TBK1-MVB12b pathway to enable paracrine cGAS-STING signalling. Nat. Microbiol. 4 (4), 701–713.

Nayak, T.K., Mamidi, P., Kumar, A., Singh, L.P.K., Sahoo, S.S., Chattopadhyay, S., Chattopadhyay, S., 2017. Regulation of viral replication, apoptosis and pro-inflammatory responses by 17-AAG during chikungunya virus infection in macrophages. Viruses 9 (1), 3.

Newton, K., Wickliffe, K.E., Maltzman, A., Dugger, D.L., Strasser, A., Pham, V.C., Lill, J.R., Roose-Girma, M., Warming, S., Solon, M., Ngu, H., Webster, J.D., Dixit, V.M., 2016. RIPK1 inhibits ZBP1-driven necroptosis during development. Nature 540 (7631), 129–133.

Ning, X., Wang, Y., Jing, M., Sha, M., Lv, M., Gao, P., Zhang, R., Huang, X., Feng, J.-M., Jiang, Z., 2019. Apoptotic caspases suppress type I interferon production via the cleavage of cGAS, MAVS, and IRF3. Mol. Cell 74 (1), 19–31.

Nogusa, S., Thapa, R.J., Dillon, C.P., Liedmann, S., Oguin, T.H., Ingram, J.P., Rodriguez, D.A., Kosoff, R., Sharma, S., Sturm, O., Verbist, K., Gough, P.J., Bertin, J., Hartmann, B.M., Sealfon, S.C., Kaiser, W.J., Mocarski, E.S., López, C.B., Thomas, P.G., Oberst, A., Green, D.R., Balachandran, S., 2016. RIPK3 activates parallel pathways of MLKL-driven necroptosis and FADD-mediated apoptosis to protect against influenza A virus. Cell Host Microbe 20 (1), 13–24.

Obregón-Henao, A., Duque-Correa, M.A., Rojas, M., García, L.F., Brennan, P.J., Ortiz, B.L., Belisle, J.T., 2012. Stable extracellular RNA fragments of *Mycobacterium tuberculosis* induce early apoptosis in human monocytes via a caspase-8 dependent mechanism. PLoS One 7 (1), e29970.

O'Donnell, M.A., Perez-Jimenez, E., Oberst, A., Ng, A., Massoumi, R., Xavier, R., Green, D.R., Ting, A.T., 2011. Caspase 8 inhibits programmed necrosis by processing CYLD. Nat. Cell Biol. 13 (12), 1437–1442.

Okamoto, K., Fujisawa, J.-I., Reth, M., Yonehara, S., 2006. Human T-cell leukemia virus type-I oncoprotein Tax inhibits Fas-mediated apoptosis by inducing cellular FLIP through activation of NF-kappaB. Genes Cells 11 (2), 177–191.

Olsson, M., Forsberg, J., Zhivotovsky, B., 2015. Caspase-2: the reinvented enzyme. Oncogene 34 (15), 1877–1882.

Omoto, S., Guo, H., Talekar, G.R., Roback, L., Kaiser, W.J., Mocarski, E.S., 2015. Suppression of RIP3-dependent necroptosis by human cytomegalovirus. J. Biol. Chem. 290 (18), 11635–11648.

Orning, P., Weng, D., Starheim, K., Ratner, D., Best, Z., Lee, B., Brooks, A., Xia, S., Wu, H., Kelliher, M.A., Berger, S.B., Gough, P.J., Bertin, J., Proulx, M.M., Goguen, J.D., Kayagaki, N., Fitzgerald, K.A., Lien, E., 2018. Pathogen blockade of TAK1 triggers caspase-8-dependent cleavage of gasdermin D and cell death. Science (New York, NY) 362 (6418), 1064–1069.

Pajuelo, D., Gonzalez-Juarbe, N., Tak, U., Sun, J., Orihuela, C.J., Niederweis, M., 2018. NAD+ depletion triggers macrophage necroptosis, a cell death pathway exploited by *Mycobacterium tuberculosis*. Cell Rep. 24 (2), 429–440.

Pan, T., Wu, S., He, X., Luo, H., Zhang, Y., Fan, M., Geng, G., Ruiz, V.C., Zhang, J., Mills, L., Bai, C., Zhang, H., 2014. Necroptosis takes place in human immunodeficiency virus type-1 (HIV-1)-infected CD4+ T lymphocytes. PLoS One 9 (4), e93944.

Pearson, J.S., Giogha, C., Ong, S.Y., Kennedy, C.L., Kelly, M., Robinson, K.S., Lung, T.W.F., Mansell, A., Riedmaier, P., Oates, C.V.L., Zaid, A., Mühlen, S., Crepin, V.F., Marches, O., Ang, C.-S., Williamson, N.A., O'Reilly, L.A., Bankovacki, A., Nachbur, U., Infusini, G., Webb, A.I., Silke, J., Strasser, A., Frankel, G., Hartland, E.L., 2013. A type III effector antagonizes death receptor signalling during bacterial gut infection. Nature 501 (7466), 247–251.

Pearson, J.S., Giogha, C., Mühlen, S., Nachbur, U., Pham, C.L.L., Zhang, Y., Hildebrand, J.M., Oates, C.V., Lung, T.W.F., Ingle, D., Dagley, L.F., Bankovacki, A., Petrie, E.J., Schroeder, G.N., Crepin, V.F., Frankel, G., Masters, S.L., Vince, J., Murphy, J.M., Sunde, M., Webb, A.I., Silke, J., Hartland, E.L., 2017. EspL is a bacterial cysteine protease effector that cleaves RHIM proteins to block necroptosis and inflammation. Nat. Microbiol. 2, 16258.

Peters, K., Chattopadhyay, S., Sen, G.C., 2008. IRF-3 activation by Sendai virus infection is required for cellular apoptosis and avoidance of persistence. J. Virol. 82 (7), 3500–3508.

Pierini, R., Juruj, C., Perret, M., Jones, C.L., Mangeot, P., Weiss, D.S., Henry, T., 2012. AIM2/ASC triggers caspase-8-dependent apoptosis in Francisella-infected caspase-1-deficient macrophages. Cell Death Differ. 19 (10), 1709–1721.

Pinkoski, M.J., Waterhouse, N.J., Heibein, J.A., Wolf, B.B., Kuwana, T., Goldstein, J.C., Newmeyer, D.D., Bleackley, R.C., Green, D.R., 2001. Granzyme B-mediated apoptosis proceeds predominantly through a Bcl-2-inhibitable mitochondrial pathway. J. Biol. Chem. 276 (15), 12060–12067.

Platnich, J.M., Chung, H., Lau, A., Sandall, C.F., Bondzi-Simpson, A., Chen, H.-M., Komada, T., Trotman-Grant, A.C., Brandelli, J.R., Chun, J., Beck, P.L., Philpott, D.J., Girardin, S.E., Ho, M., Johnson, R.P., MacDonald, J.A., Armstrong, G.D., Muruve, D.A., 2018. Shiga toxin/lipopolysaccharide activates caspase-4 and gasdermin D to trigger mitochondrial reactive oxygen species upstream of the NLRP3 inflammasome. Cell Rep. 25 (6), 1525–1536.

Qu, Y., Misaghi, S., Izrael-Tomasevic, A., Newton, K., Gilmour, L.L., Lamkanfi, M., Louie, S., Kayagaki, N., Liu, J., Kömüves, L., Cupp, J.E., Arnott, D., Monack, D., Dixit, V.M., 2012. Phosphorylation of NLRC4 is critical for inflammasome activation. Nature 490 (7421), 539–542.

Queen, J., Agarwal, S., Dolores, J.S., Stehlik, C., Satchell, K.J.F., 2015. Mechanisms of inflammasome activation by Vibrio cholerae secreted toxins vary with strain biotype. Infect. Immun. 83 (6), 2496–2506.

Ramanathan, M.P., Chambers, J.A., Pankhong, P., Chattergoon, M., Attatippaholkun, W., Dang, K., Shah, N., Weiner, D.B., 2006. Host cell killing by the West Nile Virus NS2B-NS3 proteolytic complex: NS3 alone is sufficient to recruit caspase-8-based apoptotic pathway. Virology 345 (1), 56–72.

Rathinam, V.A.K., Jiang, Z., Waggoner, S.N., Sharma, S., Cole, L.E., Waggoner, L., Vanaja, S.K., Monks, B.G., Ganesan, S., Latz, E., Hornung, V., Vogel, S.N., Szomolanyi-Tsuda, E., Fitzgerald, K.A., 2010. The AIM2 inflammasome is essential for host defense against cytosolic bacteria and DNA viruses. Nat. Immunol. 11 (5), 395–402.

Riedl, S.J., Salvesen, G.S., 2007. The apoptosome: signalling platform of cell death. Nat. Rev. Mol. Cell Biol. 8 (5), 405–413.

Robinson, N., McComb, S., Mulligan, R., Dudani, R., Krishnan, L., Sad, S., 2012. Type I interferon induces necroptosis in macrophages during infection with Salmonella enterica serovar Typhimurium. Nat. Immunol. 13 (10), 954–962.

Rogers, C., Fernandes-Alnemri, T., Mayes, L., Alnemri, D., Cingolani, G., Alnemri, E.S., 2017. Cleavage of DFNA5 by caspase-3 during apoptosis mediates progression to secondary necrotic/pyroptotic cell death. Nat. Commun. 8, 14128.

Rupper, A.C., Cardelli, J.A., 2008. Induction of guanylate binding protein 5 by gamma interferon increases susceptibility to *Salmonella enterica* serovar Typhimurium-induced pyroptosis in RAW 264.7 cells. Infect. Immun. 76 (6), 2304–2315.

Sainski, A.M., Dai, H., Natesampillai, S., Pang, Y.-P., Bren, G.D., Cummins, N.W., Correia, C., Meng, X.W., Tarara, J.E., Ramirez-Alvarado, M., Katzmann, D.J., Ochsenbauer, C., Kappes, J.C., Kaufmann, S.H., Badley, A.D., 2014. Casp8p41 generated by HIV protease kills CD4 T cells through direct Bak activation. J. Cell Biol. 206 (7), 867–876.

Sanman, L.E., Qian, Y., Eisele, N.A., Ng, T.M., van der Linden, W.A., Monack, D.M., Weerapana, E., Bogyo, M., 2016. Disruption of glycolytic flux is a signal for inflammasome signaling and pyroptotic cell death. eLife, 5, e13663.

Sarhan, J., Liu, B.C., Muendlein, H.I., Li, P., Nilson, R., Tang, A.Y., Rongvaux, A., Bunnell, S.C., Shao, F., Green, D.R., Poltorak, A., 2018. Caspase-8 induces cleavage of gasdermin D to elicit pyroptosis during Yersinia infection. Proc. Natl. Acad. Sci. U. S. A. 115 (46), E10888–E10897.

Sauer, J.-D., Witte, C.E., Zemansky, J., Hanson, B., Lauer, P., Portnoy, D.A., 2010. *Listeria monocytogenes* triggers AIM2-mediated pyroptosis upon infrequent bacteriolysis in the macrophage cytosol. Cell Host Microbe 7 (5), 412–419.

Sborgi, L., Rühl, S., Mulvihill, E., Pipercevic, J., Heilig, R., Stahlberg, H., Farady, C.J., Müller, D.J., Broz, P., Hiller, S., 2016. GSDMD membrane pore formation constitutes the mechanism of pyroptotic cell death. EMBO J. 35 (16), 1766–1778.

Scaffidi, P., Misteli, T., Bianchi, M.E., 2002. Release of chromatin protein HMGB1 by necrotic cells triggers inflammation. Nature 418 (6894), 191–195.

Shresta, S., MacIvor, D.M., Heusel, J.W., Russell, J.H., Ley, T.J., 1995. Natural killer and lymphokine-activated killer cells require granzyme B for the rapid induction of apoptosis in susceptible target cells. Proc. Natl. Acad. Sci. U. S. A. 92 (12), 5679–5683.

Skaletskaya, A., Bartle, L.M., Chittenden, T., McCormick, A.L., Mocarski, E.S., Goldmacher, V.S., 2001. A cytomegalovirus-encoded inhibitor of apoptosis that suppresses caspase-8 activation. Proc. Natl. Acad. Sci. U. S. A. 98 (14), 7829–7834.

Song, D., Meng, J., Cheng, J., Fan, Z., Chen, P., Ruan, H., Tu, Z., Kang, N., Li, N., Xu, Y., Wang, X., Shu, F., Mu, L., Li, T., Ren, W., Lin, X., Zhu, J., Fang, X., Amrein, M.W., Wu, W., Yan, L.-T., Lü, J., Xia, T., Shi, Y., 2019. *Pseudomonas aeruginosa* quorum-sensing metabolite induces host immune cell death through cell surface lipid domain dissolution. Nat. Microbiol. 4 (1), 97–111.

Sridharan, H., Ragan, K.B., Guo, H., Gilley, R.P., Landsteiner, V.J., Kaiser, W.J., Upton, J.W., 2017. Murine cytomegalovirus IE3-dependent transcription is required for DAI/ZBP1-mediated necroptosis. EMBO Rep. 18 (8), 1429–1441.

Stutz, M.D., Ojaimi, S., Ebert, G., Pellegrini, M., 2018. Is receptor-interacting protein kinase 3 a viable therapeutic target for *Mycobacterium tuberculosis* infection? Front. Immunol. 9, 1178.

Sun, L., Wang, H., Wang, Z., He, S., Chen, S., Liao, D., Wang, L., Yan, J., Liu, W., Lei, X., Wang, X., 2012. Mixed lineage kinase domain-like protein mediates necrosis signaling downstream of RIP3 kinase. Cell 148 (1–2), 213–227.

Sun, L., Wu, J., Du, F., Chen, X., Chen, Z.J., 2013. Cyclic GMP-AMP synthase is a cytosolic DNA sensor that activates the type I interferon pathway. Science 339 (6121), 786–791.

Suzuki, T., Franchi, L., Toma, C., Ashida, H., Ogawa, M., Yoshikawa, Y., Mimuro, H., Inohara, N., Sasakawa, C., Nuñez, G., 2007. Differential regulation of caspase-1 activation, pyroptosis, and autophagy via Ipaf and ASC in Shigella-infected macrophages. PLoS Pathog. 3 (8), e111.

Suzuki, S., Franchi, L., He, Y., Muñoz-Planillo, R., Mimuro, H., Suzuki, T., Sasakawa, C., Núñez, G., 2014. Shigella type III secretion protein MxiI is recognized by Naip2 to induce Nlrc4 inflammasome activation independently of Pkcδ. PLoS Pathog. 10 (2), e1003926.

Tait, S.W.G., Green, D.R., 2010. Mitochondria and cell death: outer membrane permeabilization and beyond. Nat. Rev. Mol. Cell Biol. 11 (9), 621–632.

Tan, T.Y., Chu, J.J.H., 2013. Dengue virus-infected human monocytes trigger late activation of caspase-1, which mediates pro-inflammatory IL-1β secretion and pyroptosis. J. Gen. Virol. 94 (Pt 10), 2215–2220.

Tenev, T., Bianchi, K., Darding, M., Broemer, M., Langlais, C., Wallberg, F., Zachariou, A., Lopez, J., MacFarlane, M., Cain, K., Meier, P., 2011. The Ripoptosome, a signaling platform that assembles in response to genotoxic stress and loss of IAPs. Mol. Cell 43 (3), 432–448.

Tewari, M., Quan, L.T., O'Rourke, K., Desnoyers, S., Zeng, Z., Beidler, D.R., Poirier, G.G., Salvesen, G.S., Dixit, V.M., 1995. Yama/CPP32β, a mammalian homolog of CED-3, is a CrmA-inhibitable protease that cleaves the death substrate poly(ADP-ribose) polymerase. Cell 81 (5), 801–809.

Thapa, R.J., Ingram, J.P., Ragan, K.B., Nogusa, S., Boyd, D.F., Benitez, A.A., Sridharan, H., Kosoff, R., Shubina, M., Landsteiner, V.J., Andrake, M., Vogel, P., Sigal, L.J., tenOever, B.R., Thomas, P.G., Upton, J.W., Balachandran, S., 2016. DAI senses influenza A virus genomic RNA and activates RIPK3-dependent cell death. Cell Host Microbe 20 (5), 674–681.

Thome, M., Schneider, P., Hofmann, K., Fickenscher, H., Meinl, E., Neipel, F., Mattmann, C., Burns, K., Bodmer, J.L., Schröter, M., Scaffidi, C., Krammer, P.H., Peter, M.E., Tschopp, J., 1997. Viral FLICE-inhibitory proteins (FLIPs) prevent apoptosis induced by death receptors. Nature 386 (6624), 517–521.

Upton, J.W., Kaiser, W.J., Mocarski, E.S., 2010. Virus inhibition of RIP3-dependent necrosis. Cell Host Microbe 7 (4), 302–313.

Upton, J.W., Kaiser, W.J., Mocarski, E.S., 2012. DAI/ZBP1/DLM-1 complexes with RIP3 to mediate virus-induced programmed necrosis that is targeted by murine cytomegalovirus vIRA. Cell Host Microbe 11 (3), 290–297.

Valderrama, J.A., Riestra, A.M., Gao, N.J., LaRock, C.N., Gupta, N., Ali, S.R., Hoffman, H.M., Ghosh, P., Nizet, V., 2017. Group A streptococcal M protein activates the NLRP3 inflammasome. Nat. Microbiol. 2 (10), 1425–1434.

van Opdenbosch, N., Lamkanfi, M., 2019. Caspases in cell death, inflammation, and disease. Immunity 50 (6), 1352–1364.

Varfolomeev, E., Vucic, D., 2008. (Un)expected roles of c-IAPs in apoptotic and NFkappaB signaling pathways. Cell Cycle 7 (11), 1511–1521.

Varfolomeev, E., Vucic, D., 2018. Intracellular regulation of TNF activity in health and disease. Cytokine 101, 26–32.

Vercammen, D., Brouckaert, G., Denecker, G., van de Craen, M., Declercq, W., Fiers, W., Vandenabeele, P., 1998. Dual signaling of the Fas receptor: initiation of both apoptotic and necrotic cell death pathways. J. Exp. Med. 188 (5), 919–930.

Waguia Kontchou, C., Tzivelekidis, T., Gentle, I.E., Häcker, G., 2016. Infection of epithelial cells with Chlamydia trachomatis inhibits TNF-induced apoptosis at the level of receptor internalization while leaving non-apoptotic TNF-signalling intact. Cell. Microbiol. 18 (11), 1583–1595.

Wang, X., Tan, J., Zoueva, O., Zhao, J., Ye, Z., Hewlett, I., 2014. Novel pandemic influenza A (H1N1) virus infection modulates apoptotic pathways that impact its replication in A549 cells. Microbes Infect. 16 (3), 178–186.

Wang, Y., Jia, L., Shen, J., Wang, Y., Fu, Z., Su, S.-A., Cai, Z., Wang, J.-A., Xiang, M., 2018. Cathepsin B aggravates coxsackievirus B3-induced myocarditis through activating the inflammasome and promoting pyroptosis. PLoS Pathog. 14 (1), e1006872.

Wang, J., Deobald, K., Re, F., 2019a. Gasdermin D protects from melioidosis through pyroptosis and direct killing of bacteria. J. Immunol. 202 (12), 3468–3473.

Wang, Y., Hao, Q., Florence, J.M., Jung, B.-G., Kurdowska, A.K., Samten, B., Idell, S., Tang, H., 2019b. Influenza virus infection induces ZBP1 expression and necroptosis in mouse lungs. Front. Cell. Infect. Microbiol. 9, 286.

Warren, S.E., Mao, D.P., Rodriguez, A.E., Miao, E.A., Aderem, A., 2008. Multiple Nod-like receptors activate caspase 1 during *Listeria monocytogenes* infection. J. Immunol. 180 (11), 7558–7564.

Weber, A., Kirejczyk, Z., Besch, R., Potthoff, S., Leverkus, M., Häcker, G., 2010. Proapoptotic signalling through Toll-like receptor-3 involves TRIF-dependent activation of caspase-8 and is under the control of inhibitor of apoptosis proteins in melanoma cells. Cell Death Differ. 17 (6), 942–951.

Weiss, R., Sachet, M., Zinngrebe, J., Aschacher, T., Krainer, M., Hegedus, B., Walczak, H., Bergmann, M., 2013. IL-24 sensitizes tumor cells to TLR3-mediated apoptosis. Cell Death Differ. 20 (6), 823–833.

Weng, D., Marty-Roix, R., Ganesan, S., Proulx, M.K., Vladimer, G.I., Kaiser, W.J., Mocarski, E.S., Pouliot, K., Chan, F.K.-M., Kelliher, M.A., Harris, P.A., Bertin, J., Gough, P.J., Shayakhmetov, D.M., Goguen, J.D., Fitzgerald, K.A., Silverman, N., Lien, E., 2014. Caspase-8 and RIP kinases regulate bacteria-induced innate immune responses and cell death. Proc. Natl. Acad. Sci. U. S. A. 111 (20), 7391–7396.

White, M.J., McArthur, K., Metcalf, D., Lane, R.M., Cambier, J.C., Herold, M.J., van Delft, M.F., Bedoui, S., Lessene, G., Ritchie, M.E., Huang, D.C.S., Kile, B.T., 2014. Apoptotic caspases suppress mtDNA-induced STING-mediated type I IFN production. Cell 159 (7), 1549–1562.

Wilson, N.S., Dixit, V., Ashkenazi, A., 2009. Death receptor signal transducers: nodes of coordination in immune signaling networks. Nat. Immunol. 10 (4), 348–355.

Wu, M.-F., Chen, S.-T., Yang, A.-H., Lin, W.-W., Lin, Y.-L., Chen, N.-J., Tsai, I.-S., Li, L., Hsieh, S.-L., 2013. CLEC5A is critical for dengue virus-induced inflammasome activation in human macrophages. Blood 121 (1), 95–106.

Xie, T., Peng, W., Liu, Y., Yan, C., Maki, J., Degterev, A., Yuan, J., Shi, Y., 2013. Structural basis of RIP1 inhibition by necrostatins. Structure 21 (3), 493–499.

Yamamoto, M., Sato, S., Hemmi, H., Hoshino, K., Kaisho, T., Sanjo, H., Takeuchi, O., Sugiyama, M., Okabe, M., Takeda, K., Akira, S., 2003. Role of adaptor TRIF in the MyD88-independent Toll-like receptor signaling pathway. Science (New York, NY) 301 (5633), 640–643.

Yang, J., Zhao, Y., Shi, J., Shao, F., 2013. Human NAIP and mouse NAIP1 recognize bacterial type III secretion needle protein for inflammasome activation. Proc. Natl. Acad. Sci. U. S. A. 110 (35), 14408–14413.

Yatim, N., Jusforgues-Saklani, H., Orozco, S., Schulz, O., da Barreira Silva, R., Reis e Sousa, C., Green, D.R., Oberst, A., Albert, M.L., 2015. RIPK1 and NF-κB signaling in dying cells determines cross-priming of $CD8^+$ T cells. Science (New York, NY) 350 (6258), 328–334.

Yogarajah, T., Ong, K.C., Perera, D., Wong, K.T., 2017. AIM2 inflammasome-mediated pyroptosis in enterovirus A71-infected neuronal cells restricts viral replication. Sci. Rep. 7 (1), 5845.

Zhai, D., Yu, E., Jin, C., Welsh, K., Shiau, C.-W., Chen, L., Salvesen, G.S., Liddington, R., Reed, J.C., 2010. Vaccinia virus protein F1L is a caspase-9 inhibitor. J. Biol. Chem. 285 (8), 5569–5580.

Zhao, S., Zhou, Y., Fan, Y., Gong, Y., Yang, J., Yang, R., Li, L., Zou, L., Xu, X., Li, G., Liu, S., Zhang, C., Li, G., Liang, S., 2019. Involvement of P2X4 receptor in gp120-induced pyroptosis in dorsal root ganglia. J. Neurochem. 151 (5), 584–594.

Zhou, Q., Snipas, S., Orth, K., Muzio, M., Dixit, V.M., Salvesen, G.S., 1997. Target protease specificity of the viral serpin CrmA. Analysis of five caspases. J. Biol. Chem. 272 (12), 7797–7800.

Zhu, S., Ding, S., Wang, P., Wei, Z., Pan, W., Palm, N.W., Yang, Y., Yu, H., Li, H.-B., Wang, G., Lei, X., de Zoete, M.R., Zhao, J., Zheng, Y., Chen, H., Zhao, Y., Jurado, K.A., Feng, N., Shan, L., Kluger, Y., Lu, J., Abraham, C., Fikrig, E., Greenberg, H.B., Flavell, R.A., 2017. Nlrp9b inflammasome restricts rotavirus infection in intestinal epithelial cells. Nature 546 (7660), 667–670.

Zhu, X., Wu, T., Chi, Y., Ge, Y., Wu, B., Zhou, M., Zhu, F., Ji, M., Cui, L., 2018. Pyroptosis induced by enterovirus A71 infection in cultured human neuroblastoma cells. Virology 521, 69–76.

CHAPTER SIX

A connection in life and death: The BCL-2 family coordinates mitochondrial network dynamics and stem cell fate

Megan L. Rasmussen[a], Vivian Gama[a,b,c,d,*]

[a]Department of Cell & Developmental Biology, Vanderbilt University, Nashville, TN, United States
[b]Neuroscience Program, Vanderbilt University Medical Center, Nashville, TN, United States
[c]Vanderbilt Brain Institute, Vanderbilt University Medical Center, Nashville, TN, United States
[d]Vanderbilt Center for Stem Cell Biology, Vanderbilt University, Nashville, TN, United States
*Corresponding author: e-mail address: vivian.gama@vanderbilt.edu

Contents

Abstract

The B cell CLL/lymphoma-2 (BCL-2) family of proteins control the mitochondrial pathway of apoptosis, also known as intrinsic apoptosis. Direct binding between members of the BCL-2 family regulates mitochondrial outer membrane permeabilization (MOMP) after an apoptotic insult. The ability of the cell to sense stress and translate it into a death signal has been a major theme of research for nearly three decades; however, other mechanisms by which the BCL-2 family coordinates cellular homeostasis beyond its role in initiating apoptosis are emerging. One developing area of research is understanding how the BCL-2 family of proteins regulate development using pluripotent stem cells as a model system. Understanding BCL-2 family-mediated regulation of mitochondrial homeostasis in cell death and beyond would uncover new facets of stem cell maintenance and differentiation potential.

International Review of Cell and Molecular Biology, Volume 353
ISSN 1937-6448
https://doi.org/10.1016/bs.ircmb.2019.12.005

Abbreviations

APAF-1	apoptotic protease activating factor 1
BAD	BCL-2-associated agonist of cell death
BAK1/BAK	BCL-2 antagonist killer 1
BAX	BCL-2 associated X
BCL-2	B cell chronic lymphocytic leukemia/lymphoma 2
BCL-w	BCL-2-like 2
BCL-xL	BCL-2-like 1
BFL-1/A1	BCL-2-related protein A1
BID	BCL-2 interacting domain death agonist
BIK	BCL-2 interacting killer
BIM	BCL-2 interacting mediator of cell death
BMF	BCL-2 modifying factor
BNIP3	BCL2 and adenovirus E1B 19 kDa-interacting protein 3
BOK	BCL-2 ovarian killer
CDK1	cyclin-dependent kinase 1
DRP-1	dynamin-related protein 1
ERAD	endoplasmic reticulum-associated protein degradation pathway
ERK1/2	extracellular signal-regulated kinase 1/2
ERMES	endoplasmic reticulum-mitochondria encounter structure
ESCs/hESCs	embryonic stem cells/human embryonic stem cells
ETC	electron transport chain
HRK	Harakiri
ICM	inner cell mass
IMM	inner mitochondrial membrane
IMS	inner membrane space
iPSCs	induced pluripotent stem cells
JNK	c-Jun N-terminal kinase
KLF4	Kruppel-like factor 4
MAPK	mitogen-activated protein kinase
MCL-1	myeloid cell leukemia 1
MEF	mouse embryonic fibroblast
MFN1/2	mitofusin 1/2
MOMP	mitochondrial outer membrane permeabilization
mtDNA	mitochondrial DNA
NIX	BNIP3-like (BNIP3L)
OCT4	octamer-binding transcription factor 4
OMM	outer mitochondrial membrane
OPA1	optic atrophy 1
PINK1	PTEN-induced putative kinase
PMAIP1/NOXA	phorbol-12-myristate-13-acetate-induced protein 1
PSCs/hPSCs	pluripotent stem cells (iPSCs & ESCs)/human pluripotent stem
PUMA	P53-upregulated modulator of apoptosis
SOX2	sex-determining region Y-box 2

1. Introduction

Stem cells can simulate the earliest stages of development, since they give rise to the three main tissue lineages (Thomson et al., 1998). This capacity to differentiate into ectoderm, mesoderm, and endoderm lineages is known as pluripotency. Embryonic stem cells (ESCs) and induced pluripotent stem cells (iPSCs), collectively referred to as PSCs, are the two stem cell types that harbor this ability. Adult stem cells, also known as somatic stem cells, are multipotent and can replenish dying cells in case of tissue damage, and include hematopoietic stem cells, mesenchymal stem cells, and hair follicle stem cells (reviewed in Goodell et al., 2015). Stem cells also have the capacity of self-renewal, which is the process by which the stem cell pool is maintained indefinitely. These capabilities to regenerate and to give rise to the three germ layers have propelled an entire field of research dedicated to modeling embryonic development in culture by manipulating key signaling pathways and growth factors. The first human ESC (hESC) line was derived in 1998 from the inner cell mass (ICM) of human blastocysts (Evans and Kaufman, 1981; Martin, 1981; Thomson et al., 1998), while the discoveries of reprogramming mouse and human somatic cells into iPSCs were published in 2006 and 2007, respectively (Takahashi et al., 2007; Takahashi and Yamanaka, 2006). Reprogramming was initially achieved by inducing the expression of master pluripotency transcription factors OCT4 (Octomer-binding transcription factor 4), SOX2 (SRY (sex-determining region Y)-box 2), KLF4 (Kruppel-like factor 4) and *c*-MYC (collectively known as OSKM), but other methods of attaining iPSCs have been reported (reviewed in Takahashi and Yamanaka, 2015). The ability of PSCs to self-renew and differentiate has become an efficient tool to study basic processes of human development and various aspects of human diseases, such as diabetes, cardiomyopathy, and cancer (Assady et al., 2001; Hinson et al., 2015; Smith and Tabar, 2019). During embryonic development, genomic instability is especially dangerous for the integrity of rapidly dividing cells of the ICM. Thus, not surprisingly, stem cells are capable of executing intricate programs to quickly respond to apoptotic stress and prevent the propagation of deleterious mutations. Along with the primed cell death program, a growing number of studies on the BCL-2 family have shown changes in mitochondrial dynamics and metabolic function and regulation as stem cells differentiate and as somatic cells reprogram into iPSCs (Dumitru et al., 2012; Gama and Deshmukh, 2012;

Madden et al., 2011; Prigione et al., 2011; Rasmussen et al., 2018; Rinkenberger et al., 2000). In the following chapter, we will discuss the known fundamental mechanisms involved in these changes, centering on the BCL-2 family, as well as describe areas that are open to more detailed exploration (Fig. 1). In addition, many aspects of mitochondrial biology are beginning to emerge as hallmarks of pluripotency and self-renewal (Rastogi et al., 2019; Wanet et al., 2015). The increased sensitivity to apoptosis, the changes in mitochondrial morphology and localization, and the shifting of the metabolic program all accompany reprogramming. Furthermore, cellular events such as mitochondrial biogenesis, mitochondrial trafficking and

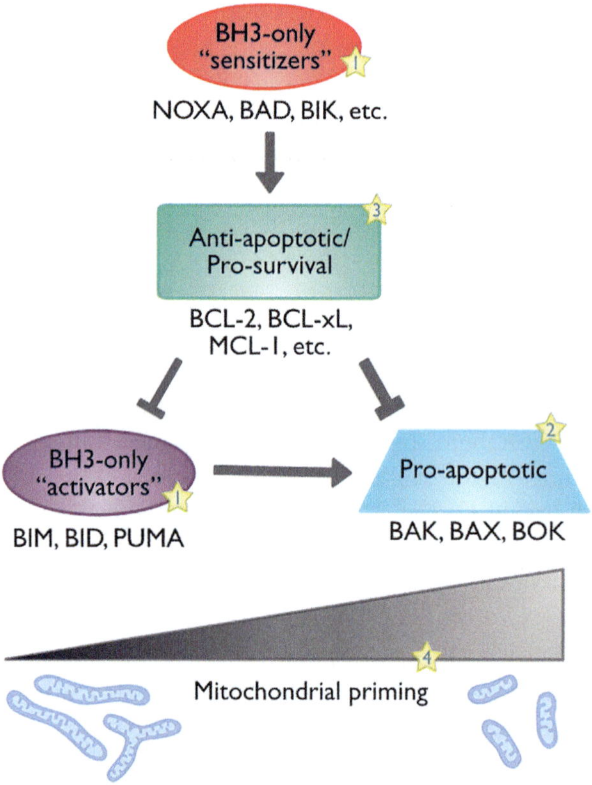

Fig. 1 The BCL-2 family regulates mitochondrial cell death and homeostasis in stem cells. This schematic depicts the canonical pathways of mitochondrial apoptosis and priming. Highlighted are the reported changes in PSC regulation of these pathways: (1) High levels of pro-apoptotic proteins. (2) BAX is maintained in an active state at the Golgi. (3) High levels of MCL-1, which is important for pluripotent maintenance and mitochondrial fission. (4) Increased fragmentation of the mitochondrial network and higher dependence on glycolytic metabolism.

motility, and mitochondrial DNA (mtDNA) transcription could also be important for reprogramming and generation of specialized tissues. Thus, the unique properties of ESCs and iPSCs make them a valuable model system to illuminate the effects of these processes on self-renewal and differentiation.

2. The BCL-2 family in stem cell death

2.1 Mitochondrial pathway of apoptosis

Caspase-dependent apoptosis occurs through both extrinsic and intrinsic pathways, which are mediated by external death ligands and mitochondrial-localized proteins, respectively (Elmore, 2007). The focus of this chapter will be on the intrinsic or mitochondrial pathway of apoptosis; the extrinsic apoptotic pathway is another form of regulated cell death that depends on detection and propagation of extracellular signals, which has been comprehensively reviewed here (Ashkenazi and Dixit, 1998; Galluzzi et al., 2018; Mehlen and Bredesen, 2011). The intrinsic pathway of apoptosis is crucial for embryonic development, tissue homeostasis, and in cellular response to irreversible perturbations. When the proteins that regulate apoptosis are dysregulated, cells can become cancerous and contribute to tumor formation, or die uncontrollably resulting in neurodegeneration. The BCL-2 family are the main regulatory proteins that control the intrinsic pathway of cell death (Fig. 2). The direct binding interactions that occur between pro-apoptotic and anti-apoptotic family members govern their activities, ultimately resulting in MOMP after receiving a death signal (Kale et al., 2018; Oltval et al., 1993). Roughly 30 family members have been identified to belong to the BCL-2 family, which is made up of structurally similar globular proteins that are defined by multiple domains, known as BCL-2 homology (BH) domains (BH1, BH2, BH3, and BH4) (Chipuk et al., 2010; Lindsten et al., 2000; Tsujimoto et al., 1984; Wang et al., 1996; Yang et al., 1996). Family members with multiple BH domains are either pro-apoptotic "effectors" (BCL-2 associated X (BAX), BCL-2 antagonist killer 1 (BAK1; known as BAK), BCL-2 ovarian killer (BOK)) or anti-apoptotic/pro-survival (BCL-2 apoptosis regulator (BCL-2), BCL-2-like 1 (known as BCL-xL), BCL-2-like 2 (BCL-w), myeloid cell leukemia (MCL-1), and BCL-2-related protein A1 (BFL-1/A1)). Another group of family members have only one BH domain, known as the BH3-only proteins. The "activator" pro-apoptotic proteins (BCL-2 interacting mediator of cell death (BIM), BCL-2 interacting domain death agonist (BID), P53-upregulated modulator of apoptosis (PUMA)) and "sensitizer" pro-apoptotic proteins

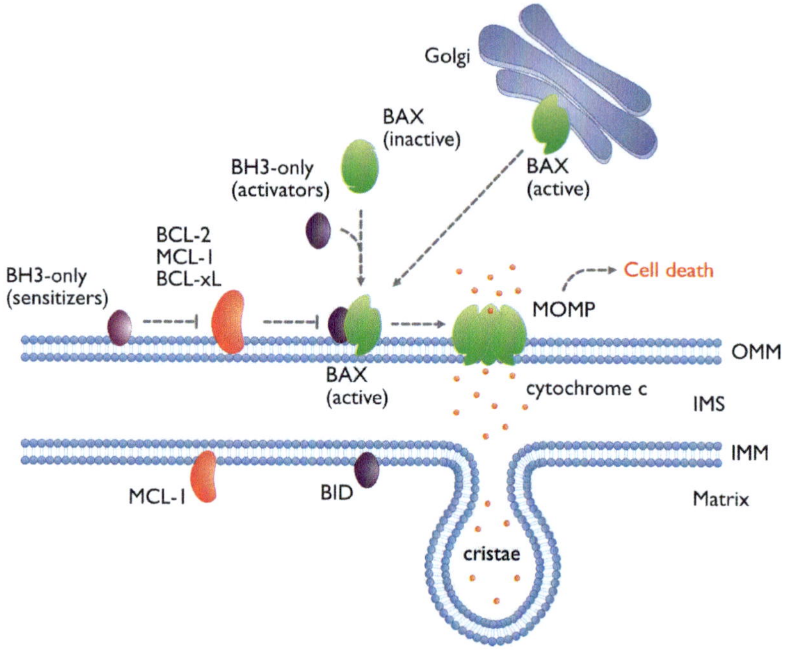

Fig. 2 The BCL-2 family regulates the mitochondrial pathway of apoptosis. When cells receive an intrinsic death signal, BAX becomes activated by the BH3-only activators, undergoes a conformational change, and translocates to the outer mitochondrial membrane (OMM). BAX (in addition to BAK and BOK) then oligomerize and form pores in the OMM, causing MOMP and the release of cytochrome c into the cytosol. BAX activation can be prevented by sequestration of the activator BH3-only proteins through the pro-survival proteins BCL-2, MCL-1, BCL-xL, etc, which are inhibited by the sensitizer BH3-only family members. In human embryonic stem cells, BAX can be maintained in an active state at the Golgi, which confers quick translocation to the OMM in case of a death signal. Additionally, MCL-1 has been found to reside at the matrix in stem cells, where it is proposed to interact with OPA1.

(phorbol-12-myristate-13-acetate-induced protein 1 (PMAIP1, best known as NOXA), BCL-2-associated agonist of cell death (BAD), BCL-2 interacting killer (BIK), BCL-2 modifying factor (BMF), and Harakiri (HRK)) work to either activate the effector proteins directly or sequester the anti-apoptotic family members from binding the effectors. After activation of BAK, BAX and/or BOK by the BH3-only proteins, they oligomerize to form pores in the outer mitochondrial membrane (OMM), initiating MOMP, reducing mitochondrial membrane potential and releasing cytochrome *c* from the inner membrane space (IMS) (Eskes et al., 2000; Goldstein et al., 2000; Kalkavan and Green, 2018; Lindsten et al., 2000; Liu et al., 1996; Moldoveanu and

Czabotar, 2019). Mechanistically, each of the effector proteins are regulated through slightly different means. While BAK is usually found anchored to the OMM, BAX resides in the cytosol in an inactive state and must undergo a conformational change to insert into the OMM through its C-terminal domain (Gavathiotis et al., 2008; Goping et al., 1998; Green and Chipuk, 2008; Hsu and Youle, 1997; Todt et al., 2015; Wolter et al., 1997). Less is known about BOK activity and regulation, but it is proposed to have the ability to initiate MOMP independent of BAK or BAX, and it is modulated by the endoplasmic-reticulum-associated protein degradation (ERAD) pathway (Llambi et al., 2016; Schulman et al., 2016, 2019; Zheng et al., 2018). Cytochrome c release results in Apoptotic protease activating factor-1 (APAF-1)/Caspase-9 assembly, known as the apoptosome (Inoue et al., 2009; Kim et al., 2005; Taylor et al., 2008; Zou et al., 1997), which activates the executioner caspases, leading to the extensive cleavage and destruction of critical cellular components (Dix et al., 2008; Fuchs and Steller, 2011; Li et al., 1997; Mahrus et al., 2008; Ramirez and Salvesen, 2018). Morphological changes occur, including cell constriction, organelle fragmentation, chromatin condensation, plasma membrane "blebbing" as the cortex ruptures, and finally the break-up of the cell into apoptotic bodies, which are the defining features of apoptosis (Kerr et al., 1972).

There are several proposed models of how the BCL-2 family execute cell death upstream of MOMP (reviewed in Shamas-Din et al., 2013): the direct activation model, the displacement model, the embedded together model, and the unified model. The latter two models, which built upon the direction activation and displacement models, highlight the complexity of the mitochondrial cell death pathway. The embedded together model proposes that the specific interactions of the BCL-2 family members are dependent on cellular equilibria, with binding events occurring based on concentration and binding affinities of each protein (Leber et al., 2007). These affinities are also dependent on proximity to the mitochondrial membrane as well as post-translational modifications. The unified model of apoptosis differs in that it proposes that the anti-apoptotic proteins have two "modes": one in binding and sequestering the activator BH3-only proteins, which can be overcome by the sensitizer BH3-only proteins, and the other binding and sequestering BAK/BAX, which results in a more robust inhibition of MOMP (Kalkavan and Green, 2018; Llambi et al., 2011). The unified model also includes the effects that the BCL-2 family proteins have on mitochondrial dynamics, thus expanding on the embedded together model.

2.2 The BCL-2 family in stem cells

The intrinsic apoptosis has been shown to be uniquely regulated in stem cells (Dumitru et al., 2012; Gama and Deshmukh, 2012; Liu et al., 2013; Madden et al., 2011; Rasmussen et al., 2018). Human PSCs (hPSCs) are extremely sensitive to DNA perturbations in a P53-dependent manner. P53 protein expression, however, was not significantly different between hPSCs and differentiated cells, pointing to another mode of action for this rapid apoptotic response. It was found that expression of the BCL-2 family in stem cells is shifted so that pro-apoptotic members are upregulated, while pro-survival members are downregulated, which serves to position them closer to the threshold of apoptosis, a property known as mitochondrial priming (Certo et al., 2006; Liu et al., 2013). BAX aides in this high mitochondrial priming state of hPSCs (Dumitru et al., 2012). BAX was found in an activated state in homeostatic conditions but localized at the Golgi until an apoptosis-initiating event occurred (e.g., etoposide-mediated DNA damage), when it would translocate to the mitochondria (Dumitru et al., 2012). How BAX is maintained at the Golgi in an active state remains elusive, as up to this point active BAX, which is identified by its N-terminal domain (Upton et al., 2007), had only been known to localize at the OMM, and furthermore only in dying cells. When hPSCs were differentiated into stem cell aggregates known as embryoid bodies, BAX was no longer active during homeostasis and retained its mitochondrial localization during apoptotic stress, indicating a stem cell-specific mechanism of BCL-2 family regulation (Dumitru et al., 2012). Maintaining BAX in an active conformation and in a high mitochondrial priming state likely allows for hPSCs to respond rapidly to cell death, thereby preventing potential mutations induced by DNA damage from persisting in the stem cell pool or in the differentiating cell lineage. Conversely, sustaining BAX in an active state would also pose significant risk for spontaneous apoptosis. Thus, isolating BAX at the Golgi could serve to safeguard hPSCs from unnecessary cell death while still allowing for its rapid induction if the cell is antagonized. At the organismal level, this sensitive response may stop the development of abnormal cells in the growing embryo. If constitutive BAX activation is also maintained by other highly proliferative cells in adult tissue, then perhaps it prevents aberrant growth of cancer cells and tumors. Dissection of the mechanisms underpinning BAX activation and localization, both at the Golgi and at the mitochondria, will be necessary to fully elucidate this apoptotic sensitivity as a defining component of stem cell homeostasis. As PSCs begin to differentiate, they quickly become resistant to DNA damage-mediated apoptosis; this was shown to be facilitated in part by a decrease in mitochondrial priming (Liu et al., 2013; Madden et al., 2011). Mitochondrial priming and apoptotic sensitivity have been exploited in recent years to identify dependencies of cancer cells on specific BCL-2 family proteins. This technique is known as BH3 profiling (Certo et al., 2006; Montero et al., 2015; Montero and Letai, 2016), and it is useful to measure the different binding affinities between the BCL-2 family members across cell types and patient-derived tumors.

The pro-apoptotic members of the BCL-2 family become activated through several upstream sensing mechanisms. As previously mentioned, PSCs are highly sensitive to DNA damage, which in turn initiates the transcription factor P53 and triggers a repair response and cell cycle arrest (Kruse and Gu, 2009). If the cell cannot resolve the damage, the BCL-2 family facilitates MOMP and the cell undergoes apoptosis in a caspase-dependent manner. Fluctuations in P53 activation have been shown to determine cell fate (Purvis et al., 2012). During reprogramming of somatic cells back to a pluripotent state, inhibition of P53 greatly improves cell survival due to the bypass of BCL-2 family-mediated cell death (Kawamura et al., 2009; Li et al., 2013). Excellent reviews covering the details of P53 function in stem cells can be found here (Bonizzi et al., 2012; Jain and Barton, 2018; Levine et al., 2016). Downstream of BCL-2 family facilitation of MOMP, activity of caspases has been shown to play dual roles in both promoting differentiation and maintaining proper distribution of differentiated cell types, as well as driving reprogramming of somatic cells (Fu et al., 2019; Fujita et al., 2008; Li et al., 2010).

3. The BCL-2 family in mitochondrial dynamics

The main players of intrinsic apoptosis, the BCL-2 family of proteins, execute their functions at the mitochondria. Beyond being the energy-producing powerhouses of the cell, mitochondria serve as major signaling organelles and constantly fuse and divide in a process known as mitochondrial dynamics (Berman et al., 2008; Chan, 2007; Friedman and Nunnari, 2014) (Fig. 3). These continuous, energy-consuming events are proposed to occur both in response to various stimuli and to relay signals throughout the cell to other organelles. Dynamic movement and fragmentation of the mitochondrial network was first documented in 1915, with various morphological changes described in detailed, hand-drawn illustrations

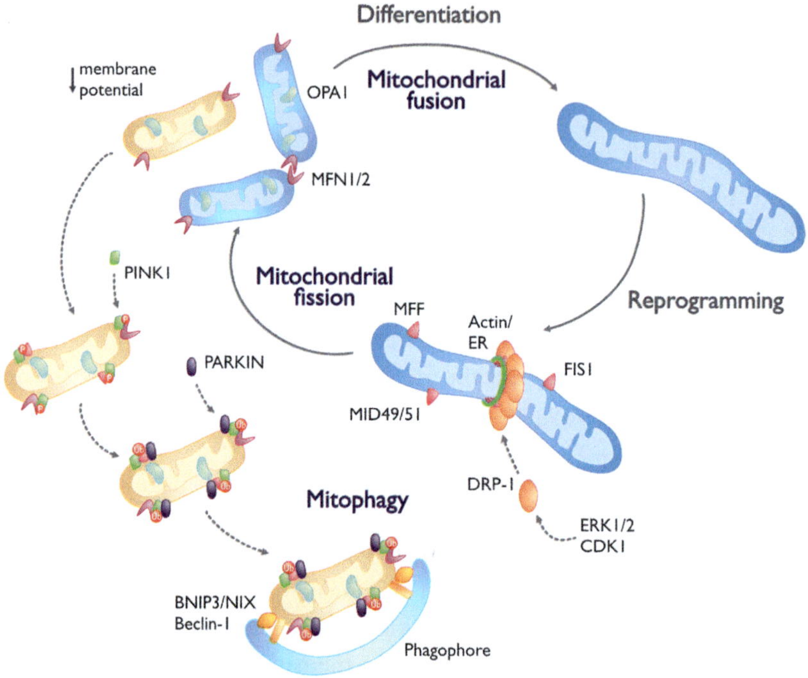

Fig. 3 The dynamic cycle of mitochondrial morphology and quality control. In homeostasis, differentiation, reprogramming, and during stress, the mitochondrial network undergoes rounds of fusion and fission. During fission, DRP-1 is activated and recruited to pre-constricted regions of mitochondria (facilitated by actin and ER tubules) by its receptors (MFF, MID49/51, FIS1). Mitochondrial fusion is achieved by the activity of MFNs, which dimerize with MFNs on adjacent mitochondria to fuse the OMM. This is followed by OPA1-mediated fusion of the inner mitochondria membrane (IMM). If mitochondria are damaged (i.e., low membrane potential), PINK1 is stabilized at the mitochondria and phosphorylates proteins on the OMM. This phosphorylation triggers Parkin-mediated ubiquitination of OMM proteins, causing recruitment of various adapter proteins and receptors associated with the phagophore. The damaged mitochondrion is then engulfed by the mature mitophagophore, which fuses with the lysosome for degradation.

(Lewis and Lewis, 1914). Yet, the notion that mitochondria are static, bean-like organelles persevered over the decades. The development of more advanced imaging technologies has allowed for improved observation and tracking of mitochondrial dynamics and motility, which has propelled the field and revealed the remarkable changes that mitochondria undergo within cells (Bereiter-Hahn and Vöth, 1994; Johnson et al., 1981; Rizzuto et al., 1996). Emerging studies have also highlighted new links between mitochondrial dynamics machinery, apoptosis and mitochondrial autophagy (mitophagy) (Fu et al., 2019; Kageyama et al., 2014; Morciano et al., 2016b; Rasmussen et al., 2018). Proper equilibrium of mitochondrial dynamics is essential for cellular and organismal homeostasis, and diseases often manifest when the proteins that regulate fission and fusion are mutated (Chan, 2007, 2020; Dorn, 2013; Westermann, 2010).

3.1 Mitochondrial fission

Mitochondrial shape changes drastically depending on the cell type, cell cycle stage, and metabolic state (Chen and Chan, 2017). Differentiated cell types, such as fibroblasts, myocytes, and endothelial cells, generally have elongated, cristae-rich mitochondria that make up complex networks, which promote mtDNA homogenization and energy generation (Kuznetsov et al., 2009). Mitochondrial fission is necessary for distribution of mitochondria to daughter cells during cell division, mitochondrial transport within cells, and isolation of damaged mitochondria for degradation by mitophagy (Ganesan et al., 2019; Nunnari et al., 1997; Pon, 2013). As differentiated cells reprogram and transition into a stem-like state, mitochondria undergo higher levels of fission and arrange to be more perinuclear (Chen and Chan, 2017; Prieto et al., 2016). The mitochondria of hESCs and iPSCs are considered more functionally immature, with globular shapes and fewer cristae, contributing to a metabolic profile that is traditionally thought to be more dependent on glycolysis for energy requirements (Chung et al., 2010; Facucho-Oliveira and St. John, 2009). Constitutive activation of the dynamin-related guanosine triphosphates (GTPases) that control fission is likely a major participating factor in maintaining high levels of fragmentation in stem cells (Chung et al., 2010; Prieto et al., 2016; Rasmussen et al., 2018).

The large dynamin-related GTPases regulate mitochondrial dynamics by inducing mitochondrial fission (division) and fusion (reviewed in Section 3.2) in a highly conserved manner (Chan, 2012; Hoppins et al., 2007; Praefcke and McMahon, 2004). Dynamin-related protein 1 (DRP-1) executes its function upon activation by several post-translational modifications, including phosphorylation, ubiquitination, sumoylation, and O-GlcNAcylation (Westermann, 2010). The enzymes that perform this activation of Dynamin-related proteins have not been completely identified, but ERK1/2 and CDK1 have been shown phosphorylate DRP-1 at Serine 616, providing insight into specific pathways that promote increased mitochondrial fragmentation (Prieto et al., 2016; Taguchi et al., 2007). Once DRP-1 is activated, it translocates from the cytosol to the OMM, where it hydrolyzes GTP and self-assembles around the mitochondria, constricting both membranes until the organelle is divided in two (Antonny et al., 2016; Francy et al., 2017; Hoppins et al., 2007; Pon, 2013). The structural domains and mechanistic details of action for DRP-1 have been described in several studies and reviews (Chappie et al., 2010; Francy et al., 2017; Gao et al., 2010; Hoppins et al., 2007; Mears et al., 2011; Meglei and McQuibban, 2009; van der Bliek and Payne, 2010); however, whether or not DRP-1 activation is differentially regulated in PSCs, which have a more fragmented mitochondrial network, has not been elucidated. DRP-1 function in mitochondrial fission is required for proper achievement of pluripotency early on in the reprogramming process (Prieto et al., 2016). As DRP-1 is also involved in the mitochondrial fragmentation observed during apoptosis (Frank et al., 2001), and iPSCs are highly sensitive to apoptotic stimuli, these cells are likely highly dependent on DRP-1 beyond its role in homeostatic conditions. It is important to note that some studies suggest the possibility that mitochondrial fission can occur in the absence of DRP-1 (Onoue et al., 2013; Stavru et al., 2013), and/or that DRP-1 requires additional proteins for severance of the mitochondria (i.e., dynamin-2) (Lee et al., 2016), a phenomenon that remains controversial (Fonseca et al., 2019; Kamerkar et al., 2018). Moving forward, additional studies into the mechanisms by which mitochondrial fission is regulated across different cell types, cell states and model systems are needed for appropriately interpreting conflicting results, and to better elucidate the intricate events coordinating mitochondrial fragmentation.

3.2 Mitochondrial fusion

As stem cells differentiate and become more mature, the mitochondrial network transforms to support the changing needs of the differentiating cell. Mitochondrial fusion is required for the proper maturation of several tissue types, and it is necessary for mtDNA homogenization and assembly of the electron transport chain (ETC) (Chan, 2012). The specific mechanisms by which proteins control mitochondrial fusion are not well understood. In vitro studies have shown that OMM fusion is a mechanistically distinct process from fusion of the inner mitochondrial membrane (IMM) (Legros et al., 2002; Meeusen et al., 2004). The large GTPases that govern the process of mitochondrial fusion are Mitofusin 1 (MFN1), Mitofusin 2 (MFN2), and Optic Atrophy 1 (OPA1) (Alexander et al., 2000; Chen et al., 2003; Delettre et al., 2000). MFN1 and MFN2 are localized to the OMM, where they facilitate outer membrane fusion by homo- or hetero-dimerizing with

MFNs on adjacent mitochondria and actively fusing the outer membranes. While the MFN paralogs are highly similar, they are not completely redundant and are both required for mitochondrial fusion (Escobar-Henriques and Joaquim, 2019). Depletion of MFN1 in mouse embryonic fibroblasts (MEFs) leads to increased fragmentation of individual mitochondria, resulting in small fragments, while MFN2 depletion results in larger fragments of aggregated mitochondria (Chen et al., 2003). Interestingly, double knockout embryos die earlier than single mutants (Chen et al., 2003) and both MFN1 and 2 levels vary in expression among different tissues (Santel et al., 2003), pointing to a non-redundant requirement for the Mitofusins during early development (Schrepfer and Scorrano, 2016).

Once OMM fusion occurs, the IMM is fused through the activity of OPA1. OPA1 regulation is complex, as it has eight different splice isoforms that lead to the formation of two proteolytically cleaved proteins, designated as long OPA1 (OPA1-L) and short OPA1 (OPA1-S), which serve several distinct and overlapping purposes (Del Dotto et al., 2017; Mishra et al., 2014). OPA1-L is anchored in the IMM with its GTPase domain exposed to IMS. It coordinates the active process of IMM fusion by forming homodimers with OPA1-L proteins on the opposite target membrane. OPA-1S is proposed to provide a more passive, structural role at the mitochondrial matrix in cristae formation and maintenance, while combinations of both long and short forms work together to fine-tune mitochondrial morphology and function (Del Dotto et al., 2018). The OPA1-S form has also been implicated in fission of the IMM (Anand et al., 2014). These studies were performed in MEFs, but it would be interesting to probe if the homeostatic balance of OPA1 isoforms is differentially regulated in stem cells and over the course of differentiation. Disproportionate levels of either mitochondrial fusion or fission results in profound pathological abnormalities. These include embryonic lethality at mid-gestation in mice deficient in any of the dynamin-related proteins (i.e., MFN1, MFN2, OPA1, and DRP1) (Alavi et al., 2007; Chen et al., 2003; Davies et al., 2007; Ishihara et al., 2009; MacVicar and Langer, 2016), as well as neurodegenerative diseases such as Charcot-Marie-Tooth syndrome and dominant optic atrophy (Alavi et al., 2007; Davies et al., 2007; Waterham et al., 2007; Züchner et al., 2004) caused by mutations in MFN2 and OPA1, respectively.

3.3 Mitochondrial dynamics and apoptosis

Mitochondrial dynamics also change during apoptosis; mitochondrial fusion is thought to protect against mitochondrial pore formation (Estaquier and

Arnoult, 2007; Jahani-Asl et al., 2010), while mitochondrial fission has been shown occur concurrently with BAK/BAX-mediated cell death (Frank et al., 2001; Youle and van der Bliek, 2012). Parallel to this, studies have shown that pro-apoptotic BAK and BAX must be inactivated for mitochondrial fusion to occur, and BAX has been shown to modulate fusion through interactions with the Mitofusins (Brooks et al., 2007; Hoppins et al., 2007, 2011; Karbowski et al., 2006). Additionally, other BCL-2 family members have recently been associated with the maintenance of mitochondrial dynamics in various cell types (Chen et al., 2011; Hardwick and Soane, 2013). In adult neurons, the anti-apoptotic protein BCL-xL is enriched in the mitochondria, where it promotes proper mitochondrial length and size, as well as localization of the mitochondria to the highly energetic synapses (Berman et al., 2008; Li et al., 2008). BCL-xL has been reported to be involved in neuronal connectivity and communication, which are required for normal brain development (Nakamura et al., 2016). Another anti-apoptotic family member, MCL-1, has been shown to also function in the regulation of mitochondrial dynamics in cancer cells and hPSCs (Morciano et al., 2016a; Rasmussen et al., 2018). MCL-1 is expressed at high levels in hPSCs, and its depletion causes mitochondrial elongation in stem cells that coincides with the loss of expression of pluripotency markers OCT4 and NANOG (Rasmussen et al., 2018). The changes in mitochondrial morphology are likely due to MCL-1's interaction with DRP-1, but the details of the state of DRP-1 activation are still unknown. Interestingly, MCL-1 was also shown to interact with OPA1, an association that likely occurs through MCL-1's matrix-localized form (Rasmussen et al., 2018). These non-apoptotic interactions further implicate MCL-1, and perhaps other BH domain proteins, in the modulation of mitochondrial dynamics and homeostasis beyond their roles in cell death. A recent study showed that BID, a pro-apoptotic BH3-only protein, is also imported into the mitochondrial matrix where it regulates cristae morphology and organization (Salisbury-Ruf et al., 2018). As BID is a known binding protein of MCL-1, this study showed these two family members interact at the matrix in cancer cells. It would be interesting to elucidate the dynamics of this interaction in stem cells. For instance, MCL-1 deficiency causes significant changes in the mitochondrial network of iPSCs where BID is highly expressed (Rasmussen et al., 2018). Whether MCL-1's remodeling of the mitochondria is dependent on BID or whether MCL-1 and BID work in concert at the matrix to maintain pluripotency/self-renewal properties in stem cells is not known. Understanding the regulation of cell fate from the perspective of mitochondrial dynamics and the BCL-2 family of proteins could shed light on the intricacies of cell signaling and organelle biology.

As previously mentioned, extensive remodeling of the mitochondrial network occurs during apoptosis. These changes result in the release of cytochrome *c* and other caspase-inducing proteins from the IMS (Galluzzi et al., 2018; Garrido et al., 2006; Goldstein et al., 2000). Mitochondrial fragmentation during apoptosis takes place through two synchronized, but independent, events: dissociation of cristae junctions, where pools of cytochrome *c* are located, and BAK/BAX oligomerization and pore formation at the outer membrane (Gao et al., 2001; Lee et al., 2004; Montessuit et al., 2010; Ow et al., 2008; Sheridan et al., 2008; Sinibaldi et al., 2013). A growing number of studies show that DRP-1 acts in response to apoptotic signals, consistent with the idea that mitochondrial fission occurs along with MOMP. DRP-1 has been shown promote BAX translocation to the mitochondria (Wang et al., 2015), and it also localizes to BAX/BAK pores (Frank et al., 2001; Tanaka and Youle, 2008) where it promotes extensive mitochondrial network fission. DRP-1 depletion prevents mitochondrial fission during apoptosis, and overexpression of the DRP-1 mutant K38A, which inhibits GTP binding by DRP-1, prevents apoptosis-induced mitochondrial fission (Estaquier and Arnoult, 2007; Smirnova et al., 2001; Sugioka et al., 2004). BAK and BAX have also been reported to localize at fission sites along with DRP-1 and MFN2, providing further evidence that MOMP is associated with the fission regulators (Hoppins et al., 2011; Karbowski et al., 2002).

OPA1 has been shown to be involved during MOMP where its disassembly is needed to facilitate cytochrome *c* release (Frezza et al., 2006; Jiang et al., 2014; Yamaguchi et al., 2008). A number of studies demonstrated that apoptosis requires OPA1-dependent cristae remodeling initiated by metalloprotease OMA1-mediated cleavage of OPA1. Furthermore, activation of OMA1 is mediated by pro-apoptotic members of the BCL-2 family. How these functions of OPA1 are regulated during apoptosis in stem cells is not clear, however, these studies point to distinct functions of OPA1 in mitochondrial fusion and in cristae remodeling during apoptosis (Arnoult et al., 2005; Frezza et al., 2006; Jiang et al., 2014; Yamaguchi et al., 2008).

3.4 Mitophagy

Concomitant with increased mitochondrial fission is the clearance of damaged mitochondria by mitophagy (Burman et al., 2017; Zhang, 2013) (Fig. 3).

Mitophagy is proposed to be an opposing process to apoptosis, and initiation of either pathway is dependent on signaling crosstalk and cell state (Pickles et al., 2018). Activation of PTEN-induced putative kinase 1 (PINK1) aids in the selective clearance of mitochondria through recruitment of the E3 ubiquitin ligase Parkin, which translocates to the mitochondria from the cytosol and targets mitochondrial membrane proteins for degradation by the autophagosome (Narendra et al., 2008). MFN1 and MFN2 are substrates for Parkin-mediated ubiquitination (Gegg et al., 2010). However, this modification targets them for proteasomal degradation rather than acting as a mitophagy signal, suggesting that elimination of mitochondrial fusion and a shift to fission occurs prior to mitophagy. BCL-2 and BCL-xL have been shown to negatively regulate mitophagy through inhibition of Beclin-1 (also known as ATG6), a BH3 domain-containing protein that regulates formation of the phagophore along with adaptor proteins such as BNIP3 (BCL2 and adenovirus E1B 19 kDa-interacting protein 3) and NIX (BNIP3-like) (Dhingra et al., 2014; Fernández et al., 2018; Maiuri et al., 2007; Pattingre et al., 2005; Vásquez-Trincado et al., 2016; Zhang and Ney, 2009). Another study in HeLa cells showed that overexpression of any of the anti-apoptotic family members (BCL-2, BCL-xL, BCL-w, MCL-1, A1) prevented PINK1/Parkin-mediated mitophagy, but this block was independent of Beclin-1 (Hollville et al., 2014).

Mitophagy can maintain cellular homeostasis and is also triggered by nutrient deprivation and other stressors; however, mitophagy can also be developmentally initiated (Pickles et al., 2018). In fact, decreased levels of mitophagy accompany increased mitochondrial fusion that occurs during stem cell differentiation, and increased mitophagy during somatic cell reprogramming promotes the switch in metabolism back to glycolytic dependence, as it reduces mitochondrial mass, mtDNA content, and ETC components that facilitate oxidative phosphorylation (Fu et al., 2019; Vazquez-Martin et al., 2012). Mitophagy is essential in post-mitotic cells, since they have lost their proliferative capacity and rely on maintaining a healthy pool of mitochondria (Evans and Holzbaur, 2019; Zhang, 2013). However, high levels of mitochondrial fission in stem cells would perhaps require high activation of mitophagy for quality control purposes (Naik et al., 2019). Supporting this, late-passage PINK1-deficient mouse iPSCs, while they expressed the pluripotency factors OCT4 and SOX2, were more prone to differentiation and showed accumulation of damaged mitochondria (Vazquez-Martin et al., 2016).

3.5 Inter-organelle contacts

Although it is known that mitochondrial fragmentation and the protein machinery that coordinate this process are essential, details of how mitochondrial membrane integrity is disrupted, yet in a controlled manner, are not well understood. Structural analysis of DRP-1 discovered that the rings formed by DRP-1 oligomers are smaller in diameter (30–50 nm) than the average diameter of mitochondria (Ingerman et al., 2005; Mears et al., 2011), suggesting there are likely other constriction mechanisms involved. Recent work has identified the endoplasmic reticulum (ER) as well as the actin cytoskeleton as key players in mitochondrial dynamics (Lewis et al., 2016; Rehklau et al., 2017; Steffen and Koehler, 2018). ER tubules can be found at sites of fission, where they form contacts with the mitochondria through the ER-mitochondria encounter structure (ERMES) complex and initiated fission even before DRP-1 constriction (Elgass et al., 2015; Lewis et al., 2016; Yang and Svitkina, 2019). Tethering of the ER to mitochondria might serve not only to facilitate mitochondrial fission, but also to regulate mitochondrial membrane biosynthesis, calcium signaling, and protein import (reviewed in Kornmann and Walter, 2010). Following ER tubule encirclement of the mitochondria, it is proposed that INF2-induced actin polymerization occurs, triggering an initial constriction mechanism to facilitate DRP-1 recruitment, assembly, and final scission of the membrane (Korobova et al., 2013; Pon, 2013). It will be interesting to determine if members of the BCL-2 family such as MCL-1 and BAK/BAX are involved in this mitochondrial fission "complex" in cells subjected to sublethal doses of apoptotic stimuli. It could also be informative to examine potential adaptations to these processes in hPSCs and cancer stem cells where the balance of mitochondrial dynamics is shifted to favor high fission activity.

Increased mitochondrial priming and mitochondrial fission are two vital components that allow for pluripotent maintenance, and these properties change upon cellular differentiation; however, the precise signaling events and protein interactions that govern these processes are yet to be elucidated. The BCL-2 family affects mitochondrial priming, but we also propose that BCL-2 proteins regulate fragmentation; reprogramming of differentiated cells into iPSCs will be a crucial tool to dissect this intriguing possibility in a temporal and cell-specific manner. Further investigation into the relationship between mitochondrial priming, fragmentation and the acquisition of pluripotency or a particular stem cell fate opens an exciting opportunity for future mechanistic studies.

4. Concluding remarks and future perspectives

As discussed throughout this chapter, the BCL-2 family of proteins regulate the mitochondrial pathway of apoptosis and are thereby at the center of several essential signaling pathways. The mitochondrial network is responsible for the regulation of apoptosis, oxidative phosphorylation, coordination of neighboring organelles, and they provide the machinery needed to generate key metabolites that serve the bioenergetic and biosynthetic needs of the cell (Fu et al., 2019). The crosstalk between the pathways of BCL-2 family-mediated cell death, mitochondrial dynamics, and mitophagy must be explored further in PSCs, as it will likely provide important details into how cells navigate the differentiation and reprogramming processes. The role of reactive oxygen species on stem cell maintenance is also an intriguing area of study, since multiple redox-sensitive pathways also control apoptosis and mitochondrial dynamics (e.g., MAPK, ERK1/2, JNK) (Bigarella et al., 2014).

Many questions still remain; for example: (1) How do the signaling pathways that converge at the mitochondria intersect to regulate cell fate? (2) How are the changes in mitochondrial network, mitochondrial turnover, and mitochondrial connectivity to other organelles regulated as cells undergo differentiation and reprogramming? (3) How do changes in mitochondrial morphology contribute to adaptations in metabolic state? (4) How do metabolite levels modulate the activity of epigenetic enzymes ultimately affecting gene expression? Answering these questions would open the opportunity to advance the fields of stem cell biology and apoptosis as well as to rationalize new therapies, as shown by the emergence of pharmacological inhibitors of the BCL-2 family.

Acknowledgments

We apologize to colleagues whose work has not been included in this review due to space limitations. We would like to thank the Gama laboratory for their experimental and conceptual contributions, which led to some of the ideas presented in this review.

This work was supported by NIH Grants: 1R35GM128915-01 (to V.G.), 1R21CA227483-01A1 (to V.G.), and AHA Grant: 19PRE34380515 (to M.R.).

References

Alavi, M.V., Bette, S., Schimpf, S., Schuettauf, F., Schraermeyer, U., Wehrl, H.F., Ruttiger, L., Beck, S.C., Tonagel, F., Pichler, B.J., Knipper, M., Peters, T., Laufs, J., Wissinger, B., 2007. A splice site mutation in the murine Opa1 gene features pathology of autosomal dominant optic atrophy. Brain 130, 1029–1042. https://doi.org/10.1093/brain/awm005.

Alexander, C., Votruba, M., Pesch, U.E.A., Thiselton, D.L., Mayer, S., Moore, A., Rodriguez, M., Kellner, U., Leo-Kottler, B., Auburger, G., Bhattacharya, S.S., Wissinger, B., 2000. OPA1, encoding a dynamin-related GTPase, is mutated in autosomal dominant optic atrophy linked to chromosome 3q28. Nat. Genet. 26, 211–215. https://doi.org/10.1038/79944.

Anand, R., Wai, T., Baker, M.J., Kladt, N., Schauss, A.C., Rugarli, E., Langer, T., 2014. The i- AAA protease YME1L and OMA1 cleave OPA1 to balance mitochondrial fusion and fission. J. Cell Biol. 204, 919–929. https://doi.org/10.1083/jcb.201308006.

Antonny, B., Burd, C., De Camilli, P., Chen, E., Daumke, O., Faelber, K., Ford, M., Frolov, V.A., Frost, A., Hinshaw, J.E., Kirchhausen, T., Kozlov, M.M., Lenz, M., Low, H.H., McMahon, H., Merrifield, C., Pollard, T.D., Robinson, P.J., Roux, A., Schmid, S., 2016. Membrane fission by dynamin: what we know and what we need to know. EMBO J. 35, 2270–2284. https://doi.org/10.15252/embj.201694613.

Arnoult, D., Grodet, A., Lee, Y.-J., Estaquier, J., Blackstone, C., 2005. Release of OPA1 during apoptosis participates in the rapid and complete release of cytochrome c and subsequent mitochondrial fragmentation. J. Biol. Chem. 280, 35742–35750. https://doi.org/10.1074/jbc.M505970200.

Ashkenazi, A., Dixit, V.M., 1998. Death receptors: signaling and modulation. Science 281, 1305–1308. https://doi.org/10.1126/science.281.5381.1305.

Assady, S., Maor, G., Amit, M., Itskovitz-Eldor, J., Skorecki, K.L., Tzukerman, M., 2001. Insulin production by human embryonic stem cells. Diabetes 50, 1691–1697. https://doi.org/10.2337/diabetes.50.8.1691.

Bereiter-Hahn, J., Vöth, M., 1994. Dynamics of mitochondria in living cells: shape changes, dislocations, fusion, and fission of mitochondria. Microsc. Res. Tech. 27, 198–219. https://doi.org/10.1002/jemt.1070270303.

Berman, S.B., Pineda, F.J., Hardwick, J.M., 2008. Mitochondrial fission and fusion dynamics: the long and short of it. Cell Death Differ. 15, 1147–1152. https://doi.org/10.1038/cdd.2008.57.

Bigarella, C.L., Liang, R., Ghaffari, S., 2014. Stem cells and the impact of ROS signaling. Development 141, 4206–4218. https://doi.org/10.1242/dev.107086.

Bonizzi, G., Cicalese, A., Insinga, A., Pelicci, P.G., 2012. The emerging role of p53 in stem cells. Trends Mol. Med. 18, 6–12. https://doi.org/10.1016/j.molmed.2011.08.002.

Brooks, C., Wei, Q., Feng, L., Dong, G., Tao, Y., Mei, L., Xie, Z.-J., Dong, Z., 2007. Bak regulates mitochondrial morphology and pathology during apoptosis by interacting with mitofusins. Proc. Natl. Acad. Sci. 104, 11649–11654. https://doi.org/10.1073/pnas.0703976104.

Burman, J.L., Pickles, S., Wang, C., Sekine, S., Vargas, J.N.S., Zhang, Z., Youle, A.M., Nezich, C.L., Wu, X., Hammer, J.A., Youle, R.J., 2017. Mitochondrial fission facilitates the selective mitophagy of protein aggregates. J. Cell Biol. 216, 3231–3247. https://doi.org/10.1083/jcb.201612106.

Certo, M., Moore, V.D.G., Nishino, M., Wei, G., Korsmeyer, S., Armstrong, S.A., Letai, A., 2006. Mitochondria primed by death signals determine cellular addiction to antiapoptotic BCL-2 family members. Cancer Cell 9, 351–365. https://doi.org/10.1016/j.ccr.2006.03.027.

Chan, D.C., 2007. Mitochondrial dynamics in disease. N. Engl. J. Med. 356, 1707–1709. https://doi.org/10.1056/NEJMp078040.

Chan, D.C., 2012. Fusion and fission: interlinked processes critical for mitochondrial health. Annu. Rev. Genet. 46, 265–287. https://doi.org/10.1146/annurev-genet-110410-132529.

Chan, D.C., 2020. Mitochondrial dynamics and its involvement in disease. Annu. Rev. Pathol. Mech. Dis. 15, 235–259. https://doi.org/10.1146/annurev-pathmechdis-012419-032711.

Chappie, J.S., Acharya, S., Leonard, M., Schmid, S.L., Dyda, F., 2010. G domain dimerization controls dynamin's assembly-stimulated GTPase activity. Nature 465, 435–440. https://doi.org/10.1038/nature09032.

Chen, H., Chan, D.C., 2017. Mitochondrial dynamics in regulating the unique phenotypes of cancer and stem cells. Cell Metab. 26, 39–48. https://doi.org/10.1016/j.cmet.2017.05.016.

Chen, H., Detmer, S.A., Ewald, A.J., Griffin, E.E., Fraser, S.E., Chan, D.C., 2003. Mitofusins Mfn1 and Mfn2 coordinately regulate mitochondrial fusion and are essential for embryonic development. J. Cell Biol. 160, 189–200. https://doi.org/10.1083/jcb.200211046.

Chen, Y., Aon, M.A., Hsu, Y.-T., Soane, L., Teng, X., McCaffery, J.M., Cheng, W.-C., Qi, B., Li, H., Alavian, K.N., Dayhoff-Brannigan, M., Zou, S., Pineda, F.J., O'Rourke, B., Ko, Y.H., Pedersen, P.L., Kaczmarek, L.K., Jonas, E.A., Hardwick, J.M., 2011. Bcl-xL regulates mitochondrial energetics by stabilizing the inner membrane potential. J. Cell Biol. 195, 263–276. https://doi.org/10.1083/jcb.201108059.

Chipuk, J.E., Moldoveanu, T., Llambi, F., Parsons, M.J., Green, D.R., 2010. The BCL-2 family reunion. Mol. Cell 37, 299–310. https://doi.org/10.1016/j.molcel.2010.01.025.

Chung, S., Arrell, D.K., Faustino, R.S., Terzic, A., Dzeja, P.P., 2010. Glycolytic network restructuring integral to the energetics of embryonic stem cell cardiac differentiation. J. Mol. Cell. Cardiol. 48, 725–734. https://doi.org/10.1016/j.yjmcc.2009.12.014.

Davies, V.J., Hollins, A.J., Piechota, M.J., Yip, W., Davies, J.R., White, K.E., Nicols, P.P., Boulton, M.E., Votruba, M., 2007. Opa1 deficiency in a mouse model of autosomal dominant optic atrophy impairs mitochondrial morphology, optic nerve structure and visual function. Hum. Mol. Genet. 16, 1307–1318. https://doi.org/10.1093/hmg/ddm079.

Del Dotto, V., Fogazza, M., Carelli, V., Rugolo, M., Zanna, C., 2018. Eight human OPA1 isoforms, long and short: what are they for? Biochim. Biophys. Acta Biomembr. 1859, 263–269. https://doi.org/10.1016/j.bbabio.2018.01.005.

Del Dotto, V., Mishra, P., Vidoni, S., Fogazza, M., Maresca, A., Caporali, L., McCaffery, J.M., Cappelletti, M., Baruffini, E., Lenaers, G., Chan, D., Rugolo, M., Carelli, V., Zanna, C., 2017. OPA1 isoforms in the hierarchical organization of mitochondrial functions. Cell Rep. 19, 2557–2571. https://doi.org/10.1016/j.celrep.2017.05.073.

Delettre, C., Lenaers, G., Griffoin, J.-M., Gigarel, N., Lorenzo, C., Belenguer, P., Pelloquin, L., Grosgeorge, J., Turc-Carel, C., Perret, E., Astarie-Dequeker, C., Lasquellec, L., Arnaud, B., Ducommun, B., Kaplan, J., Hamel, C.P., 2000. Nuclear gene OPA1, encoding a mitochondrial dynamin-related protein, is mutated in dominant optic atrophy. Nat. Genet. 26, 207–210. https://doi.org/10.1038/79936.

Dhingra, R., Margulets, V., Chowdhury, S.R., Thliveris, J., Jassal, D., Fernyhough, P., Dorn, G.W., Kirshenbaum, L.A., 2014. Bnip3 mediates doxorubicin-induced cardiac myocyte necrosis and mortality through changes in mitochondrial signaling. Proc. Natl. Acad. Sci. 111, E5537–E5544. https://doi.org/10.1073/pnas.1414665111.

Dix, M.M., Simon, G.M., Cravatt, B.F., 2008. Global mapping of the topography and magnitude of proteolytic events in apoptosis. Cell 134, 679–691. https://doi.org/10.1016/j.cell.2008.06.038.

Dorn, G.W., 2013. Mitochondrial dynamics in heart disease. Biochim. Biophys. Acta 1833, 233–241. https://doi.org/10.1016/j.bbamcr.2012.03.008.

Dumitru, R., Gama, V., Fagan, B.M., Bower, J.J., Swahari, V., Pevny, L.H., Deshmukh, M., 2012. Human embryonic stem cells have constitutively active Bax at the Golgi and are primed to undergo rapid apoptosis. Mol. Cell 46, 573–583. https://doi.org/10.1016/j.molcel.2012.04.002.

Elgass, K.D., Smith, E.A., LeGros, M.A., Larabell, C.A., Ryan, M.T., 2015. Analysis of ER– mitochondria contacts using correlative fluorescence microscopy and soft X-ray tomography of mammalian cells. J. Cell Sci. 128, 2795–2804. https://doi.org/10.1242/jcs.169136.

Elmore, S., 2007. Apoptosis: a review of programmed cell death. Toxicol. Pathol. 35, 495–516. https://doi.org/10.1080/01926230701320337.

Escobar-Henriques, M., Joaquim, M., 2019. Mitofusins: disease gatekeepers and hubs in mitochondrial quality control by E3 ligases. Front. Physiol. 10, 1–24. 517. https://doi.org/10.3389/fphys.2019.00517.

Eskes, R., Desagher, S., Antonsson, B., Martinou, J.C., 2000. Bid induces the oligomerization and insertion of Bax into the outer mitochondrial membrane. Mol. Cell. Biol. 20, 929–935. https://doi.org/10.1128/mcb.20.3.929-935.2000.

Estaquier, J., Arnoult, D., 2007. Inhibiting Drp1-mediated mitochondrial fission selectively prevents the release of cytochrome c during apoptosis. Cell Death Differ. 14, 1086–1094. https://doi.org/10.1038/sj.cdd.4402107.

Evans, C.S., Holzbaur, E.L.F., 2019. Quality control in neurons: mitophagy and other selective autophagy mechanisms. J. Mol. Biol. https://doi.org/10.1016/j.jmb.2019.06.031.

Evans, M.J., Kaufman, M.H., 1981. Establishment in culture of pluripotential cells from mouse embryos. Nature 292, 154–156. https://doi.org/10.1038/292154a0.

Facucho-Oliveira, J.M., St. John, J.C., 2009. The relationship between pluripotency and mitochondrial DNA proliferation during early embryo development and embryonic stem cell differentiation. Stem Cell Rev. Rep. 5, 140–158. https://doi.org/10.1007/s12015-009-9058-0.

Fernández, Á.F., Sebti, S., Wei, Y., Zou, Z., Shi, M., McMillan, K.L., He, C., Ting, T., Liu, Y., Chiang, W.-C., Marciano, D.K., Schiattarella, G.G., Bhagat, G., Moe, O.W., Hu, M.C., Levine, B., 2018. Disruption of the beclin 1–BCL2 autophagy regulatory complex promotes longevity in mice. Nature 558, 136–140. https://doi.org/10.1038/s41586-018-0162-7.

Fonseca, T.B., Sánchez-Guerrero, Á., Milosevic, I., Raimundo, N., 2019. Mitochondrial fission requires DRP1 but not dynamins. Nature 570, E34–E42. https://doi.org/10.1038/s41586-019-1296-y.

Francy, C.A., Clinton, R.W., Fröhlich, C., Murphy, C., Mears, J.A., 2017. Cryo-EM studies of Drp1 reveal cardiolipin interactions that activate the helical oligomer. Sci. Rep. 7, 1–12. https://doi.org/10.1038/s41598-017-11008-3.

Frank, S., Gaume, B., Bergmann-Leitner, E.S., Leitner, W.W., Robert, E.G., Catez, F., Smith, C.L., Youle, R.J., 2001. The role of dynamin-related protein 1, a mediator of mitochondrial fission, in apoptosis. Dev. Cell 1, 515–525. https://doi.org/10.1016/S1534-5807(01)00055-7.

Frezza, C., Cipolat, S., Martins de Brito, O., Micaroni, M., Beznoussenko, G.V., Rudka, T., Bartoli, D., Polishuck, R.S., Danial, N.N., De Strooper, B., Scorrano, L., 2006. OPA1 controls apoptotic cristae remodeling independently from mitochondrial fusion. Cell 126, 177–189. https://doi.org/10.1016/j.cell.2006.06.025.

Friedman, J.R., Nunnari, J., 2014. Mitochondrial form and function. Nature 505, 335–343. https://doi.org/10.1038/nature12985.

Fu, W., Liu, Y., Yin, H., 2019. Mitochondrial dynamics: biogenesis, fission, fusion, and mitophagy in the regulation of stem cell behaviors. Stem Cells Int. 2019. https://doi.org/10.1155/2019/9757201. 9757201.

Fuchs, Y., Steller, H., 2011. Programmed cell death in animal development and disease. Cell 147, 742–758. https://doi.org/10.1016/j.cell.2011.10.033.

Fujita, J., Crane, A.M., Souza, M.K., Dejosez, M., Kyba, M., Flavell, R.A., Thomson, J.A., Zwaka, T.P., 2008. Caspase activity mediates the differentiation of embryonic stem cells. Cell Stem Cell 2, 595–601. https://doi.org/10.1016/j.stem.2008.04.001.

Galluzzi, L., Vitale, I., Aaronson, S.A., Abrams, J.M., Adam, D., Agostinis, P., Alnemri, E.S., Altucci, L., Amelio, I., Andrews, D.W., Annicchiarico-Petruzzelli, M., Antonov, A.V., Arama, E., Baehrecke, E.H., Barlev, N.A., Bazan, N.G., Bernassola, F., Bertrand, M.J.M., Bianchi, K., Blagosklonny, M.V., Blomgren, K., Borner, C.,

Boya, P., Brenner, C., Campanella, M., Candi, E., Carmona-Gutierrez, D., Cecconi, F., Chan, F.K.-M., Chandel, N.S., Cheng, E.H., Chipuk, J.E., Cidlowski, J.A., Ciechanover, A., Cohen, G.M., Conrad, M., Cubillos-Ruiz, J.R., Czabotar, P.E., D'Angiolella, V., Dawson, T.M., Dawson, V.L., Laurenzi, V.D., Maria, R.D., Debatin, K.-M., DeBerardinis, R.J., Deshmukh, M., Daniele, N.D., Virgilio, F.D., Dixit, V.M., Dixon, S.J., Duckett, C.S., Dynlacht, B.D., El-Deiry, W.S., Elrod, J.W., Fimia, G.M., Fulda, S., García-Sáez, A.J., Garg, A.D., Garrido, C., Gavathiotis, E., Golstein, P., Gottlieb, E., Green, D.R., Greene, L.A., Gronemeyer, H., Gross, A., Hajnoczky, G., Hardwick, J.M., Harris, I.S., Hengartner, M.O., Hetz, C., Ichijo, H., Jäättelä, M., Joseph, B., Jost, P.J., Juin, P.P., Kaiser, W.J., Karin, M., Kaufmann, T., Kepp, O., Kimchi, A., Kitsis, R.N., Klionsky, D.J., Knight, R.A., Kumar, S., Lee, S.W., Lemasters, J.J., Levine, B., Linkermann, A., Lipton, S.A., Lockshin, R.A., López-Otín, C., Lowe, S.W., Luedde, T., Lugli, E., MacFarlane, M., Madeo, F., Malewicz, M., Malorni, W., Manic, G., Marine, J.-C., Martin, S.J., Martinou, J.-C., Medema, J.P., Mehlen, P., Meier, P., Melino, S., Miao, E.A., Molkentin, J.D., Moll, U.M., Muñoz-Pinedo, C., Nagata, S., Nuñez, G., Oberst, A., Oren, M., Overholtzer, M., Pagano, M., Panaretakis, T., Pasparakis, M., Penninger, J.M., Pereira, D.M., Pervaiz, S., Peter, M.E., Piacentini, M., Pinton, P., Prehn, J.H.M., Puthalakath, H., Rabinovich, G.A., Rehm, M., Rizzuto, R., Rodrigues, C.M.P., Rubinsztein, D.C., Rudel, T., Ryan, K.M., Sayan, E., Scorrano, L., Shao, F., Shi, Y., Silke, J., Simon, H.-U., Sistigu, A., Stockwell, B.R., Strasser, A., Szabadkai, G., Tait, S.W.G., Tang, D., Tavernarakis, N., Thorburn, A., Tsujimoto, Y., Turk, B., Berghe, T.V., Vandenabeele, P., Heiden, M.G.V., Villunger, A., Virgin, H.W., Vousden, K.H., Vucic, D., Wagner, E.F., Walczak, H., Wallach, D., Wang, Y., Wells, J.A., Wood, W., Yuan, J., Zakeri, Z., Zhivotovsky, B., Zitvogel, L., Melino, G., Kroemer, G., 2018. Molecular mechanisms of cell death: recommendations of the Nomenclature Committee on Cell Death 2018. Cell Death Differ. 25, 486–541. https://doi.org/10.1038/s41418-017-0012-4.

Gama, V., Deshmukh, M., 2012. Human embryonic stem cells: living on the edge. Cell Cycle 11, 3905–3906. https://doi.org/10.4161/cc.22233.

Ganesan, V., Willis, S.D., Chang, K.-T., Beluch, S., Cooper, K.F., Strich, R., 2019. Cyclin C directly stimulates Drp1 GTP affinity to mediate stress-induced mitochondrial hyperfission. Mol. Biol. Cell 30, 302–311. https://doi.org/10.1091/mbc.E18-07-0463.

Gao, S., von der Malsburg, A., Paeschke, S., Behlke, J., Haller, O., Kochs, G., Daumke, O., 2010. Structural basis of oligomerization in the stalk region of dynamin-like MxA. Nature 465, 502–506. https://doi.org/10.1038/nature08972.

Gao, W., Pu, Y., Luo, K.Q., Chang, D.C., 2001. Temporal relationship between cytochrome c release and mitochondrial swelling during UV-induced apoptosis in living HeLa cells. J. Cell Sci. 114, 2855–2862.

Garrido, C., Galluzzi, L., Brunet, M., Puig, P.E., Didelot, C., Kroemer, G., 2006. Mechanisms of cytochrome c release from mitochondria. Cell Death Differ. 13, 1423–1433. https://doi.org/10.1038/sj.cdd.4401950.

Gavathiotis, E., Suzuki, M., Davis, M.L., Pitter, K., Bird, G.H., Katz, S.G., Tu, H.-C., Kim, H., Cheng, E.H.-Y., Tjandra, N., Walensky, L.D., 2008. BAX activation is initiated at a novel interaction site. Nature 455, 1076–1081. https://doi.org/10.1038/nature07396.

Gegg, M.E., Cooper, J.M., Chau, K.-Y., Rojo, M., Schapira, A.H.V., Taanman, J.-W., 2010. Mitofusin 1 and mitofusin 2 are ubiquitinated in a PINK1/parkin-dependent manner upon induction of mitophagy. Hum. Mol. Genet. 19, 4861–4870. https://doi.org/10.1093/hmg/ddq419.

Goldstein, J.C., Waterhouse, N.J., Juin, P., Evan, G.I., Green, D.R., 2000. The coordinate release of cytochrome c during apoptosis is rapid, complete and kinetically invariant. Nat. Cell Biol. 2, 156–162. https://doi.org/10.1038/35004029.

Goodell, M.A., Nguyen, H., Shroyer, N., 2015. Somatic stem cell heterogeneity: diversity in the blood, skin and intestinal stem cell compartments. Nat. Rev. Mol. Cell Biol. 16, 299–309. https://doi.org/10.1038/nrm3980.

Goping, I.S., Gross, A., Lavoie, J.N., Nguyen, M., Jemmerson, R., Roth, K., Korsmeyer, S.J., Shore, G.C., 1998. Regulated targeting of BAX to mitochondria. J. Cell Biol. 143, 207–215. https://doi.org/10.1083/jcb.143.1.207.

Green, D.R., Chipuk, J.E., 2008. APOPTOSIS: stabbed in the BAX. Nature 455, 1047–1049. https://doi.org/10.1038/4551047a.

Hardwick, J.M., Soane, L., 2013. Multiple functions of BCL-2 family proteins. Cold Spring Harb. Perspect. Biol. 5, a008722. https://doi.org/10.1101/cshperspect.a008722.

Hinson, J.T., Chopra, A., Nafissi, N., Polacheck, W.J., Benson, C.C., Swist, S., Gorham, J., Yang, L., Schafer, S., Sheng, C.C., Haghighi, A., Homsy, J., Hubner, N., Church, G., Cook, S.A., Linke, W.A., Chen, C.S., Seidman, J.G., Seidman, C.E., 2015. Titin mutations in iPS cells define sarcomere insufficiency as a cause of dilated cardiomyopathy. Science 349, 982–986. https://doi.org/10.1126/science.aaa5458.

Hollville, E., Carroll, R.G., Cullen, S.P., Martin, S.J., 2014. Bcl-2 family proteins participate in mitochondrial quality control by regulating Parkin/PINK1-dependent mitophagy. Mol. Cell 55, 451–466. https://doi.org/10.1016/j.molcel.2014.06.001.

Hoppins, S., Edlich, F., Cleland, M.M., Banerjee, S., McCaffery, J.M., Youle, R.J., Nunnari, J., 2011. The soluble form of Bax regulates mitochondrial fusion via MFN2 homotypic complexes. Mol. Cell 41, 150–160. https://doi.org/10.1016/j.molcel.2010.11.030.

Hoppins, S., Lackner, L., Nunnari, J., 2007. The machines that divide and fuse mitochondria. Annu. Rev. Biochem. 76, 751–780. https://doi.org/10.1146/annurev.biochem.76.071905.090048.

Hsu, Y.-T., Youle, R.J., 1997. Nonionic detergents induce dimerization among members of the Bcl-2 family. J. Biol. Chem. 272, 13829–13834. https://doi.org/10.1074/jbc.272.21.13829.

Ingerman, E., Perkins, E.M., Marino, M., Mears, J.A., McCaffery, J.M., Hinshaw, J.E., Nunnari, J., 2005. Dnm1 forms spirals that are structurally tailored to fit mitochondria. J. Cell Biol. 170, 1021–1027. https://doi.org/10.1083/jcb.200506078.

Inoue, S., Browne, G., Melino, G., Cohen, G.M., 2009. Ordering of caspases in cells undergoing apoptosis by the intrinsic pathway. Cell Death Differ. 16, 1053–1061. https://doi.org/10.1038/cdd.2009.29.

Ishihara, N., Nomura, M., Jofuku, A., Kato, H., Suzuki, S.O., Masuda, K., Otera, H., Nakanishi, Y., Nonaka, I., Goto, Y., Taguchi, N., Morinaga, H., Maeda, M., Takayanagi, R., Yokota, S., Mihara, K., 2009. Mitochondrial fission factor Drp1 is essential for embryonic development and synapse formation in mice. Nat. Cell Biol. 11, 958–966. https://doi.org/10.1038/ncb1907.

Jahani-Asl, A., Germain, M., Slack, R.S., 2010. Mitochondria: joining forces to thwart cell death. Biochim. Biophys. Acta 1802, 162–166. https://doi.org/10.1016/j.bbadis.2009.09.006.

Jain, A.K., Barton, M.C., 2018. p53: emerging roles in stem cells, development and beyond. Development 145. https://doi.org/10.1242/dev.158360. dev158360.

Jiang, X., Jiang, H., Shen, Z., Wang, X., 2014. Activation of mitochondrial protease OMA1 by Bax and Bak promotes cytochrome c release during apoptosis. Proc. Natl. Acad. Sci. 111, 14782–14787. https://doi.org/10.1073/pnas.1417253111.

Johnson, L.V., Walsh, M.L., Bockus, B.J., Chen, L.B., 1981. Monitoring of relative mitochondrial membrane potential in living cells by fluorescence microscopy. J. Cell Biol. 88, 526–535. https://doi.org/10.1083/jcb.88.3.526.

Kageyama, Y., Hoshijima, M., Seo, K., Bedja, D., Sysa-Shah, P., Andrabi, S.A., Chen, W., Höke, A., Dawson, V.L., Dawson, T.M., Gabrielson, K.L., Kass, D.A., Iijima, M., Sesaki, H., 2014. Parkin-independent mitophagy requires Drp1 and maintains the integrity of mammalian heart and brain. EMBO J. 33, 2798–2813. https://doi.org/10.15252/embj.201488658.

Kale, J., Osterlund, E.J., Andrews, D.W., 2018. BCL-2 family proteins: changing partners in the dance towards death. Cell Death Differ. 25, 65–80. https://doi.org/10.1038/cdd.2017.186.

Kalkavan, H., Green, D.R., 2018. MOMP, cell suicide as a BCL-2 family business. Cell Death Differ. 25, 46–55. https://doi.org/10.1038/cdd.2017.179.

Kamerkar, S.C., Kraus, F., Sharpe, A.J., Pucadyil, T.J., Ryan, M.T., 2018. Dynamin-related protein 1 has membrane constricting and severing abilities sufficient for mitochondrial and peroxisomal fission. Nat. Commun. 9, 5239. https://doi.org/10.1038/s41467-018-07543-w.

Karbowski, M., Lee, Y.-J., Gaume, B., Jeong, S.-Y., Frank, S., Nechushtan, A., Santel, A., Fuller, M., Smith, C.L., Youle, R.J., 2002. Spatial and temporal association of Bax with mitochondrial fission sites, Drp1, and Mfn2 during apoptosis. J. Cell Biol. 159, 931–938. https://doi.org/10.1083/jcb.200209124.

Karbowski, M., Norris, K.L., Cleland, M.M., Jeong, S.-Y., Youle, R.J., 2006. Role of Bax and Bak in mitochondrial morphogenesis. Nature 443, 658–662. https://doi.org/10.1038/nature05111.

Kawamura, T., Suzuki, J., Wang, Y.V., Menendez, S., Morera, L.B., Raya, A., Wahl, G.M., Belmonte, J.C.I., 2009. Linking the p53 tumour suppressor pathway to somatic cell reprogramming. Nature 460, 1140–1144. https://doi.org/10.1038/nature08311.

Kerr, J.F.R., Wyllie, A.H., Currie, A.R., 1972. Apoptosis: a basic biological phenomenon with wideranging implications in tissue kinetics. Br. J. Cancer 26, 239–257. https://doi.org/10.1038/bjc.1972.33.

Kim, H.-E., Du, F., Fang, M., Wang, X., 2005. Formation of apoptosome is initiated by cytochrome c-induced dATP hydrolysis and subsequent nucleotide exchange on Apaf-1. Proc. Natl. Acad. Sci. 102, 17545–17550. https://doi.org/10.1073/pnas.0507900102.

Kornmann, B., Walter, P., 2010. ERMES-mediated ER-mitochondria contacts: molecular hubs for the regulation of mitochondrial biology. J. Cell Sci. 123, 1389–1393. https://doi.org/10.1242/jcs.058636.

Korobova, F., Ramabhadran, V., Higgs, H.N., 2013. An actin-dependent step in mitochondrial fission mediated by the ER-associated formin INF2. Science 339, 464–467. https://doi.org/10.1126/science.1228360.

Kruse, J.-P., Gu, W., 2009. Modes of p53 Regulation. Cell 137, 609–622. https://doi.org/10.1016/j.cell.2009.04.050.

Kuznetsov, A.V., Hermann, M., Saks, V., Hengster, P., Margreiter, R., 2009. The cell-type specificity of mitochondrial dynamics. Int. J. Biochem. Cell Biol. Mitochondrial Dynamics and Function in Biology and Medicine. 41, 1928–1939. https://doi.org/10.1016/j.biocel.2009.03.007. Mitochondrial Dynamics and Function in Biology and Medicine.

Leber, B., Lin, J., Andrews, D.W., 2007. Embedded together: the life and death consequences of interaction of the Bcl-2 family with membranes. Apoptosis 12, 897–911. https://doi.org/10.1007/s10495-007-0746-4.

Lee, J.E., Westrate, L.M., Wu, H., Page, C., Voeltz, G.K., 2016. Multiple dynamin family members collaborate to drive mitochondrial division. Nature 540, 139–143. https://doi.org/10.1038/nature20555.

Lee, Y., Jeong, S.-Y., Karbowski, M., Smith, C.L., Youle, R.J., 2004. Roles of the mammalian mitochondrial fission and fusion mediators Fis1, Drp1, and Opa1 in apoptosis. Mol. Biol. Cell 15, 5001–5011. https://doi.org/10.1091/mbc.e04-04-0294.

Legros, F., Lombès, A., Frachon, P., Rojo, M., 2002. Mitochondrial fusion in human cells is efficient, requires the inner membrane potential, and is mediated by mitofusins. Mol. Biol. Cell 13, 4343–4354. https://doi.org/10.1091/mbc.e02-06-0330.

Levine, A.J., Puzio-Kuter, A.M., Chan, C.S., Hainaut, P., 2016. The role of the p53 protein in stem-cell biology and epigenetic regulation. Cold Spring Harb. Perspect. Med. 6, a026153. https://doi.org/10.1101/cshperspect.a026153.

Lewis, M.R., Lewis, W.H., 1914. Mitochondria in tissue culture. Science 39, 330–333. https://doi.org/10.1126/science.39.1000.330.

Lewis, S.C., Uchiyama, L.F., Nunnari, J., 2016. ER-mitochondria contacts couple mtDNA synthesis with mitochondrial division in human cells. Science 353, aaf5549. https://doi.org/10.1126/science.aaf5549.

Li, F., He, Z., Shen, J., Huang, Q., Li, W., Liu, X., He, Y., Wolf, F., Li, C.-Y., 2010. Apoptotic caspases regulate induction of iPSCs from human fibroblasts. Cell Stem Cell 7, 508–520. https://doi.org/10.1016/j.stem.2010.09.003.

Li, H., Chen, Y., Jones, A.F., Sanger, R.H., Collis, L.P., Flannery, R., McNay, E.C., Yu, T., Schwarzenbacher, R., Bossy, B., Bossy-Wetzel, E., Bennett, M.V.L., Pypaert, M., Hickman, J.A., Smith, P.J.S., Hardwick, J.M., Jonas, E.A., 2008. Bcl-xL induces Drp1- dependent synapse formation in cultured hippocampal neurons. Proc. Natl. Acad. Sci. 105, 2169–2174. https://doi.org/10.1073/pnas.0711647105.

Li, P., Nijhawan, D., Budihardjo, I., Srinivasula, S.M., Ahmad, M., Alnemri, E.S., Wang, X., 1997. Cytochrome c and dATP-dependent formation of Apaf-1/caspase-9 complex initiates an apoptotic protease cascade. Cell 91, 479–489. https://doi.org/10.1016/s0092-8674(00)80434-1.

Li, Y., Feng, H., Gu, H., Lewis, D.W., Yuan, Y., Zhang, L., Yu, H., Zhang, P., Cheng, H., Miao, W., Yuan, W., Cheng, S.-Y., Gollin, S.M., Cheng, T., 2013. The p53–PUMA axis suppresses iPSC generation. Nat. Commun. 4, 1–9. https://doi.org/10.1038/ncomms3174.

Lindsten, T., Ross, A.J., King, A., Zong, W.X., Rathmell, J.C., Shiels, H.A., Ulrich, E., Waymire, K.G., Mahar, P., Frauwirth, K., Chen, Y., Wei, M., Eng, V.M., Adelman, D.M., Simon, M.C., Ma, A., Golden, J.A., Evan, G., Korsmeyer, S.J., MacGregor, G.R., Thompson, C.B., 2000. The combined functions of proapoptotic Bcl-2 family members bak and bax are essential for normal development of multiple tissues. Mol. Cell 6, 1389–1399. https://doi.org/10.1016/s1097-2765(00)00136-2.

Liu, J.C., Guan, X., Ryan, J.A., Rivera, A.G., Mock, C., Agarwal, V., Letai, A., Lerou, P.H., Lahav, G., 2013. High mitochondrial priming sensitizes hESCs to DNA-damage-induced apoptosis. Cell Stem Cell 13, 483–491. https://doi.org/10.1016/j.stem.2013.07.018.

Liu, X., Kim, C.N., Yang, J., Jemmerson, R., Wang, X., 1996. Induction of apoptotic program in cell-free extracts: requirement for dATP and cytochrome c. Cell 86, 147–157. https://doi.org/10.1016/S0092-8674(00)80085-9.

Llambi, F., Moldoveanu, T., Tait, S.W.G., Bouchier-Hayes, L., Temirov, J., McCormick, L.L., Dillon, C.P., Green, D.R., 2011. A unified model of mammalian BCL-2 protein family interactions at the mitochondria. Mol. Cell 44, 517–531. https://doi.org/10.1016/j.molcel.2011.10.001.

Llambi, F., Wang, Y.-M., Victor, B., Yang, M., Schneider, D.M., Gingras, S., Parsons, M.J., Zheng, J.H., Brown, S.A., Pelletier, S., Moldoveanu, T., Chen, T., Green, D.R., 2016. BOK is a non-canonical BCL-2 family effector of apoptosis regulated by ER-associated degradation. Cell 165, 421–433. https://doi.org/10.1016/j.cell.2016.02.026.

MacVicar, T., Langer, T., 2016. OPA1 processing in cell death and disease—the long and short of it. J. Cell Sci. 129, 2297–2306. https://doi.org/10.1242/jcs.159186.

Madden, D.T., Davila-Kruger, D., Melov, S., Bredesen, D.E., 2011. Human embryonic stem cells express elevated levels of multiple pro-apoptotic BCL-2 family members. PLoS One 6, e28530. https://doi.org/10.1371/journal.pone.0028530.

Mahrus, S., Trinidad, J.C., Barkan, D.T., Sali, A., Burlingame, A.L., Wells, J.A., 2008. Global sequencing of proteolytic cleavage sites in apoptosis by specific labeling of protein N termini. Cell 134, 866–876. https://doi.org/10.1016/j.cell.2008.08.012.

Maiuri, M.C., Le Toumelin, G., Criollo, A., Rain, J.-C., Gautier, F., Juin, P., Tasdemir, E., Pierron, G., Troulinaki, K., Tavernarakis, N., Hickman, J.A., Geneste, O., Kroemer, G., 2007. Functional and physical interaction between Bcl-XL and a BH3-like domain in Beclin-1. EMBO J. 26, 2527–2539. https://doi.org/10.1038/sj.emboj.7601689.

Martin, G.R., 1981. Isolation of a pluripotent cell line from early mouse embryos cultured in medium conditioned by teratocarcinoma stem cells. Proc. Natl. Acad. Sci. U. S. A. 78, 7634–7638.

Mears, J.A., Lackner, L.L., Fang, S., Ingerman, E., Nunnari, J., Hinshaw, J.E., 2011. Conformational changes in Dnm1 support a contractile mechanism for mitochondrial fission. Nat. Struct. Mol. Biol. 18, 20–26. https://doi.org/10.1038/nsmb.1949.

Meeusen, S., McCaffery, J.M., Nunnari, J., 2004. Mitochondrial fusion intermediates revealed in vitro. Science 305, 1747–1752. https://doi.org/10.1126/science.1100612.

Meglei, G., McQuibban, G.A., 2009. The dynamin-related protein Mgm1p assembles into oligomers and hydrolyzes GTP to function in mitochondrial membrane fusion. Biochemistry 48, 1774–1784. https://doi.org/10.1021/bi801723d.

Mehlen, P., Bredesen, D.E., 2011. Dependence receptors: from basic research to drug development. Sci. Signal. 4, mr2. https://doi.org/10.1126/scisignal.2001521.

Mishra, P., Carelli, V., Manfredi, G., Chan, D.C., 2014. Proteolytic cleavage of Opa1 stimulates mitochondrial inner membrane fusion and couples fusion to oxidative phosphorylation. Cell Metab. 19, 630–641. https://doi.org/10.1016/j.cmet.2014.03.011.

Moldoveanu, T., Czabotar, P.E., 2019. BAX, BAK, and BOK: a coming of age for the BCL-2 family effector proteins. Cold Spring Harb. Perspect. Biol. a036319. https://doi.org/10.1101/cshperspect.a036319, 1-19.

Montero, J., Letai, A., 2016. Dynamic BH3 profiling-poking cancer cells with a stick. Mol. Cell. Oncol. 3 e1040144https://doi.org/10.1080/23723556.2015.1040144.

Montero, J., Sarosiek, K.A., DeAngelo, J.D., Maertens, O., Ryan, J., Ercan, D., Piao, H., Horowitz, N.S., Berkowitz, R.S., Matulonis, U., Jänne, P.A., Amrein, P.C., Cichowski, K., Drapkin, R., Letai, A., 2015. Drug-induced death signaling strategy rapidly predicts cancer response to chemotherapy. Cell 160, 977–989. https://doi.org/10.1016/j.cell.2015.01.042.

Montessuit, S., Somasekharan, S.P., Terrones, O., Lucken-Ardjomande, S., Herzig, S., Schwarzenbacher, R., Manstein, D.J., Bossy-Wetzel, E., Basañez, G., Meda, P., Martinou, J.-C., 2010. Membrane remodeling induced by the dynamin-related protein Drp1 stimulates Bax oligomerization. Cell 142, 889–901. https://doi.org/10.1016/j.cell.2010.08.017.

Morciano, G., Giorgi, C., Balestra, D., Marchi, S., Perrone, D., Pinotti, M., Pinton, P., 2016a. Mcl- 1 involvement in mitochondrial dynamics is associated with apoptotic cell death. Mol. Biol. Cell 27, 20–34. https://doi.org/10.1091/mbc.E15-01-0028.

Morciano, G., Pedriali, G., Sbano, L., Iannitti, T., Giorgi, C., Pinton, P., 2016b. Intersection of mitochondrial fission and fusion machinery with apoptotic pathways: role of Mcl-1. Biol. Cell 108, 279–293. https://doi.org/10.1111/boc.201600019.

Naik, P.P., Birbrair, A., Bhutia, S.K., 2019. Mitophagy-driven metabolic switch reprograms stem cell fate. Cell. Mol. Life Sci. 76, 27–43. https://doi.org/10.1007/s00018-018-2922-9.

Nakamura, A., Swahari, V., Plestant, C., Smith, I., McCoy, E., Smith, S., Moy, S.S., Anton, E.S., Deshmukh, M., 2016. Bcl-xL is essential for the survival and function of differentiated neurons in the cortex that control complex behaviors. J. Neurosci. 36, 5448–5461. https://doi.org/10.1523/JNEUROSCI.4247-15.2016.

Narendra, D., Tanaka, A., Suen, D.-F., Youle, R.J., 2008. Parkin is recruited selectively to impaired mitochondria and promotes their autophagy. J. Cell Biol. 183, 795–803. https://doi.org/10.1083/jcb.200809125.

Nunnari, J., Marshall, W.F., Straight, A., Murray, A., Sedat, J.W., Walter, P., 1997. Mitochondrial transmission during mating in Saccharomyces cerevisiae is determined by mitochondrial fusion and fission and the intramitochondrial segregation of mitochondrial DNA. Mol. Biol. Cell 8, 1233–1242. https://doi.org/10.1091/mbc.8.7.1233.

Oltval, Z.N., Milliman, C.L., Korsmeyer, S.J., 1993. Bcl-2 heterodimerizes in vivo with a conserved homolog, Bax, that accelerates programed cell death. Cell 74, 609–619. https://doi.org/10.1016/0092-8674(93)90509-O.

Onoue, K., Jofuku, A., Ban-Ishihara, R., Ishihara, T., Maeda, M., Koshiba, T., Itoh, T., Fukuda, M., Otera, H., Oka, T., Takano, H., Mizushima, N., Mihara, K., Ishihara, N., 2013. Fis1 acts as a mitochondrial recruitment factor for TBC1D15 that is involved in regulation of mitochondrial morphology. J. Cell Sci. 126, 176–185. https://doi.org/10.1242/jcs.111211.

Ow, Y.-L.P., Green, D.R., Hao, Z., Mak, T.W., 2008. Cytochrome c: functions beyond respiration. Nat. Rev. Mol. Cell Biol. 9, 532–542. https://doi.org/10.1038/nrm2434.

Pattingre, S., Tassa, A., Qu, X., Garuti, R., Liang, X.H., Mizushima, N., Packer, M., Schneider, M.D., Levine, B., 2005. Bcl-2 antiapoptotic proteins inhibit Beclin 1-dependent autophagy. Cell 122, 927–939. https://doi.org/10.1016/j.cell.2005.07.002.

Pickles, S., Vigié, P., Youle, R.J., 2018. Mitophagy and quality control mechanisms in mitochondrial maintenance. Curr. Biol. 28, R170–R185. https://doi.org/10.1016/j.cub.2018.01.004.

Pon, L.A., 2013. Mitochondrial fission: rings around the organelle. Curr. Biol. 23, R279–R281. https://doi.org/10.1016/j.cub.2013.02.042.

Praefcke, G.J.K., McMahon, H.T., 2004. The dynamin superfamily: universal membrane tubulation and fission molecules? Nat. Rev. Mol. Cell Biol. 5, 133–147. https://doi.org/10.1038/nrm1313.

Prieto, J., León, M., Ponsoda, X., Sendra, R., Bort, R., Ferrer-Lorente, R., Raya, A., López-García, C., Torres, J., 2016. Early ERK1/2 activation promotes DRP1-dependent mitochondrial fission necessary for cell reprogramming. Nat. Commun. 7, 11124. https://doi.org/10.1038/ncomms11124.

Prigione, A., Hossini, A.M., Lichtner, B., Serin, A., Fauler, B., Megges, M., Lurz, R., Lehrach, H., Makrantonaki, E., Zouboulis, C.C., Adjaye, J., 2011. Mitochondrial-associated cell death mechanisms are reset to an embryonic-like state in aged donor-derived iPS cells harboring chromosomal aberrations. PLoS One 6, e27352. https://doi.org/10.1371/journal.pone.0027352.

Purvis, J.E., Karhohs, K.W., Mock, C., Batchelor, E., Loewer, A., Lahav, G., 2012. p53 dynamics control cell fate. Science 336, 1440–1444. https://doi.org/10.1126/science.1218351.

Ramirez, M.L.G., Salvesen, G.S., 2018. A primer on caspase mechanisms. Semin. Cell Dev. Biol. 82, 79–85. https://doi.org/10.1016/j.semcdb.2018.01.002. SI: Nuclear positioning.

Rasmussen, M.L., Kline, L.A., Park, K.P., Ortolano, N.A., Romero-Morales, A.I., Anthony, C.C., Beckermann, K.E., Gama, V., 2018. A non-apoptotic function of MCL-1 in promoting pluripotency and modulating mitochondrial dynamics in stem cells. Stem Cell Rep. 10, 684–692. https://doi.org/10.1016/j.stemcr.2018.01.005.

Rastogi, A., Joshi, P., Contreras, E., Gama, V., 2019. Remodeling of mitochondrial morphology and function: an emerging hallmark of cellular reprogramming. Cell Stress 3, 181–194. https://doi.org/10.15698/cst2019.06.189.

Rehklau, K., Hoffmann, L., Gurniak, C.B., Ott, M., Witke, W., Scorrano, L., Culmsee, C., Rust, M.B., 2017. Cofilin1-dependent actin dynamics control DRP1-mediated mitochondrial fission. Cell Death Dis. 8, e3063. https://doi.org/10.1038/cddis.2017.448.

Rinkenberger, J.L., Horning, S., Klocke, B., Roth, K., Korsmeyer, S.J., 2000. Mcl-1 deficiency results in peri-implantation embryonic lethality. Genes Dev. 6, 23–27.

Rizzuto, R., Brini, M., Giorgi, F.D., Rossi, R., Heim, R., Tsien, R.Y., Pozzan, T., 1996. Double labelling of subcellular structures with organelle-targeted GFP mutants in vivo. Curr. Biol. 6, 183–188. https://doi.org/10.1016/S0960-9822(02)00451-7.

Salisbury-Ruf, C.T., Bertram, C.C., Vergeade, A., Lark, D.S., Shi, Q., Heberling, M.L., Fortune, N.L., Okoye, G.D., Jerome, W.G., Wells, Q.S., Fessel, J., Moslehi, J., Chen, H., Roberts, L.J., Boutaud, O., Gamazon, E.R., Zinkel, S.S., 2018. Bid maintains mitochondrial cristae structure and function and protects against cardiac disease in an integrative genomics study. Elife 7, e40907. https://doi.org/10.7554/eLife.40907.

Santel, A., Frank, S., Gaume, B., Herrler, M., Youle, R.J., Fuller, M.T., 2003. Mitofusin-1 protein is a generally expressed mediator of mitochondrial fusion in mammalian cells. J. Cell Sci. 116, 2763–2774. https://doi.org/10.1242/jcs.00479.

Schrepfer, E., Scorrano, L., 2016. Mitofusins, from mitochondria to metabolism. Mol. Cell 61, 683–694. https://doi.org/10.1016/j.molcel.2016.02.022.

Schulman, J.J., Szczesniak, L.M., Bunker, E.N., Nelson, H.A., Roe, M.W., Wagner, L.E., Yule, D.I., Wojcikiewicz, R.J.H., 2019. Bok regulates mitochondrial fusion and morphology. Cell Death Differ. 26, 2682–2694. https://doi.org/10.1038/s41418-019-0327-4.

Schulman, J.J., Wright, F.A., Han, X., Zluhan, E.J., Szczesniak, L.M., Wojcikiewicz, R.J.H., 2016. The stability and expression level of Bok are governed by binding to inositol 1,4,5-trisphosphate receptors. J. Biol. Chem. 291, 11820–11828. https://doi.org/10.1074/jbc.M115.711242.

Shamas-Din, A., Kale, J., Leber, B., Andrews, D.W., 2013. Mechanisms of action of Bcl-2 family proteins. Cold Spring Harb. Perspect. Biol. 5, a008714. https://doi.org/10.1101/cshperspect.a008714.

Sheridan, C., Delivani, P., Cullen, S.P., Martin, S.J., 2008. Bax- or Bak-induced mitochondrial fission can be uncoupled from cytochrome c release. Mol. Cell 31, 570–585. https://doi.org/10.1016/j.molcel.2008.08.002.

Sinibaldi, F., Howes, B.D., Droghetti, E., Polticelli, F., Piro, M.C., Di Pierro, D., Fiorucci, L., Coletta, M., Smulevich, G., Santucci, R., 2013. Role of lysines in cytochrome c–cardiolipin interaction. Biochemistry 52, 4578–4588. https://doi.org/10.1021/bi400324c.

Smirnova, E., Griparic, L., Shurland, D.-L., van der Bliek, A.M., 2001. Dynamin-related protein Drp1 is required for mitochondrial division in mammalian cells. Mol. Biol. Cell 12, 2245–2256.

Smith, R.C., Tabar, V., 2019. Constructing and deconstructing cancers using human pluripotent stem cells and organoids. Cell Stem Cell 24, 12–24. https://doi.org/10.1016/j.stem.2018.11.012.

Stavru, F., Palmer, A.E., Wang, C., Youle, R.J., Cossart, P., 2013. Atypical mitochondrial fission upon bacterial infection. Proc. Natl. Acad. Sci. 110, 16003–16008. https://doi.org/10.1073/pnas.1315784110.

Steffen, J., Koehler, C.M., 2018. ER–mitochondria contacts: actin dynamics at the ER control mitochondrial fission via calcium release. J. Cell Biol. 217, 15–17. https://doi.org/10.1083/jcb.201711075.

Sugioka, R., Shimizu, S., Tsujimoto, Y., 2004. Fzo1, a protein involved in mitochondrial fusion, inhibits apoptosis. J. Biol. Chem. 279, 52726–52734. https://doi.org/10.1074/jbc.M408910200.

Taguchi, N., Ishihara, N., Jofuku, A., Oka, T., Mihara, K., 2007. Mitotic phosphorylation of dynamin-related GTPase Drp1 participates in mitochondrial fission. J. Biol. Chem. 282, 11521–11529. https://doi.org/10.1074/jbc.M607279200.

Takahashi, K., Tanabe, K., Ohnuki, M., Narita, M., Ichisaka, T., Tomoda, K., Yamanaka, S., 2007. Induction of pluripotent stem cells from adult human fibroblasts by defined factors. Cell 131, 861–872. https://doi.org/10.1016/j.cell.2007.11.019.

Takahashi, K., Yamanaka, S., 2015. A developmental framework for induced pluripotency. Development 142, 3274–3285. https://doi.org/10.1242/dev.114249.

Takahashi, K., Yamanaka, S., 2006. Induction of pluripotent stem cells from mouse embryonic and adult fibroblast cultures by defined factors. Cell 126, 663–676. https://doi.org/10.1016/j.cell.2006.07.024.

Tanaka, A., Youle, R.J., 2008. A chemical inhibitor of DRP1 uncouples mitochondrial fission and apoptosis. Mol. Cell 29, 409–410. https://doi.org/10.1016/j.molcel.2008.02.005.

Taylor, R.C., Cullen, S.P., Martin, S.J., 2008. Apoptosis: controlled demolition at the cellular level. Nat. Rev. Mol. Cell Biol. 9, 231–241. https://doi.org/10.1038/nrm2312.

Thomson, J.A., Itskovitz-Eldor, J., Shapiro, S.S., Waknitz, M.A., Swiergiel, J.J., Marshall, V.S., Jones, J.M., 1998. Embryonic stem cell lines derived from human blastocysts. Science 282, 1145–1147. https://doi.org/10.1126/science.282.5391.1145.

Todt, F., Cakir, Z., Reichenbach, F., Emschermann, F., Lauterwasser, J., Kaiser, A., Ichim, G., Tait, S.W., Frank, S., Langer, H.F., Edlich, F., 2015. Differential retrotranslocation of mitochondrial Bax and Bak. EMBO J. 34, 67–80. https://doi.org/10.15252/embj.201488806.

Tsujimoto, Y., Finger, L.R., Yunis, J., Nowell, P.C., Croce, C.M., 1984. Cloning of the chromosome breakpoint of neoplastic B cells with the t(14;18) chromosome translocation. Science 226, 1097–1099. https://doi.org/10.1126/science.6093263.

Upton, J.-P., Valentijn, A.J., Zhang, L., Gilmore, A.P., 2007. The N-terminal conformation of Bax regulates cell commitment to apoptosis. Cell Death Differ. 14, 932–942. https://doi.org/10.1038/sj.cdd.4402092.

van der Bliek, A.M., Payne, G.S., 2010. Dynamin subunit interactions revealed. Dev. Cell 18, 687–688. https://doi.org/10.1016/j.devcel.2010.05.001.

Vásquez-Trincado, C., García-Carvajal, I., Pennanen, C., Parra, V., Hill, J.A., Rothermel, B.A., Lavandero, S., 2016. Mitochondrial dynamics, mitophagy and cardiovascular disease. J. Physiol. 594, 509–525. https://doi.org/10.1113/JP271301.

Vazquez-Martin, A., Cufí, S., Corominas-Faja, B., Oliveras-Ferraros, C., Vellon, L., Menendez, J.A., 2012. Mitochondrial fusion by pharmacological manipulation impedes somatic cell reprogramming to pluripotency: new insight into the role of mitophagy in cell stemness. Aging 4, 393–401.

Vazquez-Martin, A., Van den Haute, C., Cufí, S., Corominas-Faja, B., Cuyàs, E., Lopez-Bonet, E., Rodriguez-Gallego, E., Fernández-Arroyo, S., Joven, J., Baekelandt, V., Menendez, J.A., 2016. Mitophagy-driven mitochondrial rejuvenation regulates stem cell fate. Aging 8, 1330–1349. https://doi.org/10.18632/aging.100976.

Wanet, A., Arnould, T., Najimi, M., Renard, P., 2015. Connecting mitochondria, metabolism, and stem cell fate. Stem Cells Dev. 24, 1957–1971. https://doi.org/10.1089/scd.2015.0117.

Wang, K., Yin, X.M., Chao, D.T., Milliman, C.L., Korsmeyer, S.J., 1996. BID: a novel BH3 domain- only death agonist. Genes Dev. 10, 2859–2869. https://doi.org/10.1101/gad.10.22.2859.

Wang, P., Wang, P., Liu, B., Zhao, J., Pang, Q., Agrawal, S.G., Jia, L., Liu, F.-T., 2015. Dynamin-related protein Drp1 is required for Bax translocation to mitochondria in response to irradiation-induced apoptosis. Oncotarget 6, 22598–22612. https://doi.org/10.18632/oncotarget.4200.

Waterham, H.R., Koster, J., van Roermund, C.W.T., Mooyer, P.A.W., Wanders, R.J.A., Leonard, J.V., 2007. A lethal defect of mitochondrial and peroxisomal fission. N. Engl. J. Med. 356, 1736–1741. https://doi.org/10.1056/NEJMoa064436.

Westermann, B., 2010. Mitochondrial fusion and fission in cell life and death. Nat. Rev. Mol. Cell Biol. 11, 872–884. https://doi.org/10.1038/nrm3013.

Wolter, K.G., Hsu, Y.-T., Smith, C.L., Nechushtan, A., Xi, X.-G., Youle, R.J., 1997. Movement of Bax from the cytosol to mitochondria during apoptosis. J. Cell Biol. 139, 1281–1292. https://doi.org/10.1083/jcb.139.5.1281.

Yamaguchi, R., Lartigue, L., Perkins, G., Scott, R.T., Dixit, A., Kushnareva, Y., Kuwana, T., Ellisman, M.H., Newmeyer, D.D., 2008. Opa1-mediated cristae opening is Bax/Bak and BH3 dependent, required for apoptosis, and independent of Bak oligomerization. Mol. Cell 31, 557–569. https://doi.org/10.1016/j.molcel.2008.07.010.

Yang, C., Svitkina, T.M., 2019. Ultrastructure and dynamics of the actin−myosin II cytoskeleton during mitochondrial fission. Nat. Cell Biol. 21, 603–613. https://doi.org/10.1038/s41556-019-0313-6.

Yang, T., Buchan, H.L., Townsend, K.J., Craig, R.W., 1996. MCL-1, a member of the BCL-2 family, is induced rapidly in response to signals for cell differentiation or death, but not to signals for cell proliferation. J. Cell. Physiol. 166, 523–536. https://doi.org/10.1002/(SICI)1097-4652(199603)166:3<523::AID-JCP7>3.0.CO;2-R.

Youle, R.J., van der Bliek, A.M., 2012. Mitochondrial fission, fusion, and stress. Science 337, 1062–1065. https://doi.org/10.1126/science.1219855.

Zhang, J., 2013. Autophagy and mitophagy in cellular damage control. Redox Biol. 1, 19–23. https://doi.org/10.1016/j.redox.2012.11.008.

Zhang, J., Ney, P.A., 2009. Role of BNIP3 and NIX in cell death, autophagy, and mitophagy. Cell Death Differ. 16, 939–946. https://doi.org/10.1038/cdd.2009.16.

Zheng, J.H., Grace, C.R., Guibao, C.D., McNamara, D.E., Llambi, F., Wang, Y.-M., Chen, T., Moldoveanu, T., 2018. Intrinsic instability of BOK enables membrane permeabilization in apoptosis. Cell Rep. 23, 2083–2094.e6. https://doi.org/10.1016/j.celrep.2018.04.060.

Zou, H., Henzel, W.J., Liu, X., Lutschg, A., Wang, X., 1997. Apaf-1, a human protein homologous to C. elegans CED-4, participates in cytochrome c-dependent activation of caspase-3. Cell 90, 405–413. https://doi.org/10.1016/s0092-8674(00)80501-2.

Züchner, S., Mersiyanova, I.V., Muglia, M., Bissar-Tadmouri, N., Rochelle, J., Dadali, E.L., Zappia, M., Nelis, E., Patitucci, A., Senderek, J., Parman, Y., Evgrafov, O., Jonghe, P.D., Takahashi, Y., Tsuji, S., Pericak-Vance, M.A., Quattrone, A., Battologlu, E., Polyakov, A.V., Timmerman, V., Schröder, J.M., Vance, J.M., 2004. Mutations in the mitochondrial GTPase mitofusin 2 cause Charcot-Marie-Tooth neuropathy type 2A. Nat. Genet. 36, 449–451. https://doi.org/10.1038/ng1341.

CPI Antony Rowe
Eastbourne, UK
May 16, 2020